1,000+ SUDOKU

Collin Deloach

Contents

Your Mission

is to solve the puzzle by filling in the empty cells with numbers from 1 to 9 without repetition in each row, column, and sub-grid.

The goal is to use logic and deduction to find the missing numbers and complete the puzzle.

						3		2
				8		4	9	1
	1				3			
2			7	4	1	9		6
6			8	9	2			7
9		1	6	3	5			4
			4				8	
8	4	6		7				
7		9						

Without repetition in each sub-grid

						3		2
				8		4	9	1
	1				3	5		
2						9		6
6			8		2	1		7
9		1				8		4
			4			7	8	
8	4	6		7		2		
7		9				6		

Without repetition in each column

Without repetition in each row

						3		2
				8		4	9	1
	1				3			
2						9		6
6			8		2			7
9		1						4
1	2	3	4	5	6	7	8	9
8	4	6		7				
7		9						

Enjoy!

EASY
Puzzles

Easy # 1

		9			5	7	8	
		6	4					2
	1	5	7	9	2	4		
			1	8		3		4
		1				2		
9		3		5	4			
		8	9	3	6	5	4	
6					7	1		
	5	7	8			6		

Easy # 2

	7	1	4					
		4	2	9		1	7	6
	9					2	8	
6	3		9	5				
		8	6		4	5		
				3	8		6	9
	8	5					2	
4	2	3		6	7	9		
					1	4	5	

Easy # 3

		6	7	2	3			4
5	1				6		3	
4		7						2
6				9			7	
	8	9				3	4	
	7			3				1
3						9		8
	5		3				2	7
7			5	8	9	1		

Easy # 4

								1
	1		8		6	4		3
	4		9		1		6	5
3		6			9			8
	2	1	3		7	6	5	
4			1			9		2
7	6		5		8		2	
1		9	6		2		8	
2								

Easy # 5

8		1	9		6		2	
	9		1		3			6
6		5		8		9		
7	1			4				
			2		7			
				3			9	4
		3		6		4		9
9			3		4		8	
	7		5		9	3		2

Easy # 6

	4		2	5			7	8
		6			1			9
	2		7	6		1	4	
7		4	3	2				
				7	8	3		2
	9	3		8	4		1	
1			5			4		
4	5			3	2		9	

Easy # 7

2		9						
4	7	8					5	9
1		5	9			3		
			8	7		9	1	
		4	2		9	5		
	5	1		4	6			
		2			3	1		5
5	4					6	9	8
						4		3

Easy # 8

		6	5					
9	1		4			3		7
	7			3		1		8
4		7		5	2			
8			7	6	4			5
			8	9		7		4
1		9		7			4	
7		2			6		1	3
					5	2		

Easy # 9

2	6		8	7			1	3
						5		
	9			3	6			
	4						9	2
5	2	9	3		7	4	8	1
8	3						5	
			6	8			3	
		6						
3	8			1	5		7	4

Easy # 10

4			1	6			5	
5					8		6	1
1		3				4		7
8					1		3	
	5	2	3		4	1	7	
	7		2					9
7		5				6		4
2	1		6					5
	4			1	5			3

Easy # 11

3	9	5	8		2			
	4			6	1			
6				3			8	2
5	6		1	7			2	8
1	8			9	5		6	3
4	5			2				6
			6	5			1	
			9		4	2	5	7

Easy # 12

				6	3	5		4
	3	6	9				2	7
	9	4	2			6		
6		3	5					2
		7				8		
2					8	9		5
		2			6	4	8	
8	1				9	2	7	
4		9	8	2				

Easy # 13

2	3			9		5		
	5				7			
			5	2	4		1	7
8				5	6		3	1
6	1						2	5
4	2		1	3				9
5	9		8	6	2			
			7				6	
		2		4			5	8

Easy # 14

		3		2				
	4		1	5			7	
	1		4	7	9		3	2
3				4			1	
1	8						6	4
	5			6				3
4	6		5	1	2		8	
	9			8	7		4	
				9		5		

Easy # 15

			1	2	8	6		4
8		4		5			1	
		2			9			
6				3	7	1		5
4		1				7		2
7		3	2	4				8
			9			4		
	4			6		3		1
9		7	5	1	4			

Easy # 16

9				8	5		1	
							2	3
6		5	4	1		7		
	6	7			4	9	3	2
5	3	9	7			1	4	
		2		4	9	8		5
7	9							
	5		8	2				1

Easy # 17

8		1				5		7
			2			1	9	
	3		1					6
2		8	6				5	
3		5	8		2	6		4
	9				4	8		1
9					6		1	
	7	4			8			
1		2				7		9

Easy # 18

			8			6	1	
	6			3				8
	8		6	1	9		2	
	4			6	5			1
6	9						8	3
2			3	9			7	
	2		1	7	4		3	
3				2			4	
	7	9			3			

Easy # 19

	5	6		4	7		8	1
		7	3	2			9	
1	3	9		8				
						1	6	5
		8				4		
6	1	3						
				1		6	2	7
	6			7	3	9		
7	9		6	5		8	1	

Easy # 20

8			1				6	3
	9		8	7	2	1		
	7					2	9	
2				5			1	
9		8				5		4
	3			8				2
	4	5					8	
		3	5	4	6		2	
7	2				8			6

Easy # 21

6	8				3	7	5	
				4			6	3
		7	6	1		2		
7		8						6
1			8		4			9
4						3		8
		9		5	6	8		
2	4			8				
	1	6	2				9	5

Easy # 22

7	2	8		3	9	6		
					1	2		5
4		5						7
				8	4		6	3
		4	3		2	5		
8	3		6	5				
6						7		4
9		1	2					
		2	7	6		1	3	9

Easy # 23

		8		3	9			7
9		3	6					8
2	7						4	9
		4	9					6
	9	2	7		4	8	1	
5					1	2		
7	3						8	2
8					3	9		1
4			8	9		7		

Easy # 24

	6		4		2			9
		1	8			2	7	6
8				6	1		3	
	9	3						4
2								8
6						5	1	
	5		2	8				1
9	7	4			5	8		
1			3		4		5	

Easy # 25

		3	4		8			1
				6		9	8	
1	9		7	2		4		
	3	7			9		2	
		2		3		5		
	8		1			3	4	
		6		5	3		1	4
	4	9		1				
3			6		4	7		

Easy # 26

					8		2	
		3	5	1			8	6
1	4			3	2			
6	3	5		8				
	1	4	2		5	3	9	
				4		5	6	7
			9	2			4	5
2	5			7	4	9		
	8		3					

Easy # 27

	8	5		1	7	2		
9		2		4	5	7		
	3		9					4
				7	6		5	2
6	7		8	5				
2					1		9	
		9	2	8		3		6
		3	7	6		1	2	

Easy # 28

	6	1			2	7	8	9
	8			1			5	
7		5	9	6				
		3	1	8				7
5								1
1				9	4	5		
				7	6	1		8
	9			4			7	
8	1	7	3			4	6	

Easy # 29

1		4						
	7		1			2		6
		6	5	3		1		4
	8	5		1	3			
9			8	7	6			1
			2	5		7	8	
3		1		8	5	9		
4		9			1		2	
						4		5

Easy # 30

	4	3		1				
9				5	2			1
2		6	4			9	5	
	3						1	8
	6		1		3		9	
1	7						2	
	2	1			8	5		7
7			2	6				4
				3		2	8	

Easy # 31

					6	8		
	2	4		7				3
	7	8			5		1	2
			9	4		7	3	
	9		7	5	3		6	
	3	7		6	8			
2	4		3			1	7	
7				1		2	9	
		5	6					

Easy # 32

	4		5		3	1	8	
3						7		5
				2	1	3		4
7		5					4	
			6	4	5			
	8					2		1
4		9	2	7				
1		8						9
	7	3	1		6		2	

Easy # 33

7			4	2				
2			9	6			4	3
	8			1	7			5
		6					2	
3		1	8	5	2	6		9
	2					8		
4			5	8			3	
1	6			4	3			7
				9	6			2

Easy # 34

1			7	4				
		5						
	7	2		5	9		3	4
	8		1			3	4	
7	2		5		4		8	1
	9	1			6		5	
9	1		2	3		8	7	
						4		
				1	8			3

Easy # 35

5						4		
	1		5	4		8	6	
		7	6	3				
9	6		1			3	7	
8	2		4		6		5	9
	5	1			3		4	2
				8	1	5		
	9	5		6	7		8	
		8						6

Easy # 36

	3	5	8					7
	1	6					4	
7			6			8		
6		7		2		1		4
	5		9		4		3	
4		3		8		7		9
		2			7			5
	7					4	9	
3					9	2	7	

Easy # 37

3		6						2
	7		2			5		
2		5	3				8	
	3	2		9		8	6	
8			6		3			7
	6	4		5		2	1	
	2				9	7		8
		9			1		2	
6						1		4

Easy # 38

	4	8	7		9	3		
	7							
3	6		8		1	7		
	1		9				6	8
8		4	2		6	5		7
9	5				7		3	
		1	5		8		7	9
							5	
		5	1		4	8	2	

Easy # 39

	5	3	6				9	1
	1	7	2			3		
				3	1	6		7
4		6	1					3
		1				5		
3					4	8		2
7		4	8	2				
		2			3	7	6	
5	3				6	2	8	

Easy # 40

8			1			7	3	
		9	4			2		
	6			5	8	1		9
2		6					5	4
	8						6	
5	9					3		2
1		8	2	6			9	
		5			3	6		
	4	7			9			3

Easy # 41

	5	3				7		
	4	7	3					6
1			7				4	
3	7			2			6	5
		6	5		3	1		
5	9			4			7	8
	2				8			7
7					2	6	1	
		5				9	8	

Easy # 42

	8		9	5			2	
4	5						7	8
			7		1			
6	1		8		5		4	9
		7		1		8		
8	4		3		9		1	5
			5		3			
1	6						5	3
	2			4	8		6	

Easy # 43

7		1	5	4			8	2
				7		1	6	3
6				9	3		5	
8	1	2						
	4						7	
						8	3	1
	6		3	5				8
9	8	5		1				
1	7			2	8	5		6

Easy # 44

			6	8				9
			7		5	3	2	4
5	7			2			8	
7	5			1	6		3	8
8	2		3	4			6	5
	8			7			9	3
3	1	7	9		4			
6				3	8			

Easy # 45

1	9	6			2			
		3	5			2		
4	5				6			7
	6	9		4		8		
	2			5			1	
		4		8		6	2	
9			1				5	6
		5			3	1		
			4			9	7	3

Easy # 46

3	6		4				1	
				2				6
		1	7	6	5	8	2	3
	8			1				
4		3	5		8	2		1
				4			7	
7	3	8	9	5	1	6		
5				7				
	9				4		8	5

Easy # 47

3	1		6		2	5		8
				4				
		2			1		6	
4		3			7		1	2
	2	6	8		3	4	9	
5	8		9			7		6
	7		4			1		
				7				
6		8	1		9		4	7

Easy # 48

	6		2	8			4	3
		2			1	9		
1			4	6		2		
	1				4			7
7	4			1			2	9
3			6				1	
		1		9	3			4
		6	7			3		
9	3			5	6		8	

Easy # 49

4	6		2			7	9	
9				7	4			8
	3	2		8				
		3				8		1
		6	8		3	9		
8		5				4		
				3		1	4	
5			4	6				2
	8	4			1		7	5

Easy # 50

		7	5	1				
		1	9	2		3	5	
	8			6	7	4		
2							1	
6		3	8	4	1	9		2
	1							8
		5	4	8			3	
	2	6		5	3	7		
				9	2	1		

Easy # 51

1		9			5	3	6	
				8		1		5
	6		1	7			2	
6	9							1
7			9		8			4
8							5	9
	4			3	1		9	
2		8		9				
	1	7	2			4		3

Easy # 52

			4		6			2
	4	9	5		2			3
		5		1				
	6				1	2		8
	5	8		6		1	9	
7		1	9				5	
				3		8		
8			6		9	7	2	
5			2		7			

Easy # 53

	1	8				7	6	
2	6				5			
		9			6			4
1					9	5	7	
	9	3	5		7	4	1	
	7	6	3					2
6			9			2		
			7				3	8
	8	2				6	5	

Easy # 54

2	8	7			6			
	3		4				6	
4		9			8	5		
8	2			9			1	
6				4				7
	9			1			8	6
		2	7			8		4
	4				3		7	
			9			3	2	5

Easy # 55

	8			5			9	
				6	3	8		1
3	9	1	2			5	6	
1				9	5	4		
5								8
		8	7	3				5
	6	7			4	1	5	9
9		5	6	1				
	1			7			3	

Easy # 56

	5	7			8	4		2
8	3		5	2				
2					9	5		3
	2				5		6	8
4								5
1	9		6				2	
3		8	2					9
				9	1		3	6
9		1	8			2	4	

Easy # 57

	5			7				
			8		1	6		
8	4		5		6	9		
1					7	3	6	
5	3			1			7	4
	7	2	4					5
		3	1		4		2	6
		5	6		2			
				9			3	

Easy # 58

		2	6	3				9
8			5		1			4
6				8	4	3		
9		1	7	6				8
	2						1	
7				4	2	9		3
		9	2	1				5
2			8		3			6
1				7	9	8		

Easy # 59

8		3	6	1				
1					9		7	3
	9	2			7		6	1
		9			8	1		6
5								2
7		8	5			9		
9	2		7			5	4	
3	5		1					9
				9	5	3		7

Easy # 60

5		3	9	4				
2		6						3
	1	9	3		2		5	
	5					2		6
			2	5	8			
9		4					1	
	4		8		9	3	6	
7						1		9
				6	4	7		5

Easy # 61

			4	1				
		6			2			4
2	9			5		8		1
7	8			6		3		
4	2	1		3		6	7	8
		9		2			4	5
5		3		7			2	9
8			2			5		
				4	6			

Easy # 62

	1	3	9	6				8
4		7			3		9	
		2			5	1		
6	5					8	2	
8								9
	2	4					6	1
		8	4			6		
	4		1			7		5
1				8	2	9	3	

Easy # 63

					6	9		
3	9	5						7
2				5	9			1
		7			5		8	3
9	3		1		2		7	6
5	1		6			2		
8			2	4				5
1						7	4	9
		3	9					

Easy # 64

9		5				8		6
	8	3			2			
6					7		9	
	6				3	2		9
1		4	2		5	7		3
5		2	7				4	
	1		9					7
			5			9	6	
2		9				4		8

Easy # 65

	9		4	7		5	6	
4						7		
		2	6	8				
1	6		9			8	2	
5	3		7		6		4	1
	4	9			8		7	3
				5	9	4		
		5						6
	1	4		6	2		5	

Easy # 66

6				5	2		9	
	2		3	8				
	8		4	7			1	3
		7						8
	1	5	6	9	8	7	4	
8						6		
7	5			3	1		2	
				4	7		8	
	3		9	6				1

Easy # 67

	2		6				5	
7		6			4			1
8	4	9			5			
	9	4		7			3	
		5		6		8		
	7			3		5	4	
			7			1	9	2
9			8			6		4
	6				2		8	

Easy # 68

				7				
		9	4		1	6		
	1	8	6		9	4	2	
7	5		1		3			8
	8			6			9	
6			8		5		1	3
	2	7	9		6	8	4	
		1	5		8	3		
				1				

Easy # 69

9	4			8	3		7	
	7	2		1	4		3	
5			2			1		
				3	6	7		4
3		6	9	4				
		7			8			2
	2		7	9		6	5	
	5		3	6			8	7

Easy # 70

7			9	6		1		3
	3				8		9	
		5	3	1			4	
4					9	3		
2		1		4		8		5
		8	5					4
	2			9	5	4		
	1		4				2	
5		3		7	2			9

Easy # 71

6		5				4		7
		7			6	3	2	
		8	7	2			5	
		9			3		4	
2	4		5		8		7	3
	8		2			1		
	7			6	2	5		
	6	2	1			7		
5		4				2		8

Easy # 72

				4	1	9		2
		2				3		6
	1		7		3		4	5
1		3					6	
			8	9	7			
	9					4		8
3	6		5		8		9	
4		8				5		
5		9	3	1				

Easy # 73

					9		5	
6	5				8	1		2
1	3			6		4		
			7	3			6	4
7			6	8	4			9
4	6			9	5			
		6		2			1	7
3		1	4				2	6
	8		9					

Easy # 74

	9		7					6
6		3				8		9
			4			1	3	
9		4	1				6	
1		7	8		4	5		2
	5				7	4		8
	6	9			8			
3		5				9		4
7					9		2	

Easy # 75

				3	7		5	8
		8	9		5	6	7	
5							4	9
4	9					8		
			1	8	9			
		6					3	7
7	6							2
	5	4	7		1	3		
8	2		3	4				

Easy # 76

			7		5		1	6
5		1	4	3		8	7	
6		3						5
4		8	3				2	
	5						8	
	3				7	5		4
9						4		2
	4	5		6	2	1		8
3	1		5		8			

Easy # 77

9	2		3	5				
8	4						9	
7		5	4		6	3		
		8					4	3
			6	2	1			
1	5					2		
		2	1		7	8		4
	7						1	5
				3	4		2	7

Easy # 78

				6		9		2
5		4			2	3	8	
	6		8	5			4	
6	7							9
4			9		6			3
8							6	1
	2			3	8		1	
	1	5	7			6		8
7		8		9				

Easy # 79

	6		2	1	9		5	
9	5		7			1		
				4	8			6
				6	7	3	2	
	9	7				5	6	
	1	6	3	9				
1			4	5				
		2			3		1	5
	3		6	8	1		4	

Easy # 80

	4		8			7		
		3		1		8		4
		9	3	4			1	
6	8		9				7	
		1	6		8	5		
	7				3		2	8
	9			3	1	4		
4		7		6		1		
		6			5		3	

Easy # 81

2	4		1	3		6		
9		8			6		5	
7					5			3
	7	6					1	5
		2				3		
3	1					7	8	
6			8					1
	2		4			5		9
		3		7	2		6	4

Easy # 82

1		2	7				5	3
	9		6					
		3		5			2	4
7	3			6	8			
4			3	9	7			6
			4	1			3	7
2	1			3		7		
					6		8	
3	8				9	2		5

Easy # 83

7			6	1	3			2
1	8				2			
		2		7				3
		7	2	8				1
8		4				2		5
3				4	9	6		
4				2		5		
			5				4	6
5			4	6	8			7

Easy # 84

1	4				5			
		5		8				4
3			9	1	4			5
		1	6	4				2
5		8				4		9
7				9	8	3		
8			2	7	1			3
2				3		8		
			8				9	7

Easy # 85

8					9	6		
	4		1	8			5	9
	5		4	3		2	1	
5		1	7	4				
				1	2	4		7
	3	5		7	4		6	
7	6			2	5		9	
		9	3					5

Easy # 86

	1				3	4	9	
2	9						6	8
	8		9	4		1		
	3				9	2		
7		1	2		8	6		9
		6	7				5	
		8		9	1		2	
1	6						8	4
	7	9	4				1	

Easy # 87

5			7	3		9		
		3	4					1
7				1			5	2
	6	8	3			2		
4			6		1			7
		2			9	6	1	
6	5			7				3
2					6	5		
		7		5	3			9

Easy # 88

	9	7		1	6			8
							6	
	6		4	3	2			7
7					1	5		4
6	8		7		9		2	3
2		5	3					6
5			1	7	8		4	
	7							
1			9	6		3	7	

Easy # 89

	4					9	3	8
		8	4	9		2	7	1
	9		2				4	
				4	7	8		
			6		1			
		4	3	5				
	6				9		1	
4	3	9		1	2	5		
5	1	2					9	

Easy # 90

			8				4	
		4		7			2	6
8	9		1	6	4			
9	2		3	4				5
4	6						9	3
7				2	9		6	1
			6	3	5		7	4
5	4			1		6		
	3				8			

Easy # 91

		5	4					9
7	8						4	
9	4		8			6		
4		8		3		7		6
	6		7		8		5	
1		7		9		2		4
		4			3		6	5
	7						1	2
3					2	4		

Easy # 92

		3	9	7		1		
				5			6	9
4	9				6	3		8
	3	4					9	
	7		4		5		2	
	5					6	4	
7		9	1				8	2
5	1			4				
		2		8	9	4		

Easy # 93

3	9		2			4		
6	4				9	5	2	
		2			6			
4	8	5						9
7		9				8		3
1						7	6	4
			5			9		
	5	4	9				3	8
		1			8		4	5

Easy # 94

1				9	2			4
4		6				8		3
			3		8			
9		2	8		7	3		6
	5			6			2	
6		4	2		3	7		9
			5		6			
3		9				2		5
2			7	3				1

Easy # 95

		1						
	4		3	1	8	6		
	3		2	5		1		9
4	7		9				5	
	5	8	1		2	7	9	
	1				3		6	4
1		2		3	5		8	
		5	6	9	7		1	
						5		

Easy # 96

9	5		2			4		7
		4	8					
	3			7		6		5
3		7		6	1			
8			3	2	7			1
			4	8		7		3
1		5		9			7	
					8	2		
7		9			3		5	6

Easy # 97

	1	7			4		6	3
		5			8			
	2	4		1		5		9
				8	6	7		5
		8	4		2	6		
9		3	1	5				
5		6		2		9	7	
			6			8		
2	7		8			3	1	

Easy # 98

			3	5		2	9	6
9	2		8			3	7	
	6		9			8		
	9							3
6	7	8				5	4	1
3							6	
		6			7		8	
	8	1			9		5	2
2	5	9		4	8			

Easy # 99

				2	4	8		
	8	9	3					
			8	6		1		3
	5	8	6			2	1	
6				5				9
	7	4			2	5	8	
4		2		3	6			
					1	3	4	
		7	5	4				

Easy # 100

7			8	2		9		
		6	4		1	5		
		8		6	5			2
1		9	3	8		6		
	7						1	
		3		5	7	2		9
9			7	1		4		
		7	6		2	8		
		1		3	9			6

Easy # 101

	4		1	7				6
1			9		3			2
7				8	4		9	
8				5	1		4	3
		1				7		
4	7		8	2				9
	1		2	3				4
9			6		7			5
2				9	5		3	

Easy # 102

			7	6				3
	7							
6		3		4	1	2		5
3		6						1
9	8	1	3		5	4	2	6
2						9		8
7		9	6	5		3		4
							1	
8				3	7			

Easy # 103

		5	3	9		2	6	
		3	2	1		5		4
1					4		7	
5	2		8	3				
				2	6		3	8
	4		9					5
8		7		6	5	4		
	5	9		8	3	7		

Easy # 104

5	4		3					
				5	8		1	
			6	4		5	7	
9	8		7				5	1
		2		8		6		
3	7				6		9	8
	3	4		6	9			
	9		5	7				
					4		2	9

Easy # 105

5	8			2			9	
		2		5	9		6	
	7				8	5		
		7			6	8		1
	4		8		1		2	
8		3	9			7		
		9	4				1	
	5		2	9		6		
	2			1			7	5

Easy # 106

3	6		4		8		9	
	9	2	6					
7								2
2		3	9		6		5	
	4	6		1		2	3	
	8		5		2	4		6
8								7
					9	5	1	
	2		7		1		6	3

Easy # 107

9	3	2		5	7		8	4
					8	5		
	8							6
	9	4	8				5	3
	7		3		6		1	
6	1				2	8	4	
3							2	
		9	1					
4	6		7	2		9	3	1

Easy # 108

1			2		7	4		
		2		6	1		8	7
	7	9		8				
	3		8			1	7	
		5		1		6		
	1	4			9		5	
				2		9	3	
8	9		4	5		7		
		1	7		3			8

Easy # 109

5	8			4	1			
					7		9	
		1	8	2			5	4
2	6	5		8				
	1	7	5		4	8	3	
				9		5	7	6
6	9			3	5	7		
	4		9					
			4	7			8	3

Easy # 110

				4			7	
		9	8		3			
		7	1		2		3	8
	6	3	2					9
9	7			1			6	2
1					6	7	8	
5	2		9		8	4		
			5		1	8		
	9			6				

Easy # 111

		9	6	4				
3			8			5		
4			9	3		6	2	
		6		5			9	1
9		3	2		4	7		5
5	2			6		4		
	5	2		1	8			4
		8			7			6
				9	6	1		

Easy # 112

9				4	5	1		
		8			3			4
	1	2		8		5		
2					4		6	7
		5	8		6	3		
6	8		9					2
		4		5		6	1	
1			6			2		
		9	4	1				5

Easy # 113

3	5	9		8	4	2	6	
					2			8
		2					1	
6		5	2			8	9	
		4	9		1	7		
	1	7			3	6		2
	9					3		
5			7					
	6	1	4	3		9	7	5

Easy # 114

	9	8		7			3	1
	4		1			9		
				2	6			
		3		1			9	2
1	2	5		8		6	4	7
4	7			6		8		
			2	5				
		6			1		2	
3	1			9		4	5	

Easy # 115

			7		4			
4							7	2
	7		9	3	6	8		5
9		3	2	6		5		7
		7				2		
2		1		8	7	6		4
7		4	6	5	3		2	
8	9							3
			4		8			

Easy # 116

7			5					
		9					5	
	5	3	1	7		4	2	8
	7	2			5		4	3
	6		9		2		1	
5	3		8			9	6	
4	2	6		8	1	3	9	
	8					2		
					6			4

Easy # 117

5	6	7			3			
	4	8			1	5		
1			7					4
8	2			9				3
	1			4			2	
9				3			8	5
2					4			7
		6	8			3	4	
			2			1	5	8

Easy # 118

	9	5	3			1		4
2			7					
	1			4		8		9
1		3		7	6			
		8	1	2	3	7		
			8	5		3		1
5		9		1			3	
					7			6
6		1			2	4	9	

Easy # 119

	3				5	1	4	
4	8		1	3				
1		6			8	7	3	
3					1		8	9
	7						1	
5	2		9					3
	5	2	8			3		7
				5	2		9	4
	4	8	3				5	

Easy # 120

		7			8		3	
				2				
4	8		3		7	5		1
2		4			6		8	7
	7	3	1		4	2	9	
5	1		9			6		3
3		1	8		9		2	6
				6				
	6		2			8		

Easy # 121

	1		3	9	5			6
6		7			2	5		
9	3							2
1				2		6		
	2	8				9	3	
		6		3				4
7							6	8
		2	4			1		5
8			2	7	6		4	

Easy # 122

8	9		2				3	
	7	3				2	5	
5				8	3		4	
7			9				1	
3		9	4		5	8		7
	6				8			4
	5		8	2				3
	8	4				5	7	
	3				6		8	2

Easy # 123

	7							1
					2	8		
2	1	8		7	9		5	6
	6	4	7				2	5
	2		5		1		9	
1	3				4	6	8	
6	4		9	3		7	1	8
		3	4					
5							4	

Easy # 124

3		9	6			1	8	
	4		2					
1				8		7	3	
	1	6		2	5			
		7	1	4	6	2		
			7	9		6	1	
	9	3		1				6
					2		5	
	5	1			4	8		3

Easy # 125

		8	6					
	5			4		9		3
2	9		7			4		5
7		5		6	1			
3			5	8	7			6
			3	2		5		7
5		1			8		9	4
9		2		5			7	
					6	1		

Easy # 126

	7	8			1		5	
1	6	4			2			
9			8					2
4		1		7				3
		2		8		6		
7				3		2		1
8					9			6
			7			5	9	4
	4		6			8	1	

Easy # 127

	9	6		8	5		4	
2	4			1	9		5	
		3	2					1
				5	7	9		4
7		5	6	9				
4					8	2		
	2		4	6			3	7
	3		5	7		4	8	

Easy # 128

3			6	8				9
8		4				3		7
			7		2			
2		1	3		8	6		4
	7			2			3	
4		3	5		6	8		2
			8		5			
1		2				5		8
9				4	3			1

Easy # 129

		2		3	9			4
9		3	5					2
7	4						1	9
		1	9					5
	9	7	4		1	2	6	
8					6	7		
4	3						2	7
2					3	9		6
1			2	9		4		

Easy # 130

9			1		2			4
				9				
5		7	8		6	3		2
	6		2		1	9	4	
		2		6		8		
	5	1	9		4		2	
2		9	6		8	7		3
				5				
8			3		9			6

Easy # 131

			7		5			
	5	6					1	
7	2		8	9	1	3		
4	3			5	2		7	8
2								3
1	6		3	8			2	9
		2	6	1	8		9	5
	7					2	3	
			2		7			

Easy # 132

	8		1					
		5		9			6	4
3		6	2				9	5
2	5			1	7			
4			5	8	2			1
			4	3			5	2
5	7				8	6		9
6	3			5		2		
					1		7	

Easy # 133

				3		9	5	7
7		8	5	6		3	1	
		5		7	4		8	
5	4	3						
	1						2	
						5	3	6
	7		4	9		8		
	5	6		2	7	1		3
3	8	4		1				

Easy # 134

			2		3			
		7				2	6	
	5		7	1	4	8		3
4		3	8	2		5		9
5								8
1		8		4	5	6		7
2		1	4	7	6		8	
	8	5				3		
			3		8			

Easy # 135

1		6				7		
2	5			6		3		
	3			8	4			5
5		9	2	1				
	1	3		7		5	8	
				4	5	1		7
8			7	9			3	
		1		3			5	4
		2				8		1

Easy # 136

1	3		2					
				8	5			3
			3	7		2		9
3	6		7				9	8
		7		6		1		
5	4				8		3	6
8		5		2	7			
4			6	5				
					9		5	2

Easy # 137

		9		3				
	5		4	8			3	
	8		1	6	9		7	5
2				7			9	
5	7						8	6
	6			5				2
1	2		3	4	5		6	
	4			9	6		5	
				1		2		

Easy # 138

5	7	3		6				
		2	7	9			3	
	8	4		1	2		6	5
						5	4	8
		6				1		
4	5	7						
2	3		4	8		6	5	
	4			2	7	3		
				5		4	9	2

Easy # 139

7								
2		5	3		7		6	
9	3		8		6		7	
1			2			5		7
	7	2	4		9	3	8	
4		3			5			6
	1		5		2		3	8
	2		6		3	1		4
								2

Easy # 140

	8		9		4		3	
		4	6	5			1	
	2			1	3	6		
	9	6	3	7			5	
1								3
	4			8	5	1	6	
		9	7	4			8	
	6			9	8	3		
	7		1		2		4	

Easy # 141

6		5			1		2	9
					8	5		
2		4		6			7	
			3	4		6		7
3			6	1	7			8
7		6		8	5			
	6			9		2		3
		1	8					
4	2		7			9		6

Easy # 142

				4				
9			7		3			1
5	3		1		9		2	7
	8	4	3		6	5		
	5			1			9	
		1	5		8	6	3	
4	2		9		1		7	5
3			8		5			6
				3				

Easy # 143

	9		6			1	8	
	1			7		4		5
		3		1		6		
	7	4		5		8		
2		9	3		7	5		4
		5		6		3	7	
		6		2		9		
9		1		3			5	
	3	2			4		6	

Easy # 144

9		8	5			6	4	
		7	9				5	
			4	3		9	8	7
		9						4
7	5	6				1	3	2
4						7		
8	9	3		1	5			
	7				6	5		
	2	5			9	3		8

Easy # 145

			7	1				6
	7							
1		6		9	3	4		5
6		1						3
2	8	3	6		5	9	4	1
4						2		8
7		2	1	5		6		9
							3	
8				6	7			

Easy # 146

				4				5
	9			8	1		6	
	5	4	2	9	6		1	
	1			6		5		
	7	6				1	3	
		5		7			8	
	3		4	1	8	6	7	
	6		9	3			2	
8				2				

Easy # 147

9	5		1		7	2		4
		7	2				1	
				3				
1		2	6				4	3
	3	8	4		9	1	7	
7	6				8	9		5
				6				
	2				3	6		
6		3	8		2		9	7

Easy # 148

6	5	4	1		2			
	9			5		1	2	
		7		9	8			
	6	9	8	3		2	1	
	8	1		4	6	9	5	
			9	6		8		
	7	6		2			9	
			4		7	6	3	2

Easy # 149

		4			9		8	
				2				
3	9		8		4	6		1
2		3			5		9	4
	4	8	1		3	2	7	
6	1		7			5		8
8		1	9		7		2	5
				5				
	5		2			9		

Easy # 150

1		8			4		6	7
2		6		8			1	
					5	9		
			3	5		1		4
5			4	9	1			2
4		1		7	2			
		3	5					
	4			1		7		6
8	6		9			3		1

Easy # 151

6	5			8	9	7		
			7	2	6		8	5
	1				3		2	6
	7	3						8
1						5	7	
7	6		8				5	
2	3		6	9	4			
		1	5	7			6	4

Easy # 152

1	9	3			6			7
		6	2	7			5	
	5		4		1	6		
	8					5		6
	2						7	
4		9					1	
		8	1		2		9	
	7			8	5	4		
5			7			3	8	2

Easy # 153

	7	4		8	3			2
		5			1	9	7	
			2	9	7	8	4	
1		2					8	
	5					2		4
	9	1	7	3	6			
	2	7	8			4		
5			4	2		7	6	

Easy # 154

2	8	6						5
					9	8		
7				6	8			1
		5			6		4	2
8	2		1		7		5	9
6	1		9			7		
4			7	3				6
		2	8					
1						5	3	8

Easy # 155

		3	2	4			8	
	2				7		6	
5			6	9		4		2
8					6	2		
1		4		8		7		3
		7	3					8
3		2		5	1			6
	4		8				1	
	1			6	3	8		

Easy # 156

7		1			6		3	
						7		2
8		6		9	2	1		
			3	2		5	9	
1			9	5	4			6
	9	2		6	8			
		4	2	8		6		7
6		7						
	5		6			3		4

Easy # 157

8			2	4	6	1	5	9
	4	1	3				8	
				5		4		
	9			8				
1		3	6		9	8		5
				3			2	
		6		2				
	7				3	6	9	
9	1	2	7	6	8			4

Easy # 158

8	6		7		5		2	1
				4				
	4		3		2		9	
		5	2		3	9		4
2				5				7
3		6	4		9	2		
	7		1		4		5	
				6				
4	2		5		7		1	8

Easy # 159

9	4		5			8		7
			6			3		
	1	3		7		5		2
	3	8	9	6				
		9	2		5	6		
				3	7	4	1	
8		1		2		9	3	
		6			9			
7		4			6		2	8

Easy # 160

	9						7	3
			3		9			
3			2	8	5	6	4	
	2	8	7	5		4	3	
		3				7		
	7	1		6	3	5	9	
	3	9	5	4	8			7
			9		6			
2	6						8	

Easy # 161

				3				
5	7		1		9		8	4
	9		8		5		1	
6		3	5		2	7		
7				1				9
		1	7		6	2		5
	5		6		7		2	
4	3		9		1		7	8
				5				

Easy # 162

		7	9		3			6
	9			1		7	3	
4			5		7	8	9	
				9		3		7
			8		4			
2		4		3				
	2	6	7		1			8
	5	1		6			7	
7			2		9	1		

Easy # 163

7	5	6	3		8			
		4		7		3	5	
	1			5	4			
	4	2	5	8		1	9	
	7	9		6	1	5	4	
			1	4			3	
	9	7		2		4		
			7		9	2	8	5

Easy # 164

6		8		2	3			9
						6		8
1		9			6		4	
			2	6			7	3
		6	9	4	7	5		
4	7			3	1			
	1		6			8		5
8		3						
5			3	7		2		6

Easy # 165

3		4		9	8	5		7
		6		5	1			
		5	2				4	
				2	3		9	4
6								3
2	4		9	7				
	8				5	3		
			7	8		4		
1		7	3	4		8		2

Easy # 166

		2			3	6	5	
				1	5			3
	3		2	4			1	7
	6		8		7		4	1
				9				
7	9		3		1		2	
9	1			5	8		7	
8			4	3				
	5	6	1			3		

Easy # 167

1	9		5	6		7	3	8
		6	1					
	4							1
6	8				1	9		3
2			4		8			5
9		1	7				4	2
7							8	
					2	3		
8	2	3		7	5		9	4

Easy # 168

9						2		
		2	1	3			6	4
6			2		7		8	
1		8			9	4		
3	4			2			9	5
		7	6			8		2
	7		3		8			9
8	6			5	2	1		
		9						8

Easy # 169

6	5							
	2		6	8			9	4
		1	9				2	5
	3	8		6	1			
9			7	3	8			2
			4	9		8	6	
7	1				9	3		
5	9			4	6		7	
							5	9

Easy # 170

3		4				9		
			2		4			
	2	5	7	6	9			1
	8	1		4	5	2	7	
	5						1	
	9	3	1	7		5	6	
5			3	9	7	6	4	
			5		2			
		2				1		5

Easy # 171

8		6		3			4	1
	9	2	1				3	5
			7					8
5		8	9	7				
9			4		1			7
				8	3	6		2
7					9			
2	3				7	4	5	
6	5			4		8		9

Easy # 172

					7	2		
3	5	1		2	9		4	7
7							8	
5		4	7				3	2
9			3		8			6
6	8				1	7		4
	3							1
8	4		9	1		5	6	3
		5	6					

Easy # 173

6			9	7	3			4
7		3			4			
	4			1				3
	7		8	3				5
4	1						3	9
2				9	1		6	
5				6			1	
			1			9		2
1			5	2	7			6

Easy # 174

1		6		2	3			7
5			8	9		4		
				7	6			8
		7					2	
3	2		7	5	4		9	1
	4					7		
7			2	3				
		1		4	5			6
8			1	6		2		9

Easy # 175

6		8			7			
4			1	6	9			7
	7			4				9
	4		7	8				6
8	3						7	5
9				3	2		1	
3				7			5	
5			3	1	8			4
			5			3		1

Easy # 176

	2	3	7					
4	6		2		8		5	
8								9
6		1	5		3		9	
	4	5		2		6	1	
	3		6		7	4		5
5								8
	7		9		1		6	4
					6	5	7	

Easy # 177

4		6	2		1			7
	7	2	6	5				
	1	3					2	
7						1	3	
			1	7	8			
	6	5						4
	9					4	6	
				3	5	9	7	
5			8		6	2		3

Easy # 178

4		7				9	1	3
							4	8
	6				4		7	2
	4	2		9	1			
	7		4		8		3	
			5	3		7	2	
7	2		6				8	
6	3							
1	5	4				3		7

Easy # 179

6	2		7			3	1	
1					2	5	7	
			3					2
		7				1	9	6
9		5				4		7
4	3	1				8		
7					6			
	1	6	9					8
	5	9			7		6	1

Easy # 180

6				2	5			
2	3	9	4		7			
	5			9			4	2
5	1		2	7			6	8
9	8			3	6		2	5
8	9			1			5	
			9		8	2	1	7
			6	5				4

Easy # 181

3							8	
				3	6			7
7		9		8	4	3		
6		7			2	5	1	
	3	1	5		8	7	9	
	9	8	6			4		2
		6	7	5		8		3
4			8	2				
	7							5

Easy # 182

2	6			1	4			
					3	2	1	
	9		5	2				
	2	9			6	7	5	
4				5				8
	7	5	4			3	6	
				6	2		7	
	8	7	1					
			7	4			3	1

Easy # 183

1			4		5		9	6
	6							5
			7	6	9		8	
	1	8	5			4	7	
	9		8		7		1	
	7	4			6	5	2	
	4		3	5	1			
2							5	
8	5		6		2			4

Easy # 184

	4	3		5		8		9
1			2		8		4	
			9			1		
2	3		4	9				
	1	5				9	3	
				2	3		5	7
		1			2			
	5		8		6			3
8		2		7		5	1	

Easy # 185

		1	3	7		6		
	6			9	4	8		
8							7	9
	8	2			6			3
	9	5		3		1	2	
4			2			7	6	
2	3							7
		9	8	4			3	
		8		6	3	2		

Easy # 186

6	3			5	2	8		
	2				7	1		
				4	1		5	
3		6		1			8	
	9	4	3		8	6	7	
	1			6		5		4
	4		1	8				
		9	2				6	
		8	4	9			1	3

Easy # 187

8	7		5		6			
1			7			6	8	
		4	8		1	7		
	9		1			2		
	4	2				3	9	
		8			2		7	
		9	3		8	1		
	3	7			4			5
			2		5		3	4

Easy # 188

7					4	2	8	
		4			5			7
	1					5	6	
1		6		3		7		5
	8		1		9		2	
9		7		4		8		1
	9	1					7	
2			7			3		
	7	3	9					8

Easy # 189

	8	1			6		3	2
2					5		1	9
6		9	1	2				
		2			1	4		6
3								1
7		5	4			2		
				5	7	9		4
9	6		2					5
5	7		6			3	2	

Easy # 190

1				9			4	2
9	5	6	4		2			
	7			1	3			
6	1		3	8			2	4
3	4			5	6		1	9
			1	6			3	
			5		7	2	6	8
7	6			2				1

Easy # 191

	1		9	5		4	6	8
5			6					9
9						1	5	7
				9	8		1	
			2		4			
	9		7	3				
4	6	3						5
2					5			4
7	5	9		4	6		3	

Easy # 192

7			9		5		4	
	2		7		4	5		3
		4		3		8		7
				1			5	6
			6		2			
1	4			9				
4		1		7		9		
2		9	4		8		6	
	3		1		9			4

Easy # 193

3			6	4			2	5
	6				8	3		
5			7	2		6		8
	2	7	4	9				
				7	2	9	6	
9		4		8	7			6
		5	3				1	
6	3			1	9			7

Easy # 194

5				1	2		4	
3			4					5
	8	2		9	5	1		
	6		2			4		
	3	5		4		2	6	
		4			1		8	
		9	1	7		8	3	
8					6			1
	2		8	3				4

Easy # 195

2	9		7	6				
		4		8		9		
	8	6			3	4	2	7
	1		8	4				2
9								8
8				7	5		9	
4	2	8	1			6	5	
		7		5		2		
				2	6		8	4

Easy # 196

5	3						6	4
			9			8	3	
		1	3					7
9	5		7			6		
1	6		5		9		7	2
		8			2		5	3
8					7	3		
	2	4			5			
3	9						4	8

Easy # 197

	5			3			2	
4	3				5			1
	8	7		6				3
	4			7			8	6
	7	8	6		2	9	1	
6	2			5			7	
7				2		1	3	
5			8				9	2
	1			9			5	

Easy # 198

5		9		6	3			
					4	3		
	6		8	5		4		7
7	8	6		4				
	9	5	3		8	1	6	
				9		7	8	2
3		8		2	9		1	
		4	6					
			1	3		9		8

Easy # 199

					7		8	
7	2	9						3
4				9	5			6
	5				1	3		4
1		2	5		3	8		7
8		6	4				2	
3			7	4				5
2						7	4	8
	7		1					

Easy # 200

3				7	2	6	9	
	5		3			4		8
		7	5			1		
	2	5					1	3
7								6
1	8					7	2	
		2			8	3		
5		4			9		6	
	3	9	6	1				7

Easy # 201

			4	5	1	7	3	
	1	3		6				4
		5			9			
	7			8	2	4	6	
	3	4				2	5	
	2	8	5	3			1	
			9			3		
3				7		8	4	
	9	2	6	4	3			

Easy # 202

		2				7		4
	9		2			8	6	
		8	9	7	4			5
	8			2		5		
4	7						1	2
		3		4			8	
3			8	6	2	1		
	5	9			3		2	
8		1				6		

Easy # 203

9		1		5	4			
					6	7		
	4		1	2		9		5
2	9	3		1				
	6	4	9		5	8	1	
				7		6	9	3
3		7		8	9		6	
		5	7					
			5	6		1		8

Easy # 204

8	9			2		5		6
7					6			
4	1				7	2	9	
				5	1	8		4
6			2		3			7
9		5	6	7				
	6	4	3				1	9
			7					5
5		8		1			2	3

Easy # 205

	8	9			3		1	7
					2	5		
	6	7		9				8
			4	2		8	3	
	2		3	5	8		6	
	3	8		1	6			
3				8		1	7	
		4	2					
7	9		5			4	8	

Easy # 206

		3	6	5		7	1	4
	5					6	8	
	1	7	3					
4	9		5	2				
		8	4		3	2		
				9	8		4	5
					7	3	2	
	8	2					6	
3	6	9		4	1	5		

Easy # 207

6			7	9		1	8	3
	9		3				4	
	3					6	9	7
				6	8			1
			9		4			
2			5	1				
3	8	2					1	
	1				7		3	
7	5	9		3	1			2

Easy # 208

	8		9		1			7
				4		1	3	
7		3	6	2			9	
	6	8			3	2		
	2			8			5	
		1	7			9	8	
	4			5	8	7		9
	3	9		7				
8			4		9		6	

Easy # 209

3	9			4		8		1
2	6				5	4	9	
5					1			
				8	6	3		2
1			4		7			5
9		8	1	5				
			5					8
	1	2	7				6	9
8		3		6			4	7

Easy # 210

2			4	5			8	
	5		7					1
4				8		5		7
	7	9	2				1	
8			9		7			6
	1				4	7	3	
1		5		9				8
9					6		4	
	2			4	8			5

Easy # 211

	3		6	5	1		8	
5					2	6	3	
		8	9	4				
7	1		2	8				
3	8						6	2
				6	7		5	8
				3	4	5		
	5	3	7					1
	4		5	9	8		7	

Easy # 212

4					7	9	1	
		7			6			4
	8					6	2	
8		2		5		4		6
	1		8		3		9	
3		4		7		1		8
	3	8					4	
9			4			5		
	4	5	3					1

Easy # 213

8			1					3
3						2	8	7
	2		3	8		9	1	5
				3	5		2	
			4		9			
	3		7	6				
7	8	3		9	1		6	
9	1	6						8
4					8			9

Easy # 214

	5					9	2	
		9		7	6			4
2			9	3				6
5		3			2		7	
1		2		9		4		8
	9		3			6		2
3				5	9			1
6			7	4		3		
	4	5					6	

Easy # 215

4			8	2	3			9
				5	4	8		
8		5	9				6	
				7	9		3	8
3	5						1	7
6	9		1	3				
	8				1	7		5
		3	2	4				
5			7	8	6			3

Easy # 216

				2				
4	1		7		3	6		8
		3	6				7	
7		6	9				8	2
	2	5	8		4	7	3	
3	9				5	4		1
	6				2	9		
9		2	5		6		4	3
				9				

Easy # 217

4			3	7			8	
6	5			1	4	9	2	
8	3	9		2				
						5	6	9
2								1
3	9	6						
				9		4	7	6
	8	4	6	5			9	2
	6			4	3			8

Easy # 218

	5		1	6		2	8	
7			8			5	4	
	2	8						
9	1			8	6			
		3	9	7	5	8		
			4	1			7	9
						1	2	
	3	2			8			4
	8	6		9	1		3	

Easy # 219

9	6			7		2	3	
				8	4			
	1		2					6
3				2		8	6	
5	8	2		9		7	1	4
	7	1		4				9
4					2		8	
			8	5				
	2	3		6			5	1

Easy # 220

		9	6	2				
6	7		9				3	
2		1		4	3			9
4		2	1		8			7
				5				
3			2		9	1		5
1			8	6		5		2
	9				2		7	6
				9	4	8		

Easy # 221

	3		1		6	9		
				2			4	6
4		9	5	8			1	
3	5				4			8
	8			3			7	
6			9				3	1
	2			7	3	1		9
1	4			9				
		3	2		1		5	

Easy # 222

2	5			1	7		6	
							2	5
6	3				5	9		
			1	5		4	7	
5			6	9	4			8
	9	4		7	3			
		3	5				8	2
7	2							
	8		7	4			5	1

Easy # 223

5	4				3	1		6
		6			4			
3	7		6			5		
2	5	1					3	
	8	3				2	7	
	9					8	5	4
		9			2		1	5
			1			3		
1		5	3				2	7

Easy # 224

		4				6		
9			8		1	3		5
					5		4	9
2			5		9	4	3	
3	4			7			5	1
	1	5	4		2			8
7	2		9					
5		3	7		6			4
		6				8		

Easy # 225

6		9						
1		2	9					8
7	5	3				2	6	
			7	6			2	1
2			5		8			6
5	1			4	3			
	2	5				6	4	3
9					5	1		2
						8		5

Easy # 226

1		6	5		8			
		3				9	4	
4	5			2	9	8	1	
6					7	4	5	
5								8
	8	4	6					9
	1	5	4	6			8	7
	6	2				5		
			7		5	2		1

Easy # 227

4	3		8		6	5		
5				2				
			4		3	7		
	7				8	4		9
9	8			6			7	5
3		5	9				6	
		3	6		1			
				9				7
		2	3		7		1	8

Easy # 228

3	6							
2			5	1			7	
		1		9	3	5	2	
8	2	3	6			7		9
4		6			9	3	1	8
	4	2	9	7		6		
	3			5	2			7
							8	1

Easy # 229

	3		1	5		7		
6	8	9			3			5
		7	2		6		3	
		4					7	3
		1				5		
2	9					6		
	4		6		1	9		
7			5			4	8	1
		5		4	7		2	

Easy # 230

8	1			5		9		
2	5				3	1	4	
					6			2
			7	8			9	5
	7		5	3	9		6	
5	9			6	2			
3			6					
	8	1	9				5	4
		5		4			7	1

Easy # 231

	9			8		7		3
7			4	2		9		
	5						8	1
				1	3		6	7
	7	2		5		1	9	
5	1		7	4				
1	2						3	
		9		6	5			2
4		7		9			1	

Easy # 232

7			8		4	6		
5		4		3	9		1	
	6					4		
	7		5			9	4	
2		8		9		3		6
	4	1			6		2	
		6					9	
	9		1	8		2		5
		5	9		7			4

Easy # 233

		3	2	9	5		1	7
			3		4			
4						3		8
2	9		8	5			7	3
	3						8	
8	6			1	3		5	4
1		2						9
			4		1			
3	4		5	7	9	8		

Easy # 234

2			1		9		6	5
	6							1
			8	6	3		2	
	1	6	9			2	7	
	8		7		5		4	
	7	2			6	5	8	
	5		4	9	7			
6							9	
9	4		6		2			8

Easy # 235

5					1	2		8
		6	9	2		1	7	
				5	3		9	
		4	1		5	7	6	
				7				
	1	3	6		9	8		
	5		2	1				
	6	1		3	4	5		
8		2	5					4

Easy # 236

		4		5	1		8	2
9		2		6				
	8		9		4	3		
5			2			1		3
		7		3		5		
4		3			8			9
		1	4		6		3	
				8		2		4
8	4		3	7		6		

Easy # 237

				8		5	4	
		6			4		3	8
3	8			5	1			2
		7			3			
	5	1	9	7	8	6	2	
			4			9		
6			8	4			5	7
8	9		7			4		
	4	5		6				

Easy # 238

3		7				6	9	8
						4	7	
	1				7	5	3	
5	7			8	9			
	3		7		4		6	
			2	6			5	3
	5	3	1				4	
	6	1						
7	2	9				3		6

Easy # 239

	5	7		2	1	3	4	
			9	5			7	
9								
	7	5					1	
6	8	1	7		4	2	5	3
	3					8	6	
								1
	6			7	9			
	9	8	5	4		7	2	

Easy # 240

4				8	7	2		
2	8		4	6		5		9
				9		4	8	1
9	4	7						
		5				3		
						9	6	4
7	9	2		5				
6		4		3	8		9	5
		8	7	1				2

Easy # 241

			7				4	3
6				1		7		
7			6	3	2			1
3				4	7	1		
5		7				9		4
		2	8	9				6
1			4	2	9			5
		5		7				9
2	9				5			

Easy # 242

3		5	8		4	9		7
				1				
6			3		2			1
	6	1	2		3		8	
		4		8		3		
	3		6		1	2	7	
8			1		5			4
				7				
5		9	4		8	1		3

Easy # 243

6	2			1	5	7		
			2	3	6		9	4
5					8		1	6
1		5					7	
	8					4		1
9	6		4					7
8	5		6	9	1			
		1	3	8			6	5

Easy # 244

		2	3	7	9		5	
		1	5	6			4	2
	5							
8		9	6			2		
	3	7	4		2	5	1	
		5			7	3		8
							2	
7	2			5	4	6		
	9		1	2	6	8		

Easy # 245

			4	5	8			1
		6	3		9	5		8
5						9		
6	1		9				3	4
8			1		4			6
4	3				5		9	7
		7						9
9		1	5		7	3		
3			2	9	6			

Easy # 246

	4		2		8			3
		3						8
8	6			7	5	9		
		4	6			8		5
2	1			5			3	7
9		8			3	1		
		5	9	2			6	1
3						5		
6			5		4		8	

Easy # 247

	6			5				
			3		9	2		
9	3		1		8	6		
2					1	3	4	
1	4			8			6	2
	9	6	4					8
		5	9		2		1	7
		9	8		7			
				4			2	

Easy # 248

	8			5		7		6
7				4				9
	9		1				5	4
5	2			9				8
8		1	2		5	4		7
3				8			2	1
6	3				9		7	
9				6				5
1		8		2			6	

Easy # 249

6			7		5			2
9	3		4		1		8	5
				6				
		1	5		7	2	6	
	5			1			4	
	7	9	6		2	5		
				9				
5	6		1		4		3	8
4			8		6			1

Easy # 250

4			8				9	
	5		6	2				
2			5	4		7	6	
	6			9		5		1
5	4		7		2		3	9
9		7		6			2	
	7	9		1	8			2
				5	6		1	
	8				3			6

Easy # 251

7				4	9		8	
	3		7	8				2
4			1		6			9
2	6		5	7				4
		3				6		
5				9	3		2	8
3			4		8			7
6				5	2		4	
	2		3	6				1

Easy # 252

		8	1		9			6
6	1			3			9	
2	9		6		4	7		
1		6		9				
			7		2			
				1		5		7
		2	3		6		5	8
	6			8			4	3
3			9		5	6		

Easy # 253

	3	6		2		4		1
	5	9			8		3	2
		8			4			
				1	5	9		6
		4	2		7	8		
1		3	4	8				
			8			1		
9	4		7			3	5	
6		1		5		7	2	

Easy # 254

		4	6		1			9
3			7		4	8	6	
	6			5		4	1	
				6		1		4
			8		3			
2		3		1				
	7	5		9			4	
	2	9	4		5			8
4			2		6	5		

Easy # 255

		1						9
3			5	6				
	4		1	9			5	8
	5	7	4				3	6
	2	8	9		5	7	1	
4	1				6	2	9	
1	7			5	3		8	
				8	4			1
8						5		

Easy # 256

5	4	9		1	6	7		
	3	8					4	
					2	5	8	
				9	3		1	7
		3	1		5	8		
1	9		7	8				
	6	2	5					
	7					4	3	
		5	4	7		2	6	1

Easy # 257

	1	3				5	8	
	5		4	1			6	
			8		7			
	7	9	5		1	4	3	
8				7				5
	3	5	2		4	1	7	
			1		2			
	6			3	5		9	
	9	7				2	1	

Easy # 258

4				9	1			3
	8	1		2				
3		5	8			7	1	
	1						3	7
	6		2		7		9	
8	7						2	
	5	6			4	9		1
				7		2	4	
7			1	5				6

Easy # 259

8					5		4	2
	4			6				
3	6	2	8	4	7	9		
				5				6
	5	3	4		2	1	7	
2				7				
		7	6	9	4	2	3	1
				1			9	
9	3		5					7

Easy # 260

			4		6			
	1	2	9	5	8			6
6		7				4		
	2	6		9	7	8	5	
	7						6	
	9	4	6	1		7	3	
		5				1		8
7			5	2	9	6	4	
			1		4			

Easy # 261

			5		4	2	9	8
5	4			8				3
			6	3			1	
4	5			7	6		3	2
8	3		2	9			4	6
	6			2	3			
3				5			2	1
7	2	5	1		9			

Easy # 262

		7	1	5	8	4	2	
		1	7	6		5		
4				2				
	4			1		7		
	7	9				3	1	
		6		3			4	
				8				6
		8		9	5	1		
	1	3	6	7	2	9		

Easy # 263

		2		4	6		9	
9	3						6	7
	6	4	5				2	
		7	6				5	
6		3	9		7	2		8
	1				8	3		
	2				4	6	8	
4	9						3	2
	7		2	6		9		

Easy # 264

9	7			6				
8				5	3	4	9	
		4	7		8			2
	5		9				2	3
1				2				5
2	8				4		7	
3			8		6	2		
	4	8	2	1				6
				4			8	9

Easy # 265

2		3				9		
	6	5	7	4	8			2
			9		2			
	5	2		7	3	8	4	
	3						2	
	7	9	2	6		3	1	
			6		9			
3			4	5	7	2	9	
		4				6		8

Easy # 266

			5	6		3		
	5							
3		6		9	7	8		2
6		3				7		
7	1	4	3		8	6	2	9
		2				1		4
4		5	6	8		9		3
							7	
		1		3	5			

Easy # 267

5	9			7		6		
4		7			2	1		
1				4				9
6				1		3		7
9	4		7		3		2	6
2		3		6				8
7				5				1
		9	1			8		5
		5		3			6	2

Easy # 268

		1	5		6	2		
7			1	8		4		
		8		9	7			5
		9		3	1	6		7
	1						8	
8		7	9	2		5		
1			2	6		7		
		2		5	3			6
		5	4		8	3		

Easy # 269

			6		1			
	8	6	3	7	4			2
5	1						4	
	2	9		1	8	3	6	
		8				2		
	5	4	2	3		7	8	
	6						2	8
8			5	4	3	1	7	
			8		6			

Easy # 270

3	2			5		6		1
				7	4			
8			6				3	
	1			6		7		3
7	9	6		2		5	4	8
5		8		4			2	
	4				6			7
			7	9				
6		1		3			8	9

Easy # 271

		7	4	2		9	1	
	5		9	7				4
4					5			6
		5			9		3	
	3	9		5		4	6	
	1		7			5		
7			3					1
5				6	1		9	
	6	1		8	7	2		

Easy # 272

		2			1	9		
7	4	9				1		
5	1	8		9	4		7	
	5		8	7				
			2		9			
				5	6		3	
	3		5	1		6	4	9
		5				8	1	3
		1	4			5		

Easy # 273

5		4						
	3			6	1	8	7	
6			7	9		2		
4	2	5	1				3	8
1	6				3	7	2	4
		6		5	9			7
	9	7	2	1			5	
						3		2

Easy # 274

			2			8		
		7			6		2	4
2		4	8				6	3
	7					1	4	5
	1	8				6	3	
6	4	2					8	
4	5				8	2		9
8	3		9			4		
		9			5			

Easy # 275

				8	3	1		
4		1		7	6			8
		8					7	
1		3			9		5	2
5	8		2		7		4	1
7	4		3			9		6
	1					2		
3			1	2		8		7
		6	7	9				

Easy # 276

		5	1		8	2		7
7						1		
			9	7	6			5
1	7		8				5	3
9			3		2			4
3	5				7		2	9
2			4	8	3			
		7						8
4		8	7		5	9		

Easy # 277

	3	6				5	9	
			9		5			
	2			1	7		3	
	1	7	5		8	9	6	
4				6				7
	6	3	7		9	8	1	
	7		8	9			2	
			4		6			
	9	1				7	4	

Easy # 278

			2	4		9	1	3
		1	9					8
	9	3	8			5		2
		9					2	
8	1	5				7	6	4
	2					1		
6		8			9	4	3	
1					5	8		
9	3	4		7	8			

Easy # 279

5							6	8
	4		5	9	1	3		7
			8		3			
3		1	7	8		2		4
		4				7		
7		9		1	4	5		6
			3		7			
9		8	1	5	6		7	
4	7							3

Easy # 280

				7			5	
		8	4		3			
		5	6		9		3	4
	1	3	9					8
8	5			6			1	9
6					1	5	4	
2	9		8		4	7		
			2		6	4		
	8			1				

Easy # 281

		9	3	6				5
5	7						4	
		3		2	5	7		
	6		7			4		2
9		1		5		8		7
3		7			2		5	
		8	5	4		2		
	3						9	4
2				9	6	3		

Easy # 282

3	1						6	5
	5				3	7	9	
	4		5	7		1		
	8				9	6		
7		6	1		4	5		9
		4	7				2	
		5		3	7		1	
	7	3	2				5	
1	6						7	4

Easy # 283

7	1	4		3				
6		9		8	5	3	1	
5			4	2		7		
						6	9	1
3								8
4	6	1						
		6		5	4			7
	5	7	6	9		1		3
				1		2	5	6

Easy # 284

		8	4	1		6		
	2	3				8	7	
			2		3			
	3	7	5		2	1	4	
4				7				9
	5	1	3		4	7	8	
			7		9			
	4	9				3	1	
		6		3	5	4		

Easy # 285

9		6		3				8
4		3	8		6		2	
	8		4		1	6		
	4	7		5				
			2		7			
				1		5	8	
		8	1		5		3	
	7		9		8	2		1
1				6		8		5

Easy # 286

5			7	4	9	8		
6	8			1	5	2		
								5
		8			1	7	3	
2		5	8		6	4		9
	3	9	4			5		
8								
		1	6	5			4	8
		3	1	8	2			7

Easy # 287

9		2		1	6	5		7
			5	9		3		
	8							
7		3				4		
2	4	6	1		9	7	3	8
		8				9		6
							5	
		9		6	5			
4		1	8	2		6		9

Easy # 288

5		7	4		6		3	
1		6		7				4
	4		5		9	6		
	5	2		8				
			3		2			
				9		8	4	
		4	9		8		7	
9				6		4		8
	2		1		4	3		9

Easy # 289

6				9		4	7	
2			3			6		5
		1		6		3		
9		4		7		5		
	8	2	1		9	7	4	
		7		3		1		9
		3		8		2		
1		8			4			3
	2	6		1				7

Easy # 290

1	2	3		6	4		5	
6	4	5						2
9					2			6
	3		1	5				
			9		6			
				3	7		8	
2			4					3
3						8	2	1
	8		3	2		6	4	7

Easy # 291

4	9					2	1	6
		8			5	3		4
						9		5
	4	3		9	2			
		9	8		1	4		
			6	7		1	3	
8		1						
3		4	1			5		
9	7	6					4	1

Easy # 292

		9		5	2			3
	3					4	5	
8			1	4				9
7		3			9		1	
6		5		1		7		8
	2		7			9		4
3				9	1			7
	7	1					4	
5			3	2		1		

Easy # 293

		8	2				9	
				6	8			
	3	6		1		2		7
	5			9		4		3
4	9	3		5		8	6	2
8		1		2			7	
2		7		4		1	5	
			9	8				
	1				2	3		

Easy # 294

		3			2		4	6
4	6			7	5		8	
				4		7		2
		9			6			
7		5	1	9	4	3		8
			2			1		
2		7		3				
	3		4	2			9	7
1	4		9			2		

Easy # 295

4	6		3					7
		5	8	1			6	2
1	2		6	4	5			
	1					3		5
5		2					7	
			9	8	6		4	3
6	9			5	2	7		
2					1		5	6

Easy # 296

2				1		7	9	
		3	8					
7	4		6			1	2	
	6	2		8	5			
	9		2	3	6		8	
			9	4		2	6	
	2	5			3		1	7
					8	5		
	7	4		2				6

Easy # 297

	3		5	8		9	1	
		9	2	3				
8						3		
4	6		7			2	9	
1	9		8		6		4	3
	5	7			2		8	1
		6						9
				7	8	5		
	8	3		6	9		2	

Easy # 298

						4		
	1			2	3			
8	3		5	9			6	2
	7						1	8
4	8	1	2		9	7	5	6
5	2						4	
2	5			6	4		9	7
			3	5			2	
		3						

Easy # 299

	5	8			7	6		3
	7				2			
	9	6		3			2	4
				4	8		5	9
	2		3		1		7	
4	6		2	7				
9	4			8		3	1	
			7				4	
5		2	1			8	6	

Easy # 300

3		7	1		6		9	
			7		3		8	
		9		4				
8					1	5	7	
1		5		6		9		8
	9	3	5					6
				5		8		
	3		6		2			
	4		3		8	1		2

Easy # 301

9	5		4			7		
	7	8		2				
2			9	7			8	4
			7			5		
	8	3	5	4	9	2	1	
		4			6			
6	9			8	3			1
				9		8	7	
		2			7		6	9

Easy # 302

3	5				9	1		
	7				6		8	
	9	8	1	4				2
4		6				7	2	
2								1
	3	7				4		8
8				2	7	9	1	
	2		3				4	
		3	8				5	6

Easy # 303

4		2			8		7	
8	6		7	3		9		
1					5			6
	5	3					1	9
		9				7		
2	1					6	3	
9			2					3
		6		9	1		8	7
	2		6			5		4

Easy # 304

3		5		2		6		
	2			3	6	9		
		7			5		3	
	7				9		5	4
		8	5		4	2		
5	1		6				7	
	6		8			4		
		3	2	6			9	
		2		4		7		3

Easy # 305

	4		6	8			3	2
		5			3	1	8	
				5	7			6
	9		3		5		2	4
				2				
3	7		4		6		1	
5			8	3				
	8	1	5			9		
4	3			7	9		5	

Easy # 306

		2		1		7		4
	6		9			1		
		4	2	6			3	
5	8		6				7	
		9	5		1	2		
	7				3		5	1
	2			4	6	3		
		7			5		4	
4		5		2		6		

Easy # 307

3		5			8	1	6	
8					9			
2		1		6			4	9
				4	5		2	3
9			6		7			8
1	4		9	8				
4	2			5		6		7
			8					4
	3	9	7			5		1

Easy # 308

		2	4	6				
	5							
4	9			5	3	6		8
1			2				8	6
9		4	5		6	2		1
3	2				7			5
2		3	9	8			1	4
							6	
				2	1	8		

Easy # 309

	7		4	3			2	
		2		5	6		8	
8						3		5
	9	8			2			4
	1	5		4		9	7	
6			9			2	3	
9		4						3
	5		8	6		4		
	8			2	4		9	

Easy # 310

			6				2	
	5				1	6		7
6	7		2			1		9
		5				7	3	8
	2	3				9	1	
1	6	7				2		
7		8			2		6	4
2		9	4				7	
	4				8			

Easy # 311

	9	1		4			8	6
			5				9	
3		7	8				2	4
	2	9	3	5				
	3		6		8		5	
				9	4	1	7	
4	7				5	6		2
	5				3			
2	1			6		9	3	

Easy # 312

	7			6	5	4		
5	4		8	1				6
						8	3	
7		1			3	2	4	8
8	6	2	1			9		3
	2	6						
3				7	1		9	4
		7	4	5			8	

Easy # 313

			8	9		1	4	2
1			2				3	
4		2	3				8	5
2						8		
5	3	1				6	9	7
		8						1
3	6				2	4		9
	1				5			3
9	2	4		7	3			

Easy # 314

	3							
8		7		1	6	4		9
			4	8		5		
9		5				2		
7	2	6	1		8	9	5	3
		3				8		6
		8		6	4			
2		1	3	7		6		8
							4	

Easy # 315

	4	6		7	2		8	3
1		8	4		3			
9							6	2
		1			5		4	6
		4				3		
6	3		1			2		
7	1							4
			5		4	8		7
4	8		6	1		5	3	

Easy # 316

1	2							
8	9	4				5		2
3	5		2				6	
			9	4		3	2	
	8		1		2		5	
	3	5		8	7			
	1				6		3	5
5		8				2	7	9
							8	6

Easy # 317

	3					6	2	
5					7	8	9	
		7			6			5
3		2		1		5		6
	9		3		4		8	
4		5		7		9		3
8			5			1		
	5	1	4					9
	4	3					5	

Easy # 318

	9		5	3				1
		3	1	2		5	6	
1					9			4
		9			5		7	
	7	5		9		1	4	
	6		3			9		
3			7					6
	4	6		8	3	2		
9				4	6		5	

Easy # 319

6			1				8	4
	1					7	5	
	8		6	5	7	9		
8				1			9	
5		7				1		2
	3			7				8
		3	8	4	1		2	
	2	8					4	
9	6				3			1

Easy # 320

	8	9			3		5	7
	5	2		7		4		6
		3			4			
				6	8	9		2
		4	7		1	3		
6		5	4	3				
			3			6		
2		6		8		1	7	
9	4		1			5	8	

Easy # 321

			5			2		6
	8	6				1	4	
4			3				8	
	4	5	2					8
	2	3	1		5	9	7	
9					3	5	1	
	3				4			7
	6	9				4	5	
8		4			1			

Easy # 322

1		4		2	3	8		
				1			2	9
	7				9	1		4
	6				4			
2	3		5	6	1		7	8
			9				5	
5		1	6				9	
9	2			7				
		7	1	9		6		2

Easy # 323

		6	1			7		9
		9		5			3	8
4				9				1
8		5		3				7
6	2		4		5		8	3
3				1		5		4
1				2				6
9	6			4		3		
2		4			8	1		

Easy # 324

	6		1	9		7	4	5
1						6	9	8
9			4					1
				1	5		6	
			2		7			
	1		8	3				
2					9			7
7	4	3						9
8	9	1		7	4		3	

Easy # 325

				7			4	3
		7	2	9		8		
9	8				3	2	6	
7		1						4
8			4		7			6
2						7		5
	9	5	1				7	2
		3		6	2	5		
1	2			4				

Easy # 326

			9					8
9	3			1		2		4
	1	2	8			5		7
7	4		5	3				
8			6		1			9
				8	9		3	2
2		5			6	9	7	
6		1		5			4	3
3					8			

Easy # 327

					6	2		3
3	7	2		4	9	6		
5		9						4
				8	4		7	1
		8	6		7	5		
7	4		5	1				
9						8		5
		4	3	7		1	6	9
8		6	2					

Easy # 328

3	9		4	1		6		8
7				2	5	4		
				3		7	9	5
6	8	9						
		1				3		
						5	6	9
2	4	6		9				
		7	5	4				6
9		3		8	6		4	7

Easy # 329

			1	9				3
		1						
9	3			4	8		7	6
3	9							8
2	8	5	3		6	7	4	9
7							2	5
1	2		9	6			3	4
						8		
5				3	1			

Easy # 330

7	5	1			4			
9			1					3
	3	2			9	7		
2	6			8				4
	9			3			6	
8				4			2	7
		5	2			4	3	
6					3			1
			6			9	7	2

Easy # 331

	8							
3	9		5		8	2		
	6	5	4		2	8		
	7		9				8	3
9		8	1		6	4		5
5	1				3		2	
		7	3		9	5	4	
		9	2		5		1	7
							9	

Easy # 332

	7		2				6	
1				7	8	5	4	
		2	1				9	3
	2	8				6		1
7								4
6		3				8	7	
2	9				5	4		
	5	1	4	6				7
	8				3		1	

Easy # 333

	9				4		8	2
5		2			6		9	3
	6	8	2	9				
		9			2	7	6	
	3						2	
	1	4	7			9		
				4	1	8	7	
1	4		6			3		9
6	8		9				4	

Easy # 334

1		8		5			4	
4	2						9	
		5		7	3			2
3	4		8	1				
	8	2		3		4	5	
				4	9		7	8
8			1	2		5		
	3						6	4
	5			6		8		9

Easy # 335

	2	6		1	9			
					7	8		
9			6	4		2	1	
2	4	3		6				
7		9	2		1	5		6
				8		7	3	2
	3	8		5	2			7
		1	8					
			1	7		6	5	

Easy # 336

2			7	1			9	
9				6	8			5
	6	1				2		
		8	9				2	4
5	4			8			1	3
6	9				4	7		
		6				4	8	
4			8	9				2
	8			7	2			1

Easy # 337

	9			6	2		3	
	2	8	5			9		6
5		4		3				
4							7	3
8			3		4			9
1	3							2
				4		2		7
2		3			7	6	1	
	1		2	8			5	

Easy # 338

	7	1				5		
					3	2	9	
6	9	2		5	7		3	
				4	5	8		6
	4		3		6		1	
5		6	1	8				
	5		2	6		7	8	3
	3	4	9					
		7				1	4	

Easy # 339

			5	1	2	8	3	
		1			6			
	2	3		4				5
	8			9	7	5	4	
	3	5				7	1	
	7	9	1	3			2	
3				8		9	5	
			6			3		
	6	7	4	5	3			

Easy # 340

1	5	6				4		2
	4	9						
	2	7	4				3	
			5	1			4	7
	6		9		4		2	
2	7			6	8			
	9				3	2	7	
						3	6	
6		2				5	8	4

Easy # 341

2				8	4			3
3		6	5			9	4	
	5	4		1				
	4						3	9
	7		1		9		8	
5	9						1	
				9		1	2	
	6	7			2	8		4
9			4	6				7

Easy # 342

		8	5	9				
7	5			8	3	6		9
	2							
1			4				9	6
5		9	2		1	7		3
2	8				9			5
							1	
8		2	6	1			3	7
				2	7	9		

Easy # 343

1		5						3
9		4	2				6	
	6		5			2		
	5	6		7		1	3	
4			8		3			9
	3	9		2		6	8	
		7			6		4	
	9				8	7		6
6						3		8

Easy # 344

8			3	1		4		
		1	6					5
3				5			8	2
	7	9	1			2		
6			7		5			3
		2			4	7	5	
7	8			3				1
2					7	8		
		3		8	1			4

Easy # 345

					6	8		4
4	9	8		3	1			6
1		5				3		
				2	3	7	9	
2			6		9			5
	3	9	5	7				
		1				5		2
3			8	9		1	6	7
6		2	4					

Easy # 346

6	1	5					2	8
4		8						
7		2	8			3		
			5	1		8	7	
		6	4		8	2		
	2	7		6	9			
		4			3	7		2
						6		3
2	6					9	8	5

Easy # 347

	8	1		6	7			
	4		9	1				
					2		6	1
4	1				8		9	5
		7		9		3		
9	5		7				8	2
5	3		6					
				8	1		5	
			5	7		6	2	

Easy # 348

		9						
4	2			1	3		6	7
			6	4			5	
7	5						8	
2	3	8	1		4	5	7	9
	9						4	3
	4			3	6			
8	1		9	2			3	4
						6		

Easy # 349

					3	1		6
2		7				9		
6	4	1		9	2			3
				5	9	8	4	
5			3		4			7
	9	4	7	8				
9			1	4		2	3	8
		2				7		5
3		5	6					

Easy # 350

9		8	6	4	1			
	1		2	8		9		6
6		4	5			3		
8						1	5	
	9	1						3
		9			8	6		1
7		6		1	9		3	
			7	2	6	5		4

Easy # 351

8	3							
1			9	2			4	
		2		6	8	9	1	
5	1	8	3			4		6
7		3			6	8	2	5
	7	1	6	4		3		
	8			9	1			4
							5	2

Easy # 352

			3			4		1
7				8			3	
3			7	1	9			8
1				4	3		8	
6	3						2	4
	9		5	2				7
8			4	9	2			6
	6			3				2
9		2			6			

Easy # 353

				4	7		9	
6	1		5			3		
9			8	3	1			6
				9	5	2		8
1		5				6		9
3		9	2	1				
2			9	7	3			4
		8			2		6	3
	3		4	6				

Easy # 354

	2	7	9		6	8		5
				3				
6					2		9	
7		3			4	6	2	
9	6		8		7		1	3
	8	5	1			9		4
	4		3					2
				4				
8		9	2		1	4	3	

Easy # 355

		3		2	7		1	5
	1		8		3	4		
8		5		6				
2			5			7		4
		9		4		2		
3		4			1			8
				1		5		3
		7	3		6		4	
1	3		4	9		6		

Easy # 356

8			4			6		3
		6	8	2	5		7	
		4				2	5	
6				4		7		
2	5						4	1
		9		5				6
	6	1				3		
	9		6	3	4	1		
7		8			9			4

Easy # 357

4	7			6		3		
	3				9			
			3	4	2		8	9
1				3	5		7	8
5	8						4	3
2	4		8	7				6
3	6		1	5	4			
			9				5	
		4		2			3	1

Easy # 358

	7	1			8	4		3
	6	3		1				5
					9		7	
			2	6		5	1	
		2	1	8	5	9		
	1	5		9	7			
	8		9					
1				4		2	3	
3		6	5			1	4	

Easy # 359

						5		9
3	5					7	8	2
		1			9	4		3
	3	4		5	7			
		5	1		8	3		
			2	6		8	4	
4		3	8			9		
5	6	2					3	8
1		8						

Easy # 360

1		3						7
			4		5		8	2
4		8	1	9		5	3	
3		5	6				2	
	4						5	
	1				2	4		3
	6	4		2	3	8		5
9	8		5		6			
5						2		9

Easy # 361

		8	6			2	1	
	7			1			6	
		1		4			9	3
	9	4		3			2	
5	8		7		4		3	9
	3			6		4	7	
8	1			7		3		
	6			5			8	
	5	7			9	6		

Easy # 362

	5					2		
	7		2		1			4
		2	6	9			3	7
	6	4			5	3		
3	9			2			8	5
		1	7			4	2	
7	4			8	2	6		
1			9		4		5	
		5					4	

Easy # 363

	9		2		1	6		
5			9			3	8	2
		6		5	4		9	
9	6					7		
		5				4		
		2					3	1
	1		6	7		5		
4	8	7			5			6
		3	4		2		7	

Easy # 364

	1	4		5	2			7
		2	8	6	3			1
						2		
1					5		9	8
2		7	1		4	3		6
3	9		6					2
		1						
9			5	1	7	8		
5			4	2		1	6	

Easy # 365

	5			2	7			
6	7		1	8			3	2
						4		
	9						5	6
4	6	5	2		8	9	1	3
1	2						4	
		7						
2	1			3	4		8	9
			7	1			2	

Easy # 366

7		5						8
	9	3		7				4
	4			6	2	9		
1		9	3	5				
4	5			8			6	9
				2	9	8		5
		6	8	1			4	
5				4		2	9	
3						5		6

Easy # 367

		2				3		6
			6		2			
6			9	4	7	1	8	
	4	9	3	7		6	1	
	6						3	
	5	3		8	6	2	7	
	2	6	7	1	4			3
			2		8			
9		8				4		

Easy # 368

		4	8	9	7			3
	5	9		6	2			7
						9		
6					5		3	1
5		1	2		9	8		6
4	3		7					9
		6						
8			6	7		2	9	
9			1	5	4	6		

Easy # 369

	7	9				8	1	
			6				7	4
5			7			3		
	9	6	3					1
	1	5	9		6	2	3	
4					2	7	9	
		4			3			7
8	2				9			
	6	7				4	8	

Easy # 370

				6		8		4
	3		8	9			5	
8		2			4	7	3	
3	2							8
9			2		6			1
6							4	2
	8	9	5			1		7
	1			7	8		2	
5		6		2				

Easy # 371

		7	8	3			9	
5		9				4		8
		8			5	1	3	
		2			1		4	
3	4		9		7		8	1
	7		3			6		
	5	3	6			8		
9		4				3		7
	8			5	3	9		

Easy # 372

		8	1		3		2	7
	3		7		5	6		
7				9			3	5
				7		3	5	
			2		8			
	8	4		5				
1	9			6				3
		3	4		7		9	
4	6		3		9	2		

Easy # 373

	4			9	6			
8	3	2	1		7			
9				8			1	7
2	9		6	5			7	1
6	1			3	2		9	8
4	2			7				9
			3		4	7	2	5
			9	2			6	

Easy # 374

			3	1			5	9
7			6					
5	1			2	9	3		
				9		5	2	8
4		9	1		5	6		3
6	8	5		7				
		6	5	4			8	7
					7			1
9	4			6	1			

Easy # 375

	2	3	5	9				8
4	1				8	6		
	7				6		9	
8		7				5	6	
2								9
	9	5				4		7
	8		4				5	
		2	3				1	6
9				7	2	8	3	

Easy # 376

6	9				1			
		1		4				3
4			7	6	3			1
		4	1	9				6
9		8				1		5
3				8	2	7		
5			8	7	9			4
8				1		5		
			5				8	7

Easy # 377

	6	1	3	7			9	
						8		3
		5		9	6			1
7	5				8	1	3	2
2	3	9	7				8	4
5			1	6		3		
9		2						
	8			5	7	4	1	

Easy # 378

3			4	5				
					2	5		1
7	5			1	6			
5		3			7	8		4
	6			4			9	
8		4	6			2		7
			8	6			1	2
9		8	1					
				7	5			8

Easy # 379

		3	8		6	5		4
5						6		
			2	5	4			7
3	7		6				8	2
4			7		2			3
2	8				5		6	1
8			9	6	3			
		1						6
6		7	5		1	8		

Easy # 380

		5	8					
	3							8
8	9		2	5		1	4	7
5	7				8	9		4
6			3		7			2
9		8	1				3	6
7	6	4		1	2		9	3
1							7	
					6	4		

Easy # 381

4	2				5	7		
							4	6
1	5			8	6		2	
			7	6		8	9	
2			8	9	3			5
	6	8		5	1			
	3		6	1			5	4
5	4							
		9	5				7	3

Easy # 382

	5							4
	4		5		3	8		
2			1	6		9	5	
5	1				9			8
	6	4		1		7	3	
7			4				2	5
	7	9		3	2			1
		5	8		1		9	
1							4	

Easy # 383

4	8	2		1	5		7	
					9	3	4	
	3	6				2		
				8	6	1		7
	6		1		4		3	
1		8	7	3				
		7				6	2	
	9	5	4					
	4		2	7		5	9	1

Easy # 384

	5		8		1	2		6
		8	4		6		1	
1				2		8		9
				3		7	6	
			7		5			
	1	3		4				
3		1		8				4
	2		3		4	1		
4		5	1		9		7	

Easy # 385

5								
	1		5	7		4		6
	4		9	2	8			5
	8	3	7				4	
9	2		6		4		5	1
	5				2	3	9	
8			1	4	7		3	
4		2		5	6		7	
								4

Easy # 386

	1	8	7			6		
		4	2	6	8	7	3	
2					4	5		
4		9		1	3			
		2				3		
			4	7		9		6
		3	8					5
	4	7	5	9	6	1		
		5			1	8	7	

Easy # 387

	2	5	3	8				
4				9				5
8		9			6	2	3	4
		7	9	4			2	
	5						9	
	9			3	1	5		
9	4	2	7			1		8
3				1				2
				2	8	9	4	

Easy # 388

3		9		5	8		4	
			4	2	3	5		9
		1			7	2		3
	7	4						5
1						4	9	
4		3	5			9		
2		7	3	8	6			
	1		9	4		3		6

Easy # 389

6	3		2			7		
7		1		9				
	9		3	7			2	1
			7			6		
1		4	6	2	3	9		8
		2			5			
3	5			1	4		8	
				3		1		7
		9			7		3	5

Easy # 390

	5		9			3		8
		6	8		5	9		
9	8		1		3			
7			5			4		
6		4				2		7
		8			4			9
			4		1		6	2
		7	2		8	5		
2		9			6		1	

Easy # 391

1			6	4	5		8	3
6			1	9				4
		3		8				
	3			6				1
7	1						6	2
9				2			3	
				5		9		
5				7	4			6
2	6		9	1	8			7

Easy # 392

		3	1	7	5		6	2
	9					3	4	
			3		9			
7	1		4	5			3	6
3								4
8	4			2	3		9	5
			9		2			
	2	1					7	
9	3		5	6	7	4		

Easy # 393

	8		3		6		5	
	6	9	5		8	3	7	
				2				
7			4		3	1		2
		5		8		7		
4		3	1		7			8
				3				
	7	6	8		5	9	2	
	4		7		1		3	

Easy # 394

5		4	2		3	9		6
				8				
8			7		6			1
	3		6		7	8	1	
		6		3		2		
	5	7	8		1		6	
2			9		8			3
				5				
6		8	3		2	4		9

Easy # 395

		3			4	2		1
4			3	7			6	5
				5	1		4	
2			9		6		5	7
				8				
8	6		4		5			3
	9		7	4				
5	8			1	9			6
1		2	5			4		

Easy # 396

8	9	3		6	1			
		5			4			1
1		7			3		9	8
	2							5
4	5	1				8	7	6
3							2	
9	3		1			2		4
5			3			1		
			2	8		9	5	3

Easy # 397

4				5	7			1
8			4	9			7	
	7	1				3		
		9	1				5	3
6	8			7			1	2
1	4				5	7		
		4				8	3	
	5			8	9			4
2			7	3				5

Easy # 398

	4		3	7	1	8	9	6
7		8	2					4
				6		7		
9				4				
	8	2	1		9	4	6	
				2				3
		1		3				
5					2	1		9
8	9	3	5	1	4		7	

Easy # 399

	5				3	7		8
3			9	1	8			6
		8	6	7				
8	9		3	4				
4	2						7	9
				9	2		3	5
				6	1	9		
9			5	8	4			7
7		4	2				8	

Easy # 400

	7	2		5			3	4
	1				3			
	6	8			1	2		5
				4	8		6	7
	3		5		9		1	
4	2		3	1				
6		3	9			8	2	
			1				4	
7	4			8		5	9	

Easy # 401

	4		2	8	6			5
8	2							3
5		9			3	6		
4				3		5		
	3	1				8	2	
		5		2				7
		3	7			4		6
9							5	1
1			3	9	5		7	

Easy # 402

4					1		2	6
3		6	2	4				
	7	2			3		5	4
		4			2	9		3
5								2
8		1	9			4		
1	8		3			5	4	
				1	8	6		9
6	3		4					1

Easy # 403

				1		9	3	
	6	3			9		4	5
5			3	8				2
6		5				3		
		8	6		1	7		
		1				6		9
7				4	3			6
3	8		2			4	7	
	1	2		6				

Easy # 404

3	1			2				5
		2		1	5			7
4					3	1		
		4			7	3	9	
6			3		9			2
	3	8	5			4		
		5	6					9
1			2	5		7		
2				9			1	4

Easy # 405

		9		7	3	6		
	2	5	4	9	6	3		
				2				5
		3		6			5	
	6	1				8	3	
	5			1		7		
7				4				
		8	2	3	7	1	6	
		6	9	8		4		

Easy # 406

	2	1	3					9
6		3		2				
	5		7	6		2	1	
			1					8
9		5	2	8	4	6		7
4					3			
	8	6		3	2		9	
				9		3		6
3					8	4	2	

Easy # 407

	1	6			2			5
	9		6		8		4	
			7		5	6		2
	8				7	1		
	7	2				9	6	
		9	4				7	
8		1	5		3			
	2		8		4		1	
4			1			8	3	

Easy # 408

				2			3	8
	6				8	2		7
2		7		3	5	1		
	4				7			
3	5		9	4	2		6	1
			8				9	
		6	2	8		4		3
9		2	4				8	
8	3			6				

Easy # 409

				9		1		
1	9		5	3	2			4
3				8	4			2
4				2			1	
6	2						4	7
	1			6				8
2			3	7				5
7			9	4	8		2	6
		8		5				

Easy # 410

8	3		9		5	7		
		6	1		3			9
9	1			4			3	
1		9		3				
			7		8			
				1		2		7
	9			6			5	4
4			3		2	9		
		8	4		9		2	6

Easy # 411

			6			7	3	
		6	3	7	2	8		
		3		1				6
		9		3	4			7
3		2				6		1
8			1	2		5		
1				8		9		
		8	7	5	9	1		
	2	5			1			

Easy # 412

	6	3	1		5	7		
7	1		6	9				
5	2							1
		7					5	2
			5	7	4			
6	9					3		
8							3	6
				2	9		8	7
		9	4		6	2	1	

Easy # 413

		8				1	7	
9					1	8	3	
	3				8			4
7	9			6			8	1
		4	1		7	9		
2	8			3			5	7
8			2				6	
	4	9	6					8
	2	5				7		

Easy # 414

		1		5	6		4	9
	9		7				2	
6			4			3	8	
2	1					5		7
		6				1		
5		9					3	2
	8	7			9			3
	5				3		1	
4	6		2	1		9		

Easy # 415

	1	2	4		3			9
4	7		8		9			3
	3							
	6		1			2	3	
3		1	5		7	4		8
	5	4			2		9	
							1	
6			2		1		8	4
1			9		4	6	5	

Easy # 416

				9		6		2
		1	2		7		8	
9	2		8	4		7		
2		8			9			5
		4		8		3		
3			6			1		8
		2		3	1		9	6
	9		5		2	8		
5		6		7				

Easy # 417

7	8	3			2		6	
		1	6	7				2
4			8		9	7		
9						4	1	
2								8
	5	6						7
		5	9		1			6
6				2	8	5		
	2		5			3	9	4

Easy # 418

9			2		1	6		
	4		9			3	2	5
		6		4	7			9
6	9					8		
		4				7		
		2					1	3
1			6	8		4		
5	7	8			4		6	
		3	7		2			8

Easy # 419

	2					9	6	
6			9	7				8
		9		4	8			1
2		7			6		4	
3		6		9		1		5
	9		7			8		6
8			4	1		7		
7				2	9			3
	1	2					8	

Easy # 420

		8		5	2			4
5	2			9	1		3	
		1	6			2		
2			1				8	
6	4			8			7	5
	8				4			6
		7			8	5		
	1		7	3			4	2
8			4	1		7		

Easy # 421

	4		5				7	
	7			6	1	5		
1		2		3	7			6
		9	1					5
7		4		5		9		1
5					6	2		
3			6	8		4		2
		1	2	4			5	
	2				9		6	

Easy # 422

7	8	2		4				
5			2	6		7		
3		1		9	5	4	8	
						3	1	8
4								9
2	3	8						
	5	7	3	1		8		4
		3		5	2			7
				8		6	5	3

Easy # 423

	8		1	2		4	3	5
		1				9	2	8
		2	3			1		
				1	4		8	
			7		5			
	1		9	6				
		7			2	5		
6	3	5				2		
1	2	9		5	3		6	

Easy # 424

		8	9	3				5
		5		4	2	1		
4	3						8	
	2		5			7		8
7		1		2		6		3
5		4			7		9	
	4						7	2
		7	2	5		8		
2				9	8	3		

Easy # 425

	3	6					9	
8		2		3			1	
1				7	4	8		
	5	8	2	6				
6	1			9			8	7
				4	8	9	6	
		7	9	5				1
	6			1		4		8
	2					6	7	

Easy # 426

	3	9	5		4			8
5		6	1		8			4
		4						
		7	9			4	3	
4	9		2		6		5	1
	5	2			3	8		
						9		
7			3		9	1		5
9			8		5	2	7	

Easy # 427

3				9			7	6
	2			3	5			
9	4	8	7		6			
8	3		5	1			6	7
5	7			4	8		3	9
			4		2	6	8	1
			3	8			5	
2	8			6				3

Easy # 428

	6		5		1		9	
			4		3	1		6
1	4				6	5		
	7				5			2
2	8						7	9
6			7				1	
		3	9				6	8
8		9	3		7			
	5		1		8		2	

Easy # 429

9			4	8			5	1
		8					9	
	5		2	9				
6		7	3				2	5
5		1	8		6	9		7
4	3				2	1		8
				3	8		4	
	6					5		
8	9			6	5			2

Easy # 430

		6	2	1	3	4		8
	4			8				
		2	6	7		1		
4				2		6		
6		5				9		2
		7		9				4
		3		5	1	2		
				3			7	
2		9	7	6	8	5		

Easy # 431

5		2	8	9		6		1
6				1	4			
							4	
8						1		6
1	2	9	5		6	8	7	3
7		3						2
	8							
			4	6				7
9		6		5	1	3		4

Easy # 432

			2	6		8		
4					7	6		
	3	6	9	1		4	5	
1	4		5	7				
	5						8	
				3	1		7	4
	7	9		4	5	3	2	
		5	6					9
		4		9	3			

Easy # 433

6					2			
	5	1	6					9
	2	9			1	6		7
7	9	3					1	
1	4						5	3
	8					2	9	4
9		7	1			5	3	
8					3	9	7	
			7					1

Easy # 434

8			9	1				
	6	4	2				1	
5		2		4	8	3		
3		5	1		2	7		
				5				
		6	8		3	9		2
		1	7	9		2		3
	7				1	4	6	
				2	4			1

Easy # 435

5	3			2				7
9	2		7		3	4		
		7	9		6		3	
	8	9		1				
			4		8			
				6		7	1	
	7		6		1	2		
		8	5		7		4	6
6				3			7	1

Easy # 436

4		5		2	1			9
9		6			4		7	
						1		6
			7	1			2	8
		9	2	8	3	4		
1	2			4	5			
6		4						
	8		4			3		7
3			1	5		6		4

Easy # 437

				1			3	7
	5		6		7	2		
3		2	9	8			6	
5	9				3			8
	8			5			4	
7			2				5	6
	1			4	5	6		2
		5	1		6		9	
6	3			2				

Easy # 438

		4			2	5	7	
						8	2	
8	7					6	1	9
7		5		8	6			
		8	4		9	7		
			1	3		9		5
3	8	1					9	7
	4	9						
	5	7	9			2		

Easy # 439

1			7	2			5	3
	8		3	1		7		
		7			8	9		
8					3		6	
3	6			8			9	7
	5		1					8
		1	6			5		
		8		9	5		3	
5	9			4	1			2

Easy # 440

3	8	5					1	
					7			5
	9			3	5		4	
1					3	2	8	
	5	8	4		9	1	7	
	3	4	7					9
	2		9	6			3	
8			5					
	4					6	5	1

Easy # 441

		6	4	1	8	5		
		4		5				6
			6			1	7	
		1		7	6			5
6		3				7		9
8			2	9		4		
	9	8			3			
3				6		9		
		5	7	8	9	3		

Easy # 442

		4	1		3			5
6			2	5		7		
9	1	5			7		2	
		3					6	4
		7				1		
2	8					5		
	7		8			4	3	9
		2		7	1			8
8			3		6	2		

Easy # 443

	6				7	5		
		3		5	2		8	
5	7			3			2	
		6			8	7		1
	4		7		1		3	
7		9	2			6		
	3			1			6	5
	5		3	2		8		
		2	4				1	

Easy # 444

7	8	6		5	4			2
3		1				6		
					9	3		8
				7	1	5	2	
1			5		8			3
	5	7	2	3				
9		4	8					
		2				1		6
8			6	2		4	5	9

Easy # 445

	1						9	
	4	6	3		5	7		
9		7	6					
4	9		7		6	2		
6		3		8		4		9
		5	2		9		6	3
					7	8		2
		9	1		8	6	4	
	5						1	

Easy # 446

7							9	8
	9			5	3	1		
		8	9	2		3		
	2	7			8			5
	8	4		9		6	1	
9			2			8	3	
		2		7	9	4		
		3	5	1			2	
1	7							3

Easy # 447

				9				3
	9	3	8	2	7	1		
		2		4	1	7		
		1		7			3	
	7	6				5	1	
	3			6		4		
		7	2	5		8		
		5	9	1	4	6	7	
4				8				

Easy # 448

				3		5	2	
7		3		2	4			9
	1				5	7		3
	8				7			
	4	2	6	8	3	9	1	
			5				6	
3		6	8				5	
1			3	5		2		8
	2	5		1				

Easy # 449

5					8		6	4
	4			2				
2	6	3	1	4	9	5		
				5				3
	5	2	3		1	6	8	
9				8				
		4	5	1	7	3	9	6
				9			1	
3	1		8					7

Easy # 450

	5		1	8				7
	1				6		4	8
	7	6				1	9	
	2				4			9
9		8	7		5	4		1
5			8				3	
	9	7				5	8	
6	8		3				1	
1				6	8		7	

Easy # 451

						6	5	
		1			5	2	4	
6	4					9	3	7
4		2		6	9			
		6	1		7	4		
			3	8		7		2
8	6	3					7	4
	2	4	7			5		
	1	7						

Easy # 452

			7				4	
3	7			9		1	2	
9		1	4			5	6	
2	6		5	3				
	4		8		9		7	
				4	7		1	3
	1	5			8	7		6
	8	9		5			3	2
	3				4			

Easy # 453

	5		7		4		6	8
			2	8	1			5
8							7	
7		8	4			5		9
2			9		6			3
9		5			8	6		2
	8							4
6			3	4	9			
3	4		8		5		2	

Easy # 454

2			1			5		6
		5	2	3	7		4	
		1				3	7	
5				1		4		
3	7						1	9
		8		7				5
	5	9				6		
	8		5	6	1	9		
4		2			8			1

Easy # 455

4					8	7		
	5						9	8
		7			4		2	6
9		5		3		8		7
	2		5		1		6	
7		1		4		5		2
3	7		1			2		
5	1						7	
		6	7					3

Easy # 456

9					2		6	
				4				
	2	3	6		9	1		8
3		4			7	9	2	
6	9		1		3		5	4
	1	8	5			6		7
1		6	2		5	7	4	
				7				
	7		4					2

Easy # 457

9					2			
	4	8	9					6
	6	2			4		9	5
5	3	6				4		
4		1				8		3
		7				6	2	1
6	5		4			3	8	
7					3	5	6	
			5					4

Easy # 458

7			1	3		2		
2	9	1		6				
5		8		4	7	6	9	
						5	8	9
6								4
1	5	9						
	7	2	5	8		9		6
				9		3	7	5
		5		7	1			2

Easy # 459

	3	6			9		8	1
		7			4		3	
1	8	9		5	3			
2							7	
7	4	3				8	5	6
	9							2
			2	8		1	9	7
	7		9			3		
9	1		3			2	4	

Easy # 460

1	7		3					
	9	8	7		6		2	
		6				5		
4		9	2		1		5	
2	8			7			4	9
	1		9		3	2		8
		2				6		
	3		5		4	8	9	
					9		3	2

Easy # 461

7		8						
1		2	8			3		
9	5	4					2	8
			4	5		8	1	
		9	7		8	2		
	2	1		9	6			
2	9					6	8	4
		7			3	1		2
						9		3

Easy # 462

		1			8		3	2
	8				7	1		
6							7	5
	5	6		4		7	1	
2			6		9			3
	1	9		8		6	2	
9	6							1
		3	1				4	
1	4		9			2		

Easy # 463

	4		2	3				7
8	7	3		6				
6	5			1	7	3		4
						7	2	6
	9						5	
7	6	1						
5		6	3	9			7	1
				5		6	4	2
4				8	2		3	

Easy # 464

	8	4	3	9		7		6
								1
				4	6		9	
4		8			5	2		
	7	3	2		1	6	4	
		6	4			1		9
	4		7	1				
2								
3		7		2	8	9	1	

Easy # 465

			6	5			2	3
	7	6	3					
				1	8		6	
	6	9	5			2	1	
5				9				7
	8	4			1	6	9	
	4		9	8				
					2	8	3	
8	1			3	5			

Easy # 466

8			3	2				9
				1		3	6	
	3	7			6	4		8
7	8						3	
	2		7		1		5	
	1						7	6
3		2	9			5	4	
	9	1		7				
5				4	3			7

Easy # 467

9		4		2				1
5					9		4	
	2			4	1			6
	5				6	7	9	
8			9		7			2
	3	9	1				5	
4			2	1			6	
	1		8					7
2				7		4		5

Easy # 468

4			8		2			9
5		2			4		8	
			5		3	4	2	
1					8	6		
7		6				9		1
		4	1					2
	9	7	3		1			
	3		9			7		4
8			2		7			6

Easy # 469

9	3			6	2			1
7			4	5			8	
				1	3			4
	1					6		
2		6	1	7	8	5		9
		8					1	
1			6	2				
	9			8	7			3
4			9	3			6	5

Easy # 470

	2	1	5	7				
8				3				1
7		3			6	2	5	8
		4	3	8			2	
	1						3	
	3			5	9	1		
3	8	2	4			9		7
5				9				2
				2	7	3	8	

Easy # 471

		5						
4		9	6		2		5	
	6	8	5		7		4	
		2	7			9		6
6	8		1		9		3	5
7		3			5	4		
	3		2		8	1	6	
	2		3		6	5		7
						3		

Easy # 472

	9	8		5				6
	5	7			2		4	9
					3	7		
			1	8		5	6	
	1		5	2	6		3	
	6	5		3	7			
		2	3					
9	8		6			4	5	
5				4		9	1	

Easy # 473

8			6		5	4	2	
2			9		8	1		5
						8		
	5	4			9	6		
7	8		4		3		5	1
		2	8			7	9	
		7						
5		3	1		6			7
	9	8	5		7			6

Easy # 474

		2		9		8		
	5	9			6		2	
8		4		5			7	
		7		2		5	1	
9		8	5		1	7		6
	1	6		7		3		
	4			1		6		7
	8		2			4	3	
		5		4		2		

Easy # 475

5			7	2		6		3
	7				9		4	
		9	3	5			7	
9					3	8		
3		8		9		4		7
		6	5					9
	9			4	6	3		
	5		8				6	
6		4		1	5			2

Easy # 476

	6		8			9		2
2			4	1		7		8
9		4						
3	1			4	6			
		8	5	3	1	2		
			7	8			1	4
						8		9
8		9		7	4			5
6		5			8		3	

Easy # 477

				7				
2					4	7		
	7	4	5		2		1	3
7	1				5	3	9	
4		5	6		3	8		1
	8	2	7				4	6
9	3		8		1	2	6	
		1	2					8
				4				

Easy # 478

3		9		2	7		4	
						9		
		1	5	9	4		6	
	2				3		8	6
	3	8	7		9	5	2	
6	1		4				9	
	9		8	3	1	2		
		2						
	5		2	4		7		9

Easy # 479

5		7		9		2		
9		3	2		5		8	
	2		3		1			5
6	3			4				
			8		6			
				1			2	4
2			1		4		9	
	6		7		2	1		8
		1		5		4		2

Easy # 480

2	6		8		5		3	
8		3	2	7				
1		5				8		
	3					1		5
			5	3	9			
7		2					6	
		4				2		6
				1	7	3		4
	7		9		2		1	8

Easy # 481

		2			6	9		8
				5	8		6	
6			2	3			7	5
9			4		7		5	3
				1				
1	7		6		5			2
5	1			8	4			7
	4		3	6				
8		9	5			6		

Easy # 482

	8		2	3			9	6
		2						3
4			9	1				
	9	5	8				4	1
	7	6	3		9	5	2	
8	2				1	7	3	
				6	8			2
6						9		
2	5			9	4		6	

Easy # 483

5				6	7		8	
	6		1	2				7
1			9		3			2
2				7	4		9	5
		9				8		
6	5		8	1				4
7			6		2			8
3				9	8		5	
	2		5	4				9

Easy # 484

		5			6			
			7	5	9	8		4
9		4		2			7	
8				3	1	7		2
4		7				1		5
1		3	5	4				9
	4			8		3		7
6		1	2	7	4			
			6			4		

Easy # 485

	7		5	2			3	8
		4	3				7	6
5	6							
	9	2		5	4			
3			1	9	2			7
			8	3		2	5	
							6	3
1	4				3	9		
6	3			8	5		1	

Easy # 486

	3				5	4		
				2	4		6	
1	7			6	3	8		
7		1		4			8	
	9	2	7		8	1	5	
	4			1		6		2
		8	2	9			4	7
	2		4	8				
		9	3				1	

Easy # 487

	3			5			4	
				1	8	9		3
9	4	8	6				1	5
		9		4	5			7
		5				3		
3			2	8		5		
2	1				7	4	5	9
5		4	1	9				
	9			2			8	

Easy # 488

	4				7			5
				6	5		1	
	3	2		1	4			9
2		3		5			9	
6	8		3		9		7	2
	5			2		6		1
9			6	8		3	5	
	6		5	9				
8			4				2	

Easy # 489

	7	2			4		3	6
	1	4		7		8		9
		8			5			
				5	3	2		8
		5	4		1	3		
9		6	7	8				
			3			5		
8		3		1		9	2	
1	2		5			6	7	

Easy # 490

7		8				2		6
2			4	7				1
			6		9			
9		3	2		7	4		8
	6			9			2	
8		2	5		4	7		9
			7		5			
1				8	2			3
3		9				5		7

Easy # 491

		2	8	7			9	1
		7	3					5
8			1	2				
1				5		4	8	
7		8	9		2	5		6
	9	5		1				2
				8	1			4
3					6	1		
9	5			4	3	2		

Easy # 492

			6		9	4	2	3
	2	9		3		5		
			5	2			7	
	1	7		6	2	8	5	
	5	2	7	4		1	3	
	9			5	7			
		5		8		3	1	
2	6	8	1		3			

Easy # 493

	5				1		8	9
	8	4				2	6	
	2		8	9				5
	1				8			4
5		7	4		2	8		6
6			7				3	
2				8	5		4	
	6	5				9	2	
8	7		9				5	

Easy # 494

	5		2		6		4	
			9		8	5		6
	9	6			5			2
	1				2	3		
	7	3				4	1	
		5	1				6	
8			4			7	5	
4		7	8		1			
	2		6		7		3	

Easy # 495

		2	1	8		3	6	
	1	7						
9			3			2	7	
8		5		1	9			
	3		4	5	8		2	
			6	3		1		8
	4	9			3			5
						7	3	
	7	3		6	1	4		

Easy # 496

	3		8	2				
		5			6		2	3
6			7	4	3			8
3		7	6	9				
9		1				2		7
				7	1	6		5
7			5	3	9			2
2	9		1			3		
				8	4		7	

Easy # 497

								5
		6		9	2			
	8	2	7	3		4	9	
		1				6	8	
6	5	8	9		3	7	4	1
	7	9				5		
	9	7		4	5	3	1	
			2	7		9		
2								

Easy # 498

				5				
	8				2			3
2		7	3		8	4	9	
	7	5			6	8		2
8	3		4		7		5	1
4		9	1			3	6	
	4	3	2		1	6		5
6			5				2	
				6				

Easy # 499

9			3	1			6	
6	4	1		7				
8	7			5	6	1	9	
						6	7	3
2								8
7	6	5						
	8	7	1	2			5	6
				8		7	3	9
	9			4	3			1

Easy # 500

			5		8	7	1	3
5	8			3				6
			9	6			4	
8	5			2	9		6	7
3	6		7	1			8	9
	9			7	6			
6				5			7	4
2	7	5	4		1			

Easy # 501

	2		6					7
8		3				6		
7		6	3				4	
6	3			5			8	4
		4	8		3	2		
1	8			7			9	6
	6				5	4		2
		8				1		9
5					9		6	

Easy # 502

4		9	2	3	7	8		
				1			2	
		1		8	5	4		
		2		9				6
3		8				9		4
6				4		3		
		4	3	2		5		
	6			7				
		3	4	5	1	6		7

Easy # 503

1		7	6		8			
6			1			7	5	
	9		3		7		2	
		5	8				3	
	7	2				1	8	
	8				9	2		
	5		9		3		1	
	4	3			5			9
			4		6	5		3

Easy # 504

	8	7	2	9		3	1	
		6		1	7			
								4
		5				6	8	
6	4	8	1		9	2	3	5
	2	1				4		
7								
			7	2		1		
	1	2		3	4	9	5	

Easy # 505

		2		6	1		4	
1	3						2	9
	8	6	9				1	
		3	8				5	
8		1	4		2	3		6
	7				6	4		
	1				7	9	6	
4	6						3	2
	2		6	9		1		

Easy # 506

	5					1		
		7	6		1	5	2	
			8	5	2		3	
3	7		1				8	6
	2		3		8		7	
6	8				5		4	1
	6		9	1	7			
	1	3	5		4	6		
		4					1	

Easy # 507

2	6			3	9			4
8			7	1			5	
				4	6			7
	4					3		
9		3	4	8	5	1		2
		5					4	
4			3	9				
	2			5	8			6
7			2	6			3	1

Easy # 508

			7	3	9	6		1
8		3		2			7	
7					6			
		5		7	4	1		8
1		4				7		3
3		9	1	8		2		
			6					4
	3			9		5		7
2		7	5	4	3			

Easy # 509

				8		7		
	9		4		1			
	7		2		3	1		4
	1	6	3					9
9		7		2		6		3
2					6	4	7	
5		3	9		4		8	
			5		2		4	
		9		6				

Easy # 510

	5	9			7		1	
7	2	3			4			
8			9					4
3		7		5				6
		4		9		2		
5				6		4		7
9					8			2
			5			1	8	3
	3		2			9	7	

Easy # 511

			5			2	8	3
9					8			4
	3		4			9	7	
5				1		6		7
		6		9		4		
3		7		5				1
	5	9			7		2	
8			9					6
7	4	3			6			

Easy # 512

9	2		3		5			8
	5						6	
3		4	7					
	9	1	8		4			6
2		8		3		9		1
4			9		7	2	8	
					9	8		7
	8						5	
7			6		1		2	9

Easy # 513

	3		6	2			1	9
7			1	3		6		
		6			7	8		
	7				1			4
4	1			7			6	8
9			3				7	
		3	4			9		
		7		8	9			1
8	9			5	3		2	

Easy # 514

		4	3	1	2		7	6
	3					9	8	
			8		6			
2	6		7	8			4	5
4								7
1	7			2	4		9	3
			6		7			
	4	7					6	
8	1		2	3	9	7		

Easy # 515

		5		3			9	7
2		9	6				3	5
	4		8					
6	5			8	1			
7			5	4	6			8
			7	2			5	6
					8		1	
5	1				4	9		3
9	2			5		6		

Easy # 516

			4	1	6	3	2	
	4				3			
	5	1		9				4
		8		4	7	2	5	
	2	7				4	1	
	1	6	2	5		9		
1				6		8	4	
			3				7	
	9	4	8	7	1			

Easy # 517

7		1				9		
3			7	9	8	1		
	5	6			3		8	
		3		4			7	
4	2						1	8
	7			8		5		
	6		8			7	9	
		7	6	2	4			5
		8				2		4

Easy # 518

	7	4	1	6	9			
9			8	4		7	1	
	1	6	2			3		
	4					9		2
7		9					3	
		7			4	1	9	
	5	1		9	7			3
			5	8	1	2	6	

Easy # 519

				3	9	4	6	
6			2		4	9		5
	4					8	2	
	8	2						6
			7	6	2			
5						3	9	
	9	5					1	
8		4	9		7			3
	6	1	3	8				

Easy # 520

3						7		6
	6		7	8			1	
		7		5	1		4	
	3	8			6			5
	2	6		7		4	9	
7			8			1	6	
	1		5	4		8		
	8			3	7		2	
4		3						1

Easy # 521

4		5			2		6	
	8	2	6	7				9
		3			1	8		
7	1					9	3	
9								6
	3	4					7	8
		9	4			7		
8				9	3	6	2	
	4		8			5		1

Easy # 522

8			9		7		5	6
			2	6	5		3	
	6							7
	8	3	7			9	2	
	5		3		2		8	
	2	9			6	7	4	
4							7	
	9		1	7	8			
3	7		6		4			9

Easy # 523

9			4	1			2	8
				7	3		4	
		7			8	6		1
5			8		7		9	2
				2				
3	8		9		4			6
1		6	7			5		
	7		1	8				
8	9			3	5			7

Easy # 524

4			7	3				5
		5		2	6			1
	1					3	2	
8		1			5		7	
9		2		7		8		4
	6		8			5		3
	8	7					3	
2			1	6		7		
1				5	7			8

Easy # 525

				1	2			
8			6			3		
3		5		4			6	7
		7		6			1	3
1	6	9		5		2	4	8
4	8			2		5		
6	7			3		8		9
		2			6			1
			1	9				

Easy # 526

	2	9	3	6		8	4	
	5			8	2			
								1
	7					9	5	
5	9	1	8		6	4	3	7
	8	3					1	
2								
			2	3			8	
	3	8		4	1	7	6	

Easy # 527

	1	5	3					8
	6		2	7		1	5	
7		3		1				
			5					4
8		6	1	4	9	7		2
9					3			
				8		3		7
	4	7		3	1		8	
3					4	9	1	

Easy # 528

5		1		3				
8				7	4	3	1	
	4		8		1			6
		9	3			1		4
2				4				7
6		4			5	2		
4			1		9		3	
	3	5	6	2				1
				8		9		5

Easy # 529

	2		9	1	8			
7							1	
3	1		4		7			2
	6	2			4	1	7	
	5		3		6		8	
	8	3	1			2	6	
8			2		1		5	4
	4							1
			6	4	5		3	

Easy # 530

					1	3		5
6		8						9
5	7	3		9	8	1		
				2	9		7	4
		2	1		7	6		
7	9		6	4				
		9	5	7		4	1	8
8						2		6
2		1	3					

Easy # 531

	1		9		8		4	3
								8
	8		2		4	1		5
5		4			9			2
	6	8	5		7	4	3	
1			8			9		6
8		9	4		6		2	
6								
7	4		3		2		6	

Easy # 532

		9		3	8		6	
	7	6		2				9
2	1							5
4	6		7	1				
9		1		5		3		6
				8	6		5	1
7							1	3
1				9		6	8	
	3		5	4		9		

Easy # 533

3	5			7		9		4
6	8				2	7	5	
2					4			
				9	8	3		6
4			7		1			2
5		9	4	2				
			2					9
	4	6	1				8	5
9		3		8			7	1

Easy # 534

4	9						8	6
	6		1	9			5	
			8		3			
2	3		6		9		4	1
		8		3		6		
6	4		7		1		3	9
			9		7			
	5			4	6		2	
3	2						9	7

Easy # 535

		4	3	1		5	2	7
7						9		6
9		2	8					
1	4		6	5				
		9	2		1	6		
				9	4		1	5
					2	8		3
6		7						4
3	1	8		4	7	2		

Easy # 536

5	3		8		2			
2		9		4	6	5	8	
	8					4	3	
7					4	9	6	
9								8
	6	8	2					4
	7	6					1	
	9	5	7	3		8		6
			9		8		4	5

Easy # 537

			9	4				3
		9						
4	3			6	8		1	5
3	4							8
2	8	7	3		5	1	6	4
1							2	7
9	2		4	5			3	6
						8		
7				3	9			

Easy # 538

		5	4		9		1	2
4			6		1	9		
	9			2			8	4
				3		1		7
			7		5			
3		9		6				
9	3			4			6	
		2	3		6			9
5	6		9		8	7		

Easy # 539

	3				4		1	
1	5	2					3	
4	8	7		1	3			2
2			8	3				
			7		9			
				6	5			3
6			4	7		3	5	1
	1					6	7	4
	7		1				9	

Easy # 540

		3		9	6		2	
8	7							1
	5	2		8				3
4	2		5	7				
3		7		1		9		2
				6	2		1	7
7				3		2	6	
5							7	9
	9		1	4		3		

Easy # 541

	2		9		3			7
		6			4			
9		4		5		2	6	
				4	7		2	5
	6	2				1	7	
4	7		8	1				
	8	7		2		9		1
			1			6		
6			4		9		8	

Easy # 542

	2	8				4		
			4		2			
3		5	6	7	9		2	
5		2		6	8	9		7
8								2
6		4	2	3		8		1
	8		7	5	6	2		4
			3		4			
		7				3	9	

Easy # 543

	6				5			1
			8	4				
9		5		2			3	6
6		4		5			9	
1	8	2		3		5	7	4
	3			8		1		2
7	1			6		9		5
				7	4			
4			5				8	

Easy # 544

				7	1		5	9
		5		2		6		
1	9	6	3			7	2	
9				6	2		8	
2								5
	5		4	1				2
	4	7			8	2	9	6
		9		4		1		
6	2		7	9				

Easy # 545

		4	7			6		9
1							8	2
2			4	1	8		7	
		8		3				7
	4	2				5	3	
6				4		8		
	6		3	5	9			8
5	3							4
8		1			4	9		

Easy # 546

8	1	3					2	
	9			3	8		6	
					5	8		
		2			3		1	4
1	8		6		9		5	2
6	3		5			9		
		1	8					
	4		9	7			3	
	6					2	8	7

Easy # 547

			3		1			
	1	2					7	
3	4		5	6	7	8		
9	8			1	4		3	5
4								8
7	2		8	5			4	6
		4	2	7	5		6	1
	3					4	8	
			4		3			

Easy # 548

2	9			5	8		7	3
	1			7	6			
	7		4			9		
				4	2	5		9
1								2
4		9	5	3				
		8			7		2	
			3	8			9	
6	3		2	9			8	4

Easy # 549

				4	9		7	5
		5		8		2		
7	9	2	1			4		8
	7			2	8			3
	8						5	
5			6	9			8	
6		4			3	8	2	7
		7		6		9		
8	2		4	7				

Easy # 550

		8	2			5		
9	2				4			6
3	7	4			5			
	4	7		9		1		
	5			2			3	
		9		1		4	5	
			9			7	6	8
7			3				2	4
		2			8	3		

Easy # 551

	3	9				7		
7			8	9			4	
4				3	1			6
		1	4				7	2
6	2			1			9	5
3	4				2	8		
2			1	4				7
	1			8	7			9
		3				2	1	

Easy # 552

		2		6	1			
							1	
9		8	3	5		6		2
		3				2		6
5	9	6	8		2	7	4	3
7		4				9		
2		5		8	6	1		7
	3							
			1	2		4		

Easy # 553

					5		7	
8	9			6	3			
		3	9	1			8	6
1	2	8		9				
	3	5	8		6	9	4	
				7		8	5	2
2	7			4	8	5		
			6	5			9	4
	6		7					

Easy # 554

4		1			8	2	7	6
6				1				9
	2	9	7	4				
		5	1	6			2	
	9						1	
	1			7	3	9		
				2	4	1	6	
7				3				2
1	6	2	5			3		4

Easy # 555

8							3	9
		6	9		8	2	4	
				5	4		8	6
3	9					6		
			1	6	9			
		2					5	4
6	7		5	3				
	8	3	4		1	5		
4	2							7

Easy # 556

		9						
8			3	1		9	6	
5			8	9	4	2		
7	5		6					1
1		4	9		3	7		6
9					8		5	2
		1	2	6	7			9
	9	3		8	1			4
						1		

Easy # 557

	6		1		7		4	
			5		3	6		7
	5	7			6			1
	9				1	2		
	8	2				4	9	
		6	9				7	
3			4			8	6	
4		8	3		9			
	1		7		8		2	

Easy # 558

5	7		9				8	
		6	2	5	8	7	3	1
				1				5
	1			9				
2		4	7		5	3		9
				2			7	
6				4				
3	4	7	5	6	1	2		
	2				9		6	3

Easy # 559

1	6	5			4	8		
3			8	6			4	
	2		5		9			6
	9					3		2
	4						5	
8		7					6	
7			9		3		8	
	8			4	5			7
		4	7			9	2	1

Easy # 560

8			4	2				7
4			8	7	1	5		6
	6			5				
		6		8				4
3		4				8		9
2				9		6		
				1			2	
9		8	2	4	5			3
1				3	7			8

Easy # 561

7						6	8	2
5			6	8				9
	6		1					
2		4	8				7	
1		7	9		5	2		6
	9				1	5		8
					6		2	
8				3	9			4
6	7	3						5

Easy # 562

		9		2	6		1	
	7		4	5			9	
1						5		2
	3	1			9			4
	8	2		4		3	7	
6			3			9	5	
3		4						5
	1			9	4		3	
	2		1	6		4		

Easy # 563

		5	1	7	9		8	
		6				7	9	
1			6			5		2
5				6		8		
7	9						6	3
		4		9				5
8		1			4			6
	5	3				2		
	4		5	2	6	3		

Easy # 564

	1			4	3			7
	4		2		8		3	
5			1	7			9	
8	9		6	1			4	
		5				8		
	6			3	5		7	9
	8			6	9			4
	5		4		7		1	
9			5	8			2	

Easy # 565

			3	6				8
		3						
6	8			7	4		1	9
8	6							4
2	4	5	8		9	1	7	6
1							2	5
3	2		6	9			8	7
						4		
5				8	3			

Easy # 566

			3					2
		5	7		4		1	
1	2			6		4		3
	1	6	5	3				
9	5						2	1
				9	8	3	5	
4		9		1			8	5
	8		4		3	2		
2					9			

Easy # 567

	6	1	2					4
2					5			
	5	4			1	2		9
9	4	3					1	
1	8						6	3
	7					5	4	8
4		9	1			6	3	
			9					1
7					3	4	9	

Easy # 568

		5		6	9		2	1
4		1		7				
	2		4		5	8		
6			1			9		8
		3		8		6		
5		8			2			4
		9	5		7		8	
				2		1		5
2	5		8	3		7		

Easy # 569

3			6					
	4	6	8	3		2	9	7
	1					6		
	2	3			6	9		4
		5	1		2	8		
6		4	7			5	1	
		7					2	
9	5	2		7	8	1	4	
					5			9

Easy # 570

		6		7				4
9	4				6			
5			3	9	4			6
		9	8	4				1
6		7				4		3
2				3	7	5		
7			1	2	9			5
			7				3	2
1				5		7		

Easy # 571

9		7	6			5	8	
5					7	6	3	
			8					7
	6					2	5	9
2	3						1	6
1	5	8					4	
6					9			
	9	5	2					4
	2	3			6	9		5

Easy # 572

	2	3		6	9			4
							9	
	9		1	7	8			3
3					6	5		1
9	4		3		2		8	7
8		5	7					9
5			6	3	4		1	
	3							
6			2	9		7	3	

Easy # 573

	1				4	9		
				2				
3		4	9		1		5	8
2	3				7	4		1
	9	1	8		3	6	2	
5		8	6				7	9
9	8		4		6	2		7
				7				
		7	2				4	

Easy # 574

8	1		6	7	4			5
2			5			4		
7					8		3	6
			1	3		5		9
1								4
9		7		8	5			
6	8		3					2
		2			6			1
3			7	9	2		5	8

Easy # 575

5				3	1			
6	2	8	9		7			
		3		8		7		9
3		2	1	4		9		7
9		1		6	2	8		3
2		5		7		3		
			6		5	4	7	2
			3	2				1

Easy # 576

		3	5			7		
4	7			8	3		9	
		2		4	7			6
7			3				2	
5	6			2			1	4
	2				6			5
2			6	3		1		
	3		1	9			6	7
		1			2	4		

Easy # 577

			9	5		1	3	
9	6		1					
				7	2		9	
4	9		5				7	3
		5		4		6		
8	2				7		4	9
	8		4	2				
					3		1	2
	7	2		1	5			

Easy # 578

					2	9		
4							6	
6	2	9		4	3		8	1
8		7	4				1	2
2			1		6			3
5	6				7	8		9
7	8		3	5		4	9	6
	1							7
		5	7					

Easy # 579

1			9	5		7	4	
		6						1
		4	1		3		8	
8		9			6			7
	7	5		1		2	6	
3			4			1		8
	3		5		8	6		
6						8		
	4	8		2	1			9

Easy # 580

	1		5	2				
	9		8	7		5	6	
		8		3	4		7	
3						1		
5	2		1	4	3		8	6
		1						5
	4		9	6		3		
	8	7		5	2		1	
				1	7		9	

Easy # 581

9				1	8			6
2			4	7			9	
	1	7				2		
		8	9				2	5
6	5			8			7	3
1	9				5	4		
		1				5	8	
	8			4	2			7
5			8	9				2

Easy # 582

8	9				1		5	
		1	7	4	5			6
6		7						4
1				2			7	
	3	2				5	6	
	7			5				9
5						2		3
7			8	3	2	9		
	8		5				4	7

Easy # 583

9	4			7	2			6
				5	9		3	
	7					2		
3	5				6	1		9
2		1	9		7	4		8
7		8	5				6	2
		9					4	
	2		6	4				
4			3	9			2	1

Easy # 584

	7				1	5	8	
6		5				3	4	2
						1	6	
5	8			6	4			
	6		7		2		5	
			3	9			2	8
	2	7						
9	3	6				2		5
	5	8	2				1	

Easy # 585

6		2		1	7	5		
	8	7		4	5	2		
	3		6					1
				5	9		7	2
9	5		8	7				
2					4		6	
		3	5	9		4	2	
		6	2	8		3		9

Easy # 586

					2	8		
	4	9		3				5
	3	8			7		6	4
			1	9		3	5	
	1		3	7	5		2	
	5	3		2	8			
4	9		5			6	3	
3				6		4	1	
		7	2					

Easy # 587

	8			6	7		5	
5	2						6	9
			3		2			
7	9		6		5		3	1
		5		3		2		
6	3		7		4		9	5
			4		6			
4	6						1	3
	1		5	9			8	

Easy # 588

3		5	1	9		8		2
	8				7	5		
			6	5		4		
8	9		2	7				
2								4
				3	9		8	7
		8		1	3			
		2	5				1	
7		1		8	2	3		6

Easy # 589

				1		6		4
	6	1	5	3				9
7			6		9		5	
5		6			1	8		
3				5				2
		2	4			5		7
	1		8		6			5
6				2	7	4	1	
4		8		9				

Easy # 590

	9					8	6	
		8	4	1	5		2	7
			8		9			
1	4		6	5			8	2
8								6
3	6			7	8		9	5
			9		7			
9	8		5	2	1	6		
	7	4					1	

Easy # 591

	5			8	4			9
							6	1
	3	4	2	9		7		
3		7			2	5	1	6
6	4	5	7			9		2
		1		2	5	8	4	
5	7							
4			8	1			9	

Easy # 592

2				8	9			4
5			2	1			9	
	9	4				6		
		1	4				8	6
7	5			9			4	3
4	2				8	9		
		2				5	6	
	8			5	1			2
3			9	6				8

Easy # 593

6		8	1	7	5			3
				9			1	
9				3	2			8
1				6		4		
3		7				8		6
		4		8				7
8			7	1				2
	4			5				
7			8	2	9	5		4

Easy # 594

			4	5		8		1
		4	6					
3		6		1	2		5	
				6		5	2	3
	5	9	2		4	1	8	
7	2	3		8				
	9		8	7		2		4
					5	6		
2		8		4	9			

Easy # 595

4	2	7		8	1		3	6
					6	8		
6							5	
2		3	6				4	8
1			4		5			9
9	5				7	6		3
	4							7
		2	9					
5	3		1	7		2	9	4

Easy # 596

1		7	6		9	4		5
		2	7		8	3		
				3				
3	2		8		7		6	
9				6				7
	7		2		3		4	8
				4				
		6	3		1	9		
5		1	9		6	7		3

Easy # 597

	8		6	3				9
4			5		1		3	
3	2	5			9	6		
1						8	4	
9								5
	6	7						3
		9	7			1	2	4
	7		1		8			6
6				9	5		7	

Easy # 598

	5	4			6	7		2
					8		5	
	3	2		4				9
			1	3		9	4	
		1	4	6	9	8		
	4	9		8	5			
4				7		1	2	
	6		8					
2		3	9			4	7	

Easy # 599

		3	8				4	5
	8			2	7	9		1
2			3					6
3		7				6	8	
	2						1	
	6	4				7		2
7					4			8
9		8	1	6			2	
5	3				9	1		

Easy # 600

8	1	6	7		9			
	4			8		6	7	
		5		6	4			
	2	4	6	9		3	5	
	3	8		1	5	4	6	
			5	4		7		
	8	3		2			4	
			8		3	9	2	6

Easy # 601

6		2			3		4	9
4		9	2				8	
	3				9			
1	7	4				8		
	2	6				7	3	
		3				4	9	2
			1				5	
	4				5	6		3
5	9		3			1		4

Easy # 602

				6			7	8
6		1		7	3	5		
	4				8	6		1
	2				1			
7	3		9	2	6		4	5
			8				9	
9		6	2				8	
		4	6	8		2		7
8	7			4				

Easy # 603

	3	4	7					5
		8	2	5	3	4		
				6	9		8	
				8	7	2		1
7		3				8		4
8		5	1	3				
	5		6	4				
		1	8	9	5	6		
2					1	5	4	

Easy # 604

			3	1	4	7		
		1					5	
	8		6		5	4	1	
7		8	5			3		6
		4	7		3	8		
6		3			1	9		5
	7	5	1		9		6	
	9					5		
		6	2	5	8			

Easy # 605

		6	9	2	7	8		3
	7			1				
		3	5	6		1		
4				8		7		
3		8				6		2
		2		3				4
		5		7	2	3		
				9			4	
9		4	1	5	3	2		

Easy # 606

	2		5	3		6		
		3		7	2		9	
		5	9		8	1		
		7		4	5	8	2	
5								3
	3	2	7	1		9		
		9	6		3	4		
	5		1	8		2		
		1		9	4		8	

Easy # 607

	2		3	8	7	5	1	
7					4		9	
	4	1	2				7	
3	8			1	5			
	9						6	
			9	2			8	5
	3				1	2	4	
	7		5					6
	1	9	4	3	6		5	

Easy # 608

6			8		7	9	1	
8			3		9			
				4			7	
2	8		4			3		
	4	1		3		7	2	
		7			1		4	5
	2			6				
			5		8			7
	5	8	1		3			2

Easy # 609

		8	2	4		7		
	1	3				8	5	
			1		3			
	3	5	9		1	4	2	
2				5				6
	9	4	3		2	5	8	
			5		6			
	2	6				3	4	
		7		3	9	2		

Easy # 610

		5			1			
7	4				5	3	1	
1	3		7			2		
3	6	8						2
4		7				5		8
5						1	7	3
		3			9		5	4
	9	1	5				3	6
			6			9		

Easy # 611

							7	
1		3	2	4		9	6	
				3	6			4
	3	1			5	8		
9		2	8		7	6		3
		6	3			7	4	
3			9	7				
	2	9		8	1	4		7
	8							

Easy # 612

		1		9	5	3	7	
5	6							
7			3	1			4	
8	7	5	6			4		9
2		6			9	5	1	8
	5			3	7			4
							8	1
	2	7	9	4		6		

Easy # 613

3	4						6	7
			1			4	8	
		7	5					3
7	1		8			3		
8	5		6		1		2	9
		2			5		1	6
5					7	9		
	7	3			6			
4	2						7	1

Easy # 614

		3		8	9			
1		8		3	4	9		2
		4	7				8	
				9	6		3	1
4								5
3	6		4	1				
	3				1	7		
9		7	8	6		3		4
			2	7		5		

Easy # 615

	2	4	1	3				8
				6			5	2
7			8		5	4		
1	7				2		3	
3				7				9
	5		4				8	7
		7	6		8			1
2	8			4				
6				9	7	8	4	

Easy # 616

	9		6		4		8	
4			2	5			7	
	3			7	8			2
2	6		8	1			5	
		7				8		
	4			9	5		2	7
6			1	4			9	
	2			6	9			8
	1		7		3		4	

Easy # 617

7		5		9	8		3	4
	2		7	3				
		9						
1			2			4		3
5	7		9		3		2	1
8		2			6			9
						3		
				2	1		4	
2	8		5	4		1		7

Easy # 618

5	1		6		8	2		
2				3				
			5		1	7		
	7				6	5		4
4	6			8			7	2
1		2	4				8	
		1	8		9			
				4				7
		3	1		7		9	6

Easy # 619

2				3				9
	3			5		7		8
	1		9				4	3
8	5			7				4
1		6	2		5	8		7
7				9			5	2
6	2				8		9	
3		1		2			7	
9				6				1

Easy # 620

8			5				9	
				7	8			
7	4			3		2		5
	1			9		4		6
4	9	6		1		5	7	8
3		8		5			2	
2		5		6			1	3
			9	8				
	3				5			4

Easy # 621

		1	2	6	3	5	4	
		2	1	9		6		
5				4				
	5			2		1		
	1	8				7	2	
		9		7			5	
				3				9
		3		8	6	2		
	2	7	9	1	4	8		

Easy # 622

2	5		9	8	4			6
9					6		2	3
1			3			9		
			5	2		4		8
7								1
8		5		6	1			
		7			5			9
3	6		2					4
5			7	4	3		1	2

Easy # 623

6		7		1	5			
		5	4	8				
					6	9	5	
	7	8			1	5	3	
9				3				1
	5	3	8			4	2	
	4	6	7					
				4	3	2		
			1	6		8		4

Easy # 624

		4		5	6			3
1	8						9	
2		3		8			4	
3	7		2	1				
	4	1		9		5	3	
				6	3		1	9
	1			4		3		6
	2						5	1
5			9	7		4		

Easy # 625

4				6				9
2		6			3	4	5	7
	5	7	2	4				
		1	6	9			7	
	7						1	
	4			5	7	3		
				2	9	1	4	
5	9	4	8			7		2
1				7				5

Easy # 626

	1						6	7
7				9	3	4		
		6	7	2		3		
2		1			6		9	
6		8		7		5		4
	7		2			6		3
		2		1	7	8		
		3	9	4				2
1	4						3	

Easy # 627

		4	1		7	9		
3		1	5		2	6		8
				9				
9	4		7		1		5	
2				5				1
	1		4		9		6	7
				6				
8		3	2		5	1		9
		5	9		3	2		

Easy # 628

			3	8			9	5
8	5			2	9	3		
	4		7					
				9		5	1	2
	6	9	8		5	7	3	
1	7	5		4				
					4		8	
		7	5	6			4	1
6	9			7	8			

Easy # 629

8			4		9	7		
	4			3			9	8
		5	1		8		4	6
				4		8		9
			6		5			
5		2		9				
7	2		8		3	6		
3	1			7			8	
		8	2		4			3

Easy # 630

3			2	4			9	6
		5			6	1		4
				5	8		2	
7			6		5		3	9
				9				
8	6		3		2			1
	5		4	6				
4		1	5			7		
6	3			8	7			5

Easy # 631

1			2		4		6	
				8		4		3
	6	3	5	9				2
5		1			3	9		
9				1				7
		4	6			2		1
8				7	1	6	2	
3		2		6				
	1		8		2			5

Easy # 632

8				6	4	9		
		5	2	1				
		3	8	9		7		2
	6							5
	2	1	5	4	6	8	7	
5							2	
9		8		2	1	5		
				5	9	3		
		4	3	7				6

Easy # 633

		9					5	
	8		5		4	9	6	
			3	9	1	8		
9		5	4			7		8
		3	7		6	2		
8		7			9	3		6
		6	2	4	7			
	4	2	9		8		3	
	9					4		

Easy # 634

6	5			9	3		8	4
	9			5	8			
	3		7			5		
				8	2	9		6
3								1
9		2	3	6				
		9			6		7	
			4	7			1	
8	7		5	2			9	3

Easy # 635

3		9		5			1	8
			1					2
	2		7		8	9		
	7	3	9	1				
5		2				3		1
				7	3	5	6	
		5	8		4		3	
2					7			
7	8			6		2		5

Easy # 636

6			5	4		7	1	
		5						
1			3	9	2	5		
2	8		4					1
9		3	7		1	6		5
5					9		8	3
		2	6	1	4			8
						1		
	9	1		5	7			4

Easy # 637

5			9	8		3	7	2
2		3	5					
		8				4		9
	7	6	8	1				
4			7		5			1
				6	4	7	8	
1		4				9		
					2	1		5
6	5	9		7	3			8

Easy # 638

7	5						6	9
	6				1	5		
4		9			2			
		6			4		5	2
3	8		2		7		4	1
2	7		1			3		
			7			6		5
		8	5				1	
5	2						9	3

Easy # 639

	9		2	4			5	
	2		9	5	8		1	6
		1		6				
1				9			2	
2	7						3	9
	4			3				1
				8		4		
9	3		4	2	6		7	
	8			7	5		9	

Easy # 640

4		3	1	6	8			
2		1	5			7		
	9		7	2		1		8
9						2	7	
	3	2						5
1		7		5	6		2	
		9			3	4		1
			2	4	1	5		7

Easy # 641

	7		1		4			3
		6	2	8			1	5
	1					7		
	2	1			5	3		
7	8			2			4	9
		9	7			1	6	
		2					7	
5	9			4	6	2		
1			3		2		5	

Easy # 642

					5	1	6	
		9	4	6				
6		3		1	2			
	9	6			3	4	8	
2				4				7
	4	8	2			3	5	
			8	2		5		1
				3	6	8		
	8	7	1					

Easy # 643

		9			5	7	8	
8	2				4	9	1	
7		4	8	9				
9					8	4		3
		1				8		
5		6	3					9
				5	6	3		7
	6	5	4				9	1
	4	7	9			5		

Easy # 644

		3	1				4	
9	6		4			3		
7	1							8
	3	1		2		8	7	
6			5		8			9
	9	8		4		5	3	
3							8	5
		9			5		2	3
	2				3	6		

Easy # 645

				9	4			6
		6	1	2	8	5		
8		5	3				2	
				6	3	1	7	
	3	8				6	5	
	6	2	7	8				
	1				7	2		5
		7	6	4	2	9		
2			9	5				

Easy # 646

			5	1	4	7		
2			9		6	4		1
		1						6
	7	2	6			5	9	
		4	7		5	2		
	9	5			1	3	6	
3						6		
7		6	1		3			9
		9	8	6	2			

Easy # 647

					6			3
9	4	3					7	
	8			9	3		5	
7					9	2	4	
	3	4	5		8	7	6	
	9	5	6					8
	2		8	1			9	
	5					1	3	7
4			3					

Easy # 648

	4				5			
	8	5		2			7	6
6			8		9	4		
				5	6	9		7
7	5						2	4
2		1	7	9				
		7	3		8			2
4	2			1		8	9	
			9				4	

Easy # 649

				7				
	2				1	4		
5		1	4		2		9	3
7	5				8	1		2
	4	2	3		5	6	7	
9		3	6				8	4
4	3		1		6	7		8
		8	7				1	
				8				

Easy # 650

1	5			7				8
6					1	5		
		7		5	8			3
		6			3	1	9	
4			1		9			7
	1	2	8			6		
5			7	8		3		
		8	4					9
7				9			5	6

Easy # 651

1					5		3	
4	3				9	2	6	
9	6	2		8	3			
		7					1	
3	5	1				4	8	6
	9					7		
			7	6		1	9	2
	2	9	3				5	7
	1		9					3

Easy # 652

8	9		5		4			3
3		6	8					
	2						6	
	6	9	3		8			7
5		8		1		6		9
4			7		6	5	8	
	4						2	
					3	7		1
6			2		1		9	8

Easy # 653

3		9		6	5	4		8
	7							
			4	3		1		
8		1				2		
9	2	5	6		3	8	1	7
		7				3		5
		3		5	4			
							4	
2		6	7	9		5		3

Easy # 654

		8		7			9	2
3	2		9	6	8			
			1					6
8	7		5	4			3	
5	6						2	8
	9			2	6		5	4
2					1			
			2	8	7		1	5
4	8			3		2		

Easy # 655

7	1			2		9	4	
	4				3			
2	6				1	5		8
				3	8	4	6	
	3		1		7		8	
	5	9	2	4				
6		7	3				5	2
			8				3	
	8	4		7			9	6

Easy # 656

			9	6			5	7
5	6			3	7	9		
1			4					
				7		5	3	2
8		7	6		5	4		9
4	2	5		1				
					1			6
		4	5	8			2	1
7	8			4	6			

Easy # 657

	7	9	2		1			6
			9		7			8
		6		4				
	8				2	5		9
	2	5		1		6	8	
6		7	5				1	
				5		8		
7			1		3			
4			7		8	2	3	

Easy # 658

		5	8				7	
3	9		7		1	8		5
				5				
2		9	1				5	3
	3	6	9		4	1	8	
8	4				5	7		6
				8				
4		7	3		6		2	9
	6				7	3		

Easy # 659

3			5		7	2	6	
6			9		3	8		7
						3		
	7	2			9	5		
1	3		2		4		7	8
		6	3			1	9	
		1						
7		4	8		5			1
	9	3	7		1			5

Easy # 660

9	3			7		5		
5	7				2	3		1
					6		4	
			8	6			5	2
6			2	4	5			9
2	5			1	9			
	8		6					
7		3	4				8	5
		2		5			1	3

Easy # 661

3			5	4		6	2	
	6	5		2	3			
					1	9		
6	4	8		5				
1		3	6		2	7		5
				9		1	8	6
		2	9					
			2	1		5	7	
	8	9		7	6			1

Easy # 662

7	1		5		2	8	3	
				3				
		3	8					5
	4	7	2				1	3
1		6	7		9	2		8
9	8				3	5	6	
6					5	1		
				8				
	9	5	1		6		7	4

Easy # 663

2				6		3	4	
		1	7	8			6	
9						2		1
				3	4	7		2
6	2			7			1	4
8		4	9	2				
5		2						7
	6			1	3	4		
	4	9		5				6

Easy # 664

	3			1	5			7
		5	3			4	1	6
1			8		4		2	
2		7					8	
	4						3	
	1					9		5
	5		7		8			9
6	2	8			9	3		
9			4	3			5	

Easy # 665

2		4	7		6	3	5	
				4				
	6				2			4
	4	3			7	8		5
7	2		1		5		3	9
6		9	4			2	1	
3			6				9	
				2				
	8	5	9		3	1		6

Easy # 666

					5	3		2
		2	4	7				
	5	1		6	2			
1		7			6	2		9
	3			9			6	
2		9	7			4		8
			6	5		7	4	
				4	9	8		
4		5	1					

Easy # 667

		5	7	4		9	3	
4							5	
	3		1	5				
6		2	8			3	1	
9		3	4		2	6		5
	8	7			1	4		9
				8	4		7	
	2							3
	5	4		2	3	1		

Easy # 668

		7	1	3			6	
		6		9	5	8		
3	9							7
5			6			4	7	
	4	8		5		2	3	
	6	9			4			1
9							5	4
		4	5	6		7		
	5			1	7	3		

Easy # 669

5	3			4	6	2		
7					8		3	9
			2	9	3		5	4
2		8					4	
	7					5		2
8	9		3	6	1			
3	2		4					5
		7	5	2			1	3

Easy # 670

	7		9	8		5	4	
		9			2			7
	4		3	5			2	9
3		5	8	1				
				3	5	9		1
8	1			2	3		9	
4			7			6		
	9	7		6	1		3	

Easy # 671

			5			6		
3		4	7	6	9			
	9			8		3		7
8		9	1	2				4
6		1				9		3
7				3	6	2		1
9		2		4			3	
			3	9	8	1		5
		3			5			

Easy # 672

6			5	2			4	
	2		1					9
5				9		6		3
	8	7	2				3	
1			7		9			5
	3				4	9	7	
7		6		5				2
3					7		6	
	5			6	2			4

Easy # 673

4		5	9	6			8	
	2	1			5			9
		7			3	4		
3	6					8		7
	8						9	
7		2					4	6
		8	2			6		
2			4			1	3	
	4			8	7	9		5

Easy # 674

				9	6	8		
	6	1		5	7	4		
		5	3	2			7	
	8							2
1		7	2	3	8	9		6
6							8	
	2			1	4	3		
		8	9	6		7	5	
		4	5	8				

Easy # 675

	3			2			7	
		4		8			1	3
		7	5			8	2	
	8	6		7			4	
5	4		6		8		3	2
	9			4		6	5	
	1	9			7	3		
4	5			6		1		
	7			1			8	

Easy # 676

		3					8	
				9	2	7		
2		4		3	8			5
7		9			5		1	2
8	1		2		3		4	6
3	6		9			5		8
4			7	2		8		1
		8	5	4				
	2					4		

Easy # 677

	2		1		4			
	5		9		3	1		4
				8				5
7	4		3			2		
5		2		9		3		7
		9			7		5	1
2				7				
3		6	2		1		8	
			6		9		1	

Easy # 678

9						6		
		2	6		4	3		9
			1	9	8			2
6	9		4				2	7
1			7		3			5
7	2				9		3	1
3			5	4	7			
5		4	9		2	1		
		9						4

Easy # 679

		8	5	4		1	6	7
	1					5	4	8
	4		1				2	
				8	6	7		
			4		2			
		3	9	7				
	7				5		1	
3	6	1					7	
4	9	5		1	7	3		

Easy # 680

6			7	9		1		
7	4						6	3
1					5	9		7
5					7	4		
	2	1	4		6	3	7	
		3	2					8
2		7	9					1
3	1						9	6
		6		7	1			4

Easy # 681

	9	8				5		
1	4		5		8			2
	5	2	4	3				
2						9	8	
			8	2	7			
	3	4						1
				9	3	2	6	
3			7		4		5	9
		6				4	1	

Easy # 682

	4				3	2		
3		8			7			
1	9						6	3
		1			4		7	6
4	5		7		6		2	1
6	3		5			8		
9	8						3	7
			6			9		5
		3	4				8	

Easy # 683

	9	2	7	4		6		8
								4
				9	6		3	
2		9			3	1		
	3	1	9		4	8	6	
		4	5			7		3
	2		1	3				
9								
1		6		2	8	3	7	

Easy # 684

			5			2	7	4
1			7			8		9
	5				8		3	
	6			9		7	2	
		4		8		5		
	7	5		6			9	
	4		3				8	
7		8			4			2
3	2	1			9			

Easy # 685

8		3		2	7	6		
						8		3
6		4			3		5	
			2	3		7	1	
3			6	5	1			9
	1	5		7	4			
	4		3			9		8
7		8						
		9	7	1		3		2

Easy # 686

	5	7		6	9		2	
						9	3	
	2	3			5			8
			8	9			1	6
		2	6	1	4	5		
6	9			5	7			
1			5			4	8	
	3	5						
	4		9	7		3	5	

Easy # 687

4		9			6		5	1
		6			8			
1		3		5		8	2	
				2	4	9	3	
		8	5		7	6		
	2	1	8	6				
	3	2		4		7		5
			6			2		
8	9		7			1		4

Easy # 688

		6		1	7		4	
	8		9	3			6	
4						3		1
	5	4			6			9
	2	1		9		5	8	
7			5			6	3	
5		9						3
	4			6	9		5	
	1		4	7		9		

Easy # 689

2	3	4				7		6
5	7		6				9	
1	6							
			3	4		5	6	
	2		1		6		7	
	5	7		2	8			
							2	9
	1				9		5	7
7		2				6	8	3

Easy # 690

2						3		9
			9		4			
		7	2	5	8		4	1
4	8		1	9			6	7
	7						1	
1	5			8	7		2	3
5	9		8	2	3	1		
			4		1			
7		1						4

Easy # 691

	2	1			3			
7		6				2		5
5					4		7	
	5				1	3		7
8		9	3		6	4		1
6		3	4				9	
	8		7					4
3		7				9		2
			6			7	5	

Easy # 692

	9	3	1			8		
	6	7	9	3	4			
4			2	7		6	9	
	7					4		1
6		4					8	
	5	9		4	6			8
			5	2	9	1	3	
		6			7	9	4	

Easy # 693

				4	8		6	
8			6	2	5			7
6	4		7			1		
				3	7	5		6
5		4				9		3
1		7	9	5				
		6			9		3	4
4			3	6	1			5
	5		2	8				

Easy # 694

							6	
7			1		6		8	4
6			3		4	7	2	
	2	4			1		3	
9		6	2		5	4		8
	7		6			1	9	
	6	1	4		9			3
4	5		8		3			9
	9							

Easy # 695

		9			8			3
4	6	2		5	3			
3		7			2		6	4
	1							9
8	9	3				4	7	5
2							1	
6	2		3			1		8
			1	4		6	9	2
9			2			3		

Easy # 696

		4				7		2
7			6	1			8	
		2		8			9	5
				5	9	2		6
	2	8		6		9	7	
9		1	4	2				
4	9			3		8		
	8			7	5			9
2		3				6		

Easy # 697

9					6		1	3
5			3	1			9	
3		2				5		8
6					3		2	
	9	4	2		5	3	8	
	8		4					7
8		9				1		5
	5			3	9			2
4	3		1					9

Easy # 698

	1		2					4
	4	8		9	3	6	1	
	7			1	5			
				2	8	4		9
		7				8		
4		2	9	6				
			6	3			4	
	6	5	8	4		2	3	
3					1		8	

Easy # 699

5			1	2		6		8
	2	9		6				
6		8	9				4	
			8				3	
	4	5	6	3	7	2	1	
	7				9			
	9				3	7		6
				4		9	2	
3		2		9	6			4

Easy # 700

		7	8		3			5
	6				4			
8	4			9		6	7	
				4	5	7		9
	7	6				5	2	
4		5	1	2				
	5	1		7			8	2
			2				6	
6			4		8	1		

Easy # 701

	1	5		3			4	
	9	3	4		5			8
4			9		6	5		
9		2		7				
			8		2			
				6		7		4
		4	6		7			3
2			1		4	8	6	
	6			5		4	7	

Easy # 702

		6	2			8		5
7				5				6
		4		8			7	9
8		3		6				4
4	2		3		8		5	7
1				4		3		2
2	4			3		9		
6				9				8
9		1			6	7		

Easy # 703

9			6			4		
1				4	8			
4	3			1	9		2	8
				8	5	1	3	
	9						7	
	1	5	9	3				
6	8		4	5			9	1
			2	6				7
		1			3			6

Easy # 704

4		8			7	9	6	
				1		4		7
	6		4	3			5	
6	8							4
3			8		1			2
1							7	8
	2			9	4		8	
5		1		8				
	4	3	5			2		9

Easy # 705

		7		2	6			5
8			4	3		9		1
6			9	8				
	3					8		
1	2		7	5	8		3	4
		8					7	
				4	3			8
2		3		9	1			6
9			5	7		1		

Easy # 706

9		2		1				
6	8		5			9		
	1		8	9			5	2
			9			6		
2		7	6	5	8	1		3
		5			4			
8	4			2	7		3	
		1			9		8	4
				8		2		9

Easy # 707

	5	1		7				2
	7	2			4	3		5
					8		6	
			9	8		4	2	
		8	4	6	2	1		
	2	4		3	1			
	9		8					
5		7	6			2	9	
4				2		5	3	

Easy # 708

		6	5	4				8
	4		8	2		3	5	
8					6			9
	6				5	1		
	5	1		6		9	8	
		3	4				6	
4			1					3
	3	9		7	4		2	
6				9	3	5		

Easy # 709

		9				5		7
3			7	9		8	4	6
6		8	3					
	4	1	9	2				
5			4		3			2
				1	5	4	9	
					6	2		3
1	3	7		4	8			9
2		5				7		

Easy # 710

	1	2	3	7	9			
			2				9	
9				8		7	5	
	5	1	6	9		4		
	7	9				6	1	
		8		5	1	3	7	
	9	4		3				7
	6				2			
			7	6	4	9	8	

Easy # 711

8	9			7			6	
	7				2	3		
		5		3	6		8	
		9			3	1		4
	6		7		4		2	
7		4	5			9		
	5		3	8		6		
		8	4				9	
	3			6			4	8

Easy # 712

6		7						1
			1		7			
5	4		9	3	8	7		
7	5			9	6		3	8
	6						7	
1	9		7	4			2	6
		6	3	5	9		1	7
			4		1			
3						8		4

Easy # 713

			3		9			
	5	2				3	1	
		1	8	2		7		
	6	9	1		2	5	8	
3				9				1
	1	5	4		8	9	2	
		7		5	1	6		
	9	6				2	4	
			2		4			

Easy # 714

		1		9		5		4
	4		5			3		
		2	1	4			9	
8	5		2				3	
		9	8		5	6		
	3				1		7	5
	2			1	9	4		
		8			6		1	
4		3		8		9		

Easy # 715

7	9			5				8
			7		9	6	2	5
			4	8			1	
9	7			3	4		8	6
5	8		6	2			9	4
	4			6	8			
3	6	7	1		2			
8				7			6	1

Easy # 716

				4	2	3		
	1				3		8	2
3			1	5		9		4
8			7		9	4		5
				6				
6		9	3		4			1
4		6		2	7			9
2	8		4				3	
		7	5	3				

Easy # 717

8				1				4
				4	9	2	6	
2	6	4	7			1		9
	2			8	1	3		
	3						2	
		7	2	6			4	
9		2			5	4	8	6
	4	3	8	9				
6				2				3

Easy # 718

		4	7		8	1		9
			5	1	9			6
1						8		
4	6		8				7	5
9			6		5			4
5	7				1		8	2
		2						8
7			3	8	4			
8		6	1		2	7		

Easy # 719

7						4		
	8		7	4		9	3	
		2	3	1				
6	3		8			1	2	
9	5		4		3		7	6
	7	8			1		4	5
				9	8	7		
	6	7		3	2		9	
		9						3

Easy # 720

5	3	2		7	4	1		
7	1	4						2
8					2			7
		3	5	1				
			8		7			
				3	6	9		
2			4					3
3						2	9	5
		9	3	2		4	7	6

Easy # 721

5	6			4	3		1	
	9	2	1			3		
1			9	6				
6	4		5		8		2	
				7				
	3		6		1		7	5
				1	4			8
		1			6	2	9	
	5		8	9			6	7

Easy # 722

					9		3	5
	5		1	7				
	2	9		6	5			
2	7				6		5	8
		3		8		6		
5	8		7				1	4
			6	9		1	7	
				1	8		4	
1	9		2					

Easy # 723

		1		7	2			4
	7	4	1	5		9		8
				9		3	7	1
2	1	9						
8								6
						1	5	9
4	9	2		8				
1		5		6	7	8	9	
7			2	3		4		

Easy # 724

7			5			6	8	2
		8	4		2		3	
	5			8	7	1		
1		3					4	
	2						5	
	8					7		9
		9	2	5			7	
	7		1		4	9		
4	3	6			9			5

Easy # 725

1	6		8	5		2		
				6		4	1	
		9	1		2			8
	1	8			6		7	
		5		8		3		
	3		4			9	8	
6			7		1	8		
	7	4		2				
		1		3	9		4	6

Easy # 726

4		9			3	5	2	
5	1	2		6	4			
3					2	1		
	4					2		
6	8	7				9	1	3
		1					4	
		3	9					1
			3	7		6	5	2
	5	6	2			3		8

Easy # 727

5		6			3	1	2	8
8				6				9
	1	9	2	5				
		4	6	8			1	
	9						6	
	6			2	7	9		
				1	5	6	8	
2				7				1
6	8	1	4			7		5

Easy # 728

	5	3	4				8	2
4	7			9				
		8		2	3	9		
7						6		9
5			9		7			8
1		9						3
		1	3	5		4		
				7			3	6
3	9				6	1	2	

Easy # 729

4		7	8					6
	2	9		3	6	8		
	8		7	9				
	9	3	2		1	4		
				5				
		6	9		8	5	2	
				8	3		1	
		2	1	7		9	5	
8					9	7		4

Easy # 730

9	3		6	1				
	8	6			2	3	1	9
		1		8		7		
	4		8	7				3
3								4
1				9	3		2	
		4		3		9		
7	1	9	5			6	3	
				6	7		4	1

Easy # 731

6	1							
5			2	4			8	
		4		3	6	2	5	
9	5	6	1			8		3
7		1			3	6	4	9
	7	5	3	8		1		
	6			2	5			8
							9	4

Easy # 732

1		4			3			
	3			9				4
2			8	1	4			3
	1		6	4				7
3	9						4	8
5				8	9		2	
9			7	5	1			2
7				2			9	
			9			8		5

Easy # 733

8			4				6	
				5				
	2	3	6		8	1		4
4		6	9			5	1	
7	5		1		3		8	6
	9	8			7	2		3
5		9	7		4	8	3	
				9				
	4				5			9

Easy # 734

			1					9
9	2			5		8		3
	6	7	3			5		4
4	9		7	1				
7			8		3			1
				9	5		2	6
6		5			1	4	8	
2		4		8			9	7
1					7			

Easy # 735

4		8	1		5			
		1	8				6	4
	3		2		4		9	
6			5				2	
9	4						5	8
	5				3			9
	6		3		2		8	
2	7				6	3		
			7		1	2		6

Easy # 736

8					1		6	5
1	2		5	8	9			3
7			3			9		
			2	6		3		4
2								9
4		8		1	3			
		7			5			2
6			8	4	7		3	1
5	1		6					7

Easy # 737

		8	2	7				1
9				8	1	4		
2			3		6			7
7				1	5	3		9
	3						4	
8		9	4	2				5
1			8		7			4
		7	9	5				3
6				3	4	9		

Easy # 738

4		1		8			2	
8		2			5	6	4	
					7			9
			3	7		5		2
		7	5	9	2	1		
2		5		6	1			
3			7					
	4	8	9			2		3
	5			2		4		6

Easy # 739

1						9	3	8
	9		3	8		4	1	6
8			1					2
				9	6		4	
			8		2			
	5		7	4				
4					3			1
7	3	8		1	4		5	
6	1	5						4

Easy # 740

		1		4	8		6	
8		2	7	6				3
						9	5	
3	2				7	5	9	1
1	9	8	3				7	6
	1	3						
5				7	1	8		4
	8		4	5		6		

Easy # 741

7	6					2	9	3
						7	4	
		5			4	1	6	
6		1		7	2			
		7	5		3	6		
			9	8		3		1
	1	6	3			4		
	5	3						
8	7	9					3	6

Easy # 742

		9	6		3			
				8			7	
		4	9		7		2	3
	9	5	8					6
2	8			6			5	7
7					2	1	8	
9	1		2		6	5		
	5			4				
			1		9	7		

Easy # 743

2			3	7		8		
1	9	7			8		3	
		5	9		6			7
		6					2	5
		8				9		
3	4					7		
4			6		2	3		
	8		4			5	6	1
		3		8	9			4

Easy # 744

		3		5				7
5			1	4	7			3
4	2				3			
		5	3	2				4
2		9				3		6
7				9	8	1		
			6				9	1
6			9	1	2			5
9				3		6		

Easy # 745

9			4			7		5
		8		7		4		
7				2		1	3	
2		1		3		5		
	6	9	8		2	3	1	
		3		4		8		2
	9	7		8				3
		4		6		9		
8		6			1			4

Easy # 746

		4		7	2	1		
		8	4	3			2	
1	2							9
3			1			9	7	
	8	6		2		5	1	
	4	1			7			2
4							9	8
	7			8	3	4		
		5	2	9		7		

Easy # 747

7			4					
		8	7	6		5		
		9				3	7	6
	2	3	6					9
	9	4	5		8	7	3	
5					4	6	8	
9	1	7				8		
		6		1	5	2		
					7			3

Easy # 748

3		2	9					1
6		5	2	3	4			
	4		7	6		2		5
		6					9	4
4	5					1		
2		8		4	5		1	
			8	7	2	3		9
5					6	4		2

Easy # 749

5		7		9	4	3		
				5			9	8
	6				8	5		7
	2				7			
9	4		1	2	5		6	3
			8				1	
1		5	2				8	
8	9			6				
		6	5	8		2		9

Easy # 750

5		2		4			8	
	9	8			7		6	
		7		8		3		
		9		5		2	4	
2		5	4		3	6		1
	4	3		7		5		
		6		1		7		
	7		2			1	3	
	5			3		8		6

Easy # 751

							6	8
7	1			2	8		5	
6	5				1	3		
			3	8		2	4	
5			2	4	9			1
	8	2		1	7			
		4	1				3	9
	9		8	7			1	6
1	6							

Easy # 752

9			2	3				5
8						3	4	7
	6		7					
5		8	1				2	
7		6	8		2	4		1
	4				5	9		6
					1		7	
6	5	7						4
2				5	7			8

Easy # 753

	9	5	1	4		7	8	
								2
		6		8	5			
		3				6	9	
6	2	9	8		4	1	7	3
	1	8				2		
			5	1		8		
5								
	8	1		7	2	4	3	

Easy # 754

		1			9	7	8	
	2	9	7	1	6	5		
		3	5					6
			2	8		4		5
		2				6		
1		4		9	5			
3					7	2		
		8	1	4	3	9	5	
	9	7	8			3		

Easy # 755

		6	2	7		3		
		4				5	7	1
8			5					
	4	3	9					2
	8	5	4		2	9	1	
1					3	8	6	
					9			5
3	5	8				1		
		2		3	5	4		

Easy # 756

6	7			2	8		3	4
			3	6			1	
		5						
4	1						9	
7	8	9	2		6	1	4	5
	5						6	8
						3		
	6			8	3			
9	2		5	7			8	6

Easy # 757

	6	9	4		7	3	1	
				8				
		7	1		9	4		
3			5		1		2	8
	4			7			3	
5	1		2		3			7
		5	3		2	1		
				1				
	9	3	7		4	8	6	

Easy # 758

4		1			9			
	6				1			8
	5	3				1	7	
3					6	7	9	
	2	6	9		7	3	8	
	1	7	2					4
	4	5				9	1	
1			6				4	
			7			2		5

Easy # 759

	9		4	8		7	3	
		2	3	7				6
3					5			4
	6				4	3		
	1	7		6		5	2	
		5	2				6	
7			6					1
1				4	2	6		
	2	3		9	1		4	

Easy # 760

6				7		5	4	
		4	1	2			6	
9						8		7
				8	5	4		3
4	2			9			8	6
8		9	4	1				
2		8						5
	6			3	9	2		
	4	1		6				8

Easy # 761

1				7	2			
							6	
2		8	5	4		7		9
3						8		1
8	1	6	7		4	9	3	5
7		5						6
5		7		9	6	3		4
	2							
			2	5				7

Easy # 762

			5	3		1		4
				1	9			6
3	1		2					
9	8		4				6	1
		7		9		5		
4	2				5		9	8
					3		8	7
8			1	4				
2		3		5	8			

Easy # 763

6			8			9		
		2	3	5				
5			2	6		3	1	
		3		9			2	4
2		6	1		5	7		9
9	1			3		5		
	9	1		4	8			5
				2	3	4		
		8			7			3

Easy # 764

		6	5	7			8	
	8				6		4	
7			8	2		1		5
6					5	3		
5		3		6		4		8
		1	7					6
1		4		9	7			2
	7		3				1	
	6			4	1	5		

Easy # 765

7		1	8			3	6	
3				6		2	7	
	4		5					
	3	8		5	9			
		2	3	4	8	5		
			2	1		8	3	
					5		9	
	1	7		3				8
	9	3			4	6		7

Easy # 766

4		5	9		7		3	
9		1		8		5		
	6		1		5			9
1	9			5				
			3		4			
				1			2	3
8			5		2		9	
		9		6		7		8
	4		8		9	2		6

Easy # 767

1					2		8	
7		9				2		5
	4	2			6			
	9				1	5		6
3		1	6		5	9		8
2		5	3				4	
			5			3	7	
4		7				6		2
	2		1					4

Easy # 768

	9							2
					9	1		
4	7	8		1	5		9	3
	4	3	9				1	7
	5		7		2		6	
2	6				8	9	3	
3	2		5	8		4	7	6
		4	6					
7							8	

Easy # 769

	7		9		6		3	
4			7	8			2	
	8			1	4			9
	1			5	7		6	4
		7				8		
8	4		1	3			9	
7			3	6			4	
	3			9	5			6
	9		2		8		5	

Easy # 770

6			5	9			3	
		4				6		1
		1		3			8	7
				7	8	1		5
	1	3		5		8	6	
8		9	4	1				
4	8			2		3		
1		2				5		
	3			6	7			8

Easy # 771

	9		7		3	6		
		6		2	8		9	
2			9			1	5	7
9	6					4		
		2				8		
		7					1	3
8	5	4			2			6
	3		6	4		2		
		1	8		7		4	

Easy # 772

5			6	7			4	
	8	7				5		
4				8	2			9
		2	4				5	1
9	1			2			7	3
8	4				1	6		
1			2	4				5
		8				1	2	
	2			6	5			7

Easy # 773

9			7				3	
	2		9	4				
6			3	2		5	8	
	6			9		8		5
5	7		6		8		1	4
2		4		5			9	
	9	8		1	4			6
				6	9		4	
	5				3			1

Easy # 774

		9						1
5			4		1	6		9
			7	9	6	8		
	8	5	1			7	4	
		6	8		7	5		
	4	7			9	3	1	
		4	2	1	5			
8		1	9		3			4
3						1		

Easy # 775

			5	9			3	2
1		5	3					
				6	4			5
5		7	9			2		6
	9			7			1	
4		8			6	5		7
8			7	4				
					2	4		3
6	4			3	9			

Easy # 776

		8			1			
	9	1		3		6		8
6			9		5		4	
				1	4		3	6
8		6				7		4
4	1		2	7				
	8		1		9			2
2		4		6		9	7	
			7			8		

Easy # 777

				1	8	6		
6			3	7	4			9
9		4	2				7	
				6	2		5	3
4	2						9	6
7	6		5	4				
	3				5	9		7
5			6	8	7			1
		7	1	9				

Easy # 778

9		1		5	7			
					3			7
	5		4	1		8		3
5	4	8		3				
1	9		7		4		5	2
				9		6	4	8
4		7		6	9		2	
3			5					
			2	7		4		9

Easy # 779

	7		3					4
	4	3		1	5	2		
1		2		6	4	7		
				9	6		5	1
4	9		1	5				
		5	9	8		4		7
		4	5	3		9	6	
8					7		2	

Easy # 780

				6	5		9	
3					9	5		4
		9	3	8		6	7	
		4	1		7	8	6	
				2				
	7	2	9		6	3		
	2	6		5	1	7		
4		5	6					9
	1		8	9				

Easy # 781

		7	4	2		5		8
	8				5			9
		9	8	3		7	2	
4	2		3	1				
				4	2		8	1
	9	8		6	1	4		
7			9				6	
3		1		5	4	8		

Easy # 782

7			1	4	2			6
		6		7				2
4	3				6			
		7	6	3				4
3		5				6		9
2				5	8	1		
			9				5	1
5				6		9		
9			5	1	3			7

Easy # 783

	4			3	2			5
	5	3	4	7			9	8
				9		3	1	4
2	9	4						
8								6
						7	4	9
5	2	9		8				
4	7			6	3	9	8	
3			2	1			5	

Easy # 784

1	2	5		7				
	4		5	9		2		
	6	3		8	4	7		1
						6	1	3
	7						8	
6	5	1						
4		2	6	3		1	7	
		6		4	5		2	
				1		9	6	4

Easy # 785

7				8		1	6	
4			9				3	8
	1			3			4	
2	8			4			7	
	7	9	2		8	3	1	
	5			7			9	2
	4			6			8	
5	6				4			1
	9	7		2				6

Easy # 786

	5				3	7		1
		3	8	4	7	9		
7			9	1				
	8	7	3	6				
	2	6				8	1	
				8	2	5	3	
				9	4			8
		8	5	7	6	1		
6		1	2				7	

Easy # 787

				2	5			1
3		1	7					
			1	6			7	8
1		4	6			8		2
	6			4			3	
5		9			2	1		4
2	5			7	6			
					8	5		7
9			4	5				

Easy # 788

2	9		3			8		1
	6			1		4		9
		8	7					
6		1		4	5			
7			6	3	1			5
			8	7		1		6
					7	3		
5		9		2			1	
1		2			6		9	4

Easy # 789

8			6	7				
	9	6		8	3		7	2
		4						
	5		1			7	2	
7	6		4		5		3	9
	4	8			7		6	
						5		
4	8		2	5		3	9	
				4	9			7

Easy # 790

	6		5			3		
		1		9		5		6
		2	1	6			9	
8	5		2				3	
		9	8		5	4		
	3				1		7	5
	2			1	9	6		
6		3		8		9		
		8			4		1	

Easy # 791

		6					7	
4	2	9		5	8	6	1	
					6			5
1		2	6			5	9	
		8	9		7	3		
	7	3			4	1		6
2			3					
	1	7	8	4		9	3	2
	9					4		

Easy # 792

	8			2	7		3	
3		7						9
	1		8	4		7		
4			3			2	9	
	5	1		7		3	6	
	3	8			2			7
		2		1	4		8	
8						9		1
	6		7	9			2	

Easy # 793

3				4	1	8		
4		1	6			3		
	8	2				1	9	
9			1			6		
2	1		8		9		7	3
		5			7			2
	4	8				2	3	
		3			4	7		1
		9	3	1				8

Easy # 794

9	4	6	5		8			
		3		9		5	4	
	7			4	3			
	3	2	4	8		7	1	
	9	1		6	7	4	3	
			7	3			5	
	1	9		2		3		
			9		1	2	8	4

Easy # 795

	9	2			1		5	
4		3		7			2	
		1		2		8		
		9		4		3	7	
3		4	7		8	5		6
	7	8		1		4		
		5		6		1		
	4			8		2		5
	1		3			6	8	

Easy # 796

8				6				9
3		9		7			5	
6	7				1		8	
5				8			4	7
9		6	7		4	1		5
1	4			5				2
	9		8				2	3
	3			4		5		1
7				3				8

Easy # 797

		6	5	4			1	
	7	8		6	1	9		
				2	7	3		
	3							4
8		1	4	5	3	2		7
7							3	
		9	6	3				
		3	2	7		1	6	
	4			8	9	5		

Easy # 798

		3	1	8			6	7
7								
		9	3	7	2			5
	9	4	6			8		
2		8	7		1	6		4
		7			3	5	9	
8			5	6	4	7		
								8
1	7			3	8	2		

Easy # 799

	8	2		5	3		9	4
		9	6	4			3	
4	7	3		8				
						6	8	3
		1				2		
5	3	8						
				2		9	6	8
	9			7	6	4		
8	2		4	1		3	5	

Easy # 800

6	1		8	2		4		
		5	4		9		1	
				3		6		9
5		8			6			2
		2		5		7		
9			1			5		4
4		6		1				
	5		3		4	8		
		3		7	5		4	1

Easy # 801

	3	6		4	7			1
							3	
	5		9	3	1			2
4					6	2		8
6	8		7		3		9	4
5		2	1					3
3			8	6	5		4	
	4							
9			4	1		3	7	

Easy # 802

1		2	7				5	4
		4		5			2	3
	6		8					
7	4			8	9			
3			4	6	7			8
			3	1			4	7
					8		9	
2	1			4		7		
4	9				6	2		5

Easy # 803

	2		3	5		8		1
3		1		8	2			
					6			4
9	1	5		3				
2	6		1		8		3	7
				4		9	1	6
8			4					
			8	6		7		3
4		9		7	1		6	

Easy # 804

	8				5		1	
2	1	9					5	
5	4	7		1	2			9
7			4	9				
			8		1			
				7	6			3
3			7	5		1	6	2
	7					3	4	5
	5		2				7	

Easy # 805

8	9				7	6	2	
					4			8
1	6			9		3		
			5	1			3	9
	5		9	7	3		4	
9	3			4	8			
		9		2			5	6
7			4					
	1	6	3				9	2

Easy # 806

2			3		7		4	
8	6	7			4	9		
	4		5	9				2
1						4	2	
5								9
	8	3						7
9				1	2		3	
		2	9			5	6	1
	1		7		5			8

Easy # 807

9				3	6			4
4		7	1			2	6	
	1	6		8				
	6						4	2
	5		8		2		3	
1	2						8	
				2		8	9	
	7	5			9	3		6
2			6	7				5

Easy # 808

	8		6	3		4		
	1	9		2	4	8		3
3	4	5		9				
						9	6	4
	7						1	
2	9	4						
				1		6	8	9
9		1	3	7		2	4	
		8		5	6		3	

Easy # 809

4		1		2	3			
					7		1	2
8			9	1				
1	8				4		5	9
		3		9		6		
5	9		3				7	4
				4	1			5
6	5		2					
			5	3		2		7

Easy # 810

	7		8	4		6	1	
		8			2			7
	1		5	6			2	8
5		6	4	3				
				5	6	8		3
4	3			2	5		8	
1			7			9		
	8	7		9	3		5	

Easy # 811

8						2		4
4	3		8		1			
	1	9		2	5	3		8
	6				2	9		5
	9						8	
5		8	1				2	
9		3	6	4		8	5	
			9		8		3	2
6		5						7

Easy # 812

					3	1		
2							6	
6	3	1		2	4		9	5
9		8	2				5	3
3			5		6			4
7	6				8	9		1
8	9		4	7		2	1	6
	5							8
		7	8					

Easy # 813

	2	8		4				
4			1	2			8	7
1	5		7			2		
			2			5		
	8	3	5	7	1	4	9	
		7			6			
		4			2		6	1
6	1			8	3			9
				1		8	2	

Easy # 814

		3		4	6		8	
6		8	2	5				4
						9	2	
3	5				9	8	7	2
2	7	4	5				1	9
	4	7						
9				3	5	1		8
	3		8	6		2		

Easy # 815

								6
				2	9	3		
	2	7	5	3			8	9
2	7				1		4	
	5	8	4		6	2	9	
	9		2				6	3
5	8			4	7	6	3	
		2	8	6				
4								

Easy # 816

		7			4		6	2
5	4			7	6		1	
				1		5		7
		2			7			
9		1	6	4	2	8		5
			3			4		
7		5		6				
	9		8	5			3	6
3	6		7			1		

Easy # 817

9					1			
		2	7		4		9	
7	1			5		2		8
				1	2	8	4	
1		8				9		5
	3	5	8	4				
5		9		3			7	4
	8		6		7	5		
			4					9

Easy # 818

4	5			9	7		2	1
3				1	8			
1			6			4		
				6	5	9	4	
	3						5	
	6	4	9	2				
		7			1			5
			2	7				4
2	8		5	4			6	7

Easy # 819

				7	6		8	3
	3	2	1				7	
	7	6	2			1		4
7	6		8					1
	4						9	
1					9		2	8
9		5			2	4	1	
	1				7	9	3	
3	2		9	1				

Easy # 820

3					2	1		6
9	6		1	3				
	1	4			9	7		3
	3				1		8	9
7								1
5	2		8				3	
2		5	9			3	7	
				2	5		6	8
6		9	3					2

Easy # 821

8	2				4	1		3
	4				5			
3	6			1		7	5	
				7	8	6	2	
	5		1		9		4	
	3	7	5	4				
	7	6		8			9	1
			4				7	
5		2	9				3	8

Easy # 822

9		2				3		1
			5				7	9
	8		9			4		
2		5	4				1	
1		8	2		5	6		4
	7				6	9		2
		7			4		9	
6	3				2			
5		9				7		3

Easy # 823

8	5		4		2		7	9
				3				
1			8		6			3
	3	1	6		8	4		
	2			4			8	
		8	1		3	9	6	
4			3		5			2
				9				
5	7		2		4		3	8

Easy # 824

			2			9	5	
	2		1	5	3		4	
	1			4				2
	5			9	2			4
2	7						9	8
3			6	8			1	
7				2			8	
	4		9	3	8		7	
	3	8			7			

Easy # 825

			4	2	8	5		1
6					3	8		2
5		8		1	7		4	
4	3					1		
		6					5	4
	6		5	4		9		8
8		4	1					5
3		2	8	7	9			

Easy # 826

	9						5	
1			5		2		3	7
					6	8		2
9			8		1	4	7	
4		7		2		1		3
	1	3	6		7			8
6		1	7					
7	3		4		9			6
	5						1	

Easy # 827

			3			1	8	
6		8				4		2
	2		7					6
2		3	1				6	
1		7	4		3	5		9
	5				7	3		4
7					2		9	
8		5				2		3
	6	2			4			

Easy # 828

	4	6	8		3		1	
		9				5		
5	1		4					
6		5	1		4		7	
4	8			2			6	5
	3		7		5	4		8
					1		2	7
		3				9		
	5		9		2	6	4	

Easy # 829

	8		3			9		
5	9			6	8		7	1
	6			9	7			
				7	4	6		5
8								2
6		4	8	5				
			1	3			2	
7	3		9	4			6	8
		6			5		3	

Easy # 830

	4		1	5		2	8	3
3							9	6
9	2		7					
5		4	6	8				
	9		2		5		6	
				9	4	5		8
					2		7	1
6	3							4
1	7	5		4	3		2	

Easy # 831

	9		8		6	2		
			5					3
2		3		7			6	5
	7	2	9	5				
4		9				3		2
				4	1	9	5	
6	4			2		1		9
3					4			
		1	6		5		3	

Easy # 832

					6			3
	7		8	1		2		6
4		1		7	3			
7	8	2		6				
1	4		3		8		7	9
				4		5	8	2
			9	3		8		4
8		3		5	4		9	
6			7					

Easy # 833

3				9	5		6	
6	8			7		3		
	2	7				4		
	6	1	8	2				
2		3		4		6		9
				5	6	2	4	
		8				9	2	
		2		3			5	6
	9		4	1				3

Easy # 834

		8			4			5
	6	5				8	7	
7	3				9			
8					3	5	9	
	1	2	9		6	3	4	
	9	6	4					1
			6				5	8
	5	9				7	1	
2			5			4		

Easy # 835

	6				7			3
	5	9		1	6			8
				4	3		1	
9		5		3			8	
4	2		5		8		7	9
	3			9		4		1
	4		3	8				
8			4	2		5	3	
2			6				9	

Easy # 836

	1		8		4			6
7	6	8			9	3		
2			3	6			9	
	4					2		1
	9						8	
3		5					6	
	3			9	8			5
		9	5			4	1	7
5			4		2		3	

Easy # 837

	9			1			7	
				5	2	4		9
4	7	2	8				5	1
		4		7	1			6
		1				9		
9			3	2		1		
3	5				6	7	1	4
1		7	5	4				
	4			3			2	

Easy # 838

5	1		8	2		3	7	4
	4							2
		3	7					
7	5				2	6		1
8			4		5			7
3		1	6				4	9
					6	9		
6							5	
4	3	2		9	8		1	6

Easy # 839

	5		9			3		7
	9	3	5		1			
4			6		3			2
		7	1					6
3		2				9		1
1					4	2		
7			4		6			9
			8		5	7	6	
8		6			7		4	

Easy # 840

7			3					
	4	3	2	7		9	6	5
	1					3		
	9	7			3	6		4
		8	1		9	2		
3		4	5			8	1	
		5					9	
6	8	9		5	2	1	4	
					8			6

Easy # 841

		8			2		9	1
			1			6		
	1	9	6				4	2
8						7	5	9
7		6				2		4
9	2	1						6
5	9				6	1	3	
		3			5			
4	6		3			9		

Easy # 842

2			8					
8	6			5	3	4		
			2	4			5	9
				8		3	6	4
7		4	3		2	9		5
6	1	3		9				
9	3			2	7			
		7	9	1			2	3
					4			8

Easy # 843

4		9		6	1		8	
		5	3					
			8	4		1		9
				1		7	9	6
	1	2	4		9	8	3	
7	9	3		5				
2		1		3	4			
					5	4		
	3		9	2		5		7

Easy # 844

		2	1	6		9		7
7			4	2				
	6							2
	3	8	5			7		4
	9	7	6		8	3	2	
5		1			4	6	9	
8							7	
				5	6			1
2		6		8	7	4		

Easy # 845

3		9	5	1		6	2	4
6						1		
	2		4					
9		4			1	3	7	
		5	6		9	4		
	3	2	7			8		6
					7		8	
		7						9
2	1	6		8	5	7		3

Easy # 846

6				4	1	2		
	9	8					5	
2		3		9			6	
	7	2	3	8				
8	6			5			2	4
				1	2	5	8	
	8			6		1		2
	3					8	4	
		4	5	7				6

Easy # 847

		8		7		4		
5		7			8			3
	6	1		9				7
		5		6		1		9
	1	6	9		4	3	2	
9		4		8		6		
6				4		7	3	
8			1			2		4
		3		2		8		

Easy # 848

	8		6			9		
3		5		4	9			8
6	9			5	2			3
				1	4	5	2	
	1	9	5	2				
9			2	6			4	1
2			1	7		8		9
		7			8		3	

Easy # 849

	2	7	8			5		
				5	2	9		7
	4	5	9				3	2
6		9	2					5
		2				4		
5					6	1		8
4	5				9	8	1	
7		6	1	8				
		8			5	7	9	

Easy # 850

						4		
8	2		7	4		9	5	
				8	5			3
	8	2			3		1	
3	1		8		4		9	5
	4		6			3	7	
2			1	3				
	5	1		2	9		3	7
		8						

Easy # 851

			3		4		9	1
		5	1		6	7		
1		8			9		4	
		6			3			8
9		3				1		5
5			7			3		
	7		8			2		6
		9	6		7	8		
8	6		4		2			

Easy # 852

					1		6	4
	7	4		6	3			
	5		9	4				
5	4				7		9	2
		3		9		8		
9	2		3				7	1
				7	4		2	
			2	3		6	1	
2	8		6					

Easy # 853

		3	7		8		2	
5	2			6				3
7	6		3		2	1		
	4	7		9				
			1		4			
				8		3	9	
		4	5		3		1	8
8				2			3	9
	3		8		9	6		

Easy # 854

	5	1	8			9		2
			2	3		5	4	1
		4	5					8
		5					2	
8	4	9				6	7	3
	2					4		
4					9	8		
5	1	3		6	8			
7		8			5	3	1	

Easy # 855

	6				2		8	5
	7	8	5	6				
1		5			7		6	3
		6			5	9	7	
	3						5	
	4	2	9			6		
4	2		7			3		6
				2	4	8	9	
7	8		6				2	

Easy # 856

	6	9		1			4	
			4	9		3		
			2		6	9	7	1
	3	8		2	9	4	5	
	9	4	3	7		1	8	
9	5	2	8		1			
		6		4	3			
	4			5		8	1	

Easy # 857

				2		6		8
6	3			8	5		2	
		8			3		5	4
		4			8			
1		2	5	3	4	7		6
			9			3		
9	5		8			2		
	1		7	6			9	5
8		6		5				

Easy # 858

	5				3	7		1
				9		5	6	
3		6		5	1			9
	7				5			
	9	8	1	3	7	6	2	
			4				3	
8			2	6		1		4
	6	5		1				
1		4	5				9	

Easy # 859

3	5			8	7			9
	6		2			1		
9		2		1	3			7
				7	4	9	3	
	7	4	5	3				
2			9	5		4		6
		9			8		2	
6			7	4			9	8

Easy # 860

9			2					
		8	9	3		7		
		6				5	9	3
	1	5	3					6
	6	2	7		8	9	5	
7					2	3	8	
6	4	9				8		
		3		4	7	1		
					9			5

Easy # 861

8				4	5		1	
	9	1		6			5	
	6				3			4
9					4	7		2
	5		6		7		3	
7		6	8					9
1			7				9	
	4			5		1	7	
	8		4	1				5

Easy # 862

						2		
1		2		3	5		4	
		9	7	2	4		8	
	3				1		6	8
	1	6	5		2	7	3	
8	9		4				2	
	2		6	1	9	3		
	7		3	4		5		2
		3						

Easy # 863

7			9			4		
	2	4				7	1	
			8				3	2
	8	7	3					4
	9	3	1		8	6	5	
5					9	1	8	
4	7				1			
	5	2				8	7	
		9			7			6

Easy # 864

	3		6	5			1	
		6		2	1		8	
7						6		3
	7	5			3			2
	4	3		6		8	9	
6			5			1	3	
8		7						1
	1		2	8		5		
	5			7	6		4	

Easy # 865

3	5		8	2		6	4	
						2		
				3	4			1
	3	5			1		9	
1	9		3		2		6	4
	2		7			1	8	
5			9	1				
		3						
	4	9		5	6		1	8

Easy # 866

	2	7			3		6	9
	1	3		2		4		8
		4			5			
				5	6	7		4
		5	3		1	6		
8		9	2	4				
			6			5		
4		6		1		8	7	
1	7		5			9	2	

Easy # 867

	1	4		5		7		9
9			4		8		2	
		2			1			
				1	9		8	7
7		1				5		2
5	6		7	8				
			8			2		
	7		3		4			5
2		5		6		8	4	

Easy # 868

	4		7	8	3			2
	9						8	3
		7	9			1	4	
		4		9			2	
3		8				6		9
	5			3		4		
	7	2			5	9		
4	6						1	
5			4	1	9		6	

Easy # 869

		5		6			8	
	9		7	2	8		5	
8	2				5			
		2	3	8			1	
	5	6				8	7	
	4			7	6	9		
			6				4	7
	6		1	4	2		9	
	1			9		6		

Easy # 870

		8			4	3	5	
	7		2	5			4	9
				8	1			2
	6		4		8		9	7
				9				
4	1		7		2		3	
8			5	4				
7	4			1	6		8	
	5	3	8			6		

Easy # 871

	8	5	3				1	2
		1		2	5	4		
3	7			4				
7						9		4
8			4		7			1
6		4						5
				7			5	9
		6	5	8		3		
5	4				9	6	2	

Easy # 872

7			8	6		5		2
5		2						
	3		2			7		9
8	4			2	6			
		1	4	3	7	2		
			9	8			4	3
1		5			2		9	
						8		5
2		6		4	8			1

Easy # 873

2		7	5	6	8			
	8			4		3		6
			2			8		
7		3	9	8				1
8		6				7		9
4				3	7	6		5
		9			2			
1		8		5			6	
			6	9	1	4		8

Easy # 874

		9			5			8
3		2			9		5	6
5	8	6		1	2			
	2							5
4	7	1				9	8	3
8							2	
			9	4		5	6	1
1	6		5			7		9
9			3			8		

Easy # 875

			7	4	1	2		9
1		9		3			7	
		4			6			
2				5	8	7		3
9		7				8		4
8		5	4	9				1
			6			9		
	9			2		5		7
6		8	3	7	9			

Easy # 876

	3	1					9	
	2	9	6	7				
7		5	3		8			6
1						6	3	
			8	2	4			
	7	4						2
2			4		5	3		1
				6	3	5	2	
	5					7	4	

Easy # 877

		7		5	2		4	3
	5		1				9	
1			7			8	6	
2	1					7		9
		5				4		
8		9					5	2
	6	1			3			4
	2				8		7	
7	3		4	9		5		

Easy # 878

1			5			4		
	5	4	3					6
	3	2					5	
3		5		9		6		2
	6		2		3		1	
2		8		4		5		7
	2					7	8	
5					9	1	6	
		9			7			5

Easy # 879

	7	6					4	
4			2	7		1		
1				6	8			9
	8		1			4		5
9		5		8		7		3
6		1			5		2	
5			8	1				4
		8		2	4			7
	6					8	5	

Easy # 880

	4	8			6	1		
6	9	2			7			
5			4					7
9	6			8				3
	7			4			2	
8				3			7	6
4					5			2
			8			5	1	9
		9	2			6	4	

Easy # 881

		1		8		4		
	8	6			3	1	2	7
2	4		7	6				
	5		8	1				2
4								8
8				7	9		4	
				2	6		8	1
1	2	8	5			6	9	
		7		9		2		

Easy # 882

4			8	5	2	6		
	8		6				3	7
5						2		4
	2			9				6
	4	8				9	1	
3				8			2	
1		9						8
2	5				8		7	
		3	9	1	7			2

Easy # 883

		8	5	1	7	3		6
	3			6				
		5	8	4		1		
3				5		8		
8		9				2		5
		4		2				3
		7		9	1	5		
				7			4	
5		2	4	8	6	9		

Easy # 884

4	9		7					2
				6			4	
		2	3	4	8	1	9	6
1				2				
	7	9	8		1	6	2	
				7				3
9	3	1	5	8	2	4		
	8			3				
5					7		8	1

Easy # 885

		3						
	5		8	2		3		6
	7		5	3	4	9		
7	1		6				2	
	2	4	3		8	1	6	
	3				5		9	7
		2	9	6	1		3	
3		8		5	2		4	
						2		

Easy # 886

7			1				9	
	5		2	7				8
	2			8		7	1	
1		6	5					9
	8		6		1		4	
9					2	1		3
	9	7		6			8	
5				2	8		7	
	6				4			2

Easy # 887

					7			8
2	1			9	6			
		6	2	4			9	1
3	4	1		2				
6		7	1		9	2		5
				8		1	3	7
8	3			5	1	7		
			9	7			5	2
9			8					

Easy # 888

2			3					
		4	8	6		9		
		7				3	6	1
	7	9	5					8
	2	3	7		8	5	1	
1					9	2	4	
9	3	2				1		
		8		9	3	7		
					5			3

Easy # 889

5							9	6
7		4	1		3			
	1	9		8	6		4	3
		7			2		1	9
		1				3		
9	3		7			6		
1	4		9	7		2	3	
			2		1	4		8
8	7							1

Easy # 890

4					2	3		
				1				
	8	2	3		4		9	7
8	1				5	2	4	
3		4	9		8	6		1
	7	9	6				3	5
9	3		2		6	1	5	
				5				
		5	1					2

Easy # 891

	7		1				2	
4	6	3			2			
9		1			6			5
	3	6		9			8	
		2		1		4		
	9			8		2	6	
3			4			1		6
			9			5	3	7
	1				7		4	

Easy # 892

			4		6			2
	4	7	5		2			8
		5		3				
	6				3	2		1
	5	1		6		3	7	
9		3	7				5	
				8		1		
1			6		7	9	2	
5			2		9			

Easy # 893

7			5	8				6
8		3				7		1
			1		4			
4		9	7		8	5		3
	1			4			7	
3		7	2		5	8		4
			8		2			
9		4				2		8
6				3	7			9

Easy # 894

9	7			2	8	5		3
1	8	5		9				
	3		4	5				8
						8	4	9
	6						7	
8	9	2						
3				1	4		5	
				7		9	3	4
7		9	5	6			8	2

Easy # 895

	8	9		2			6	
	2				5			7
1				7	6		9	
8					7	4		3
	6		2		4		5	
4		2	1					8
	1		7	9				6
9			4				8	
	7			6		9	4	

Easy # 896

	6	5		4	9	1	8	
			7	5		6		
7								
	5	6				9		
2	9	3	6		1	5	4	8
		8				2	3	
								9
		2		6	7			
	3	7	5	1		4	6	

Easy # 897

	9	3			1		7	4
4					9		1	8
5		8	7	4				
		9			5	4		7
2								3
1		5	2			9		
				9	2	8		1
8	2		4					9
9	3		1			2	6	

Easy # 898

	3		7					
		1	3	2		6		
		9				8	2	3
5		8	2				9	
9		7	6		1	3		8
	6				7	2		1
4	9	3				1		
		2		4	6	5		
					3		8	

Easy # 899

			3					7
4		7		5			9	3
	6		2		9	4		
	5	4	6	3				
1		6				7		4
				1	8	6	3	
		8	9		3		7	
9	1			4		8		6
7					1			

Easy # 900

3			8	1	5		2	9
8			3	6				1
		9		2				
	9			8				3
4	3						8	7
6				7			9	
				5	6			
5				4	1			8
7	8		6	3	2			4

Easy # 901

	2			6	3	7		
				5		2	3	4
4	5		7	8		1	9	
9	1	4						
		8				5		
						3	4	1
	4	5		9	1		2	7
7	6	1		4				
		2	3	7			1	

Easy # 902

	9		7				4	2
			6			7	5	8
		6			2	1		
		3		4		8		7
5				2				6
6		7		3		4		
		5	1			2		
9	1	8			4			
2	7				5		8	

Easy # 903

	3	7			9	6		
2			3					5
9	1	4			5			
1	9			7				8
	5			3			4	
7				8			5	9
			7			2	6	1
3					2			4
		1	4			9	3	

Easy # 904

		1		3	2	9		5
7	6							
	3		9	8				4
6	7	4	2			1	5	
	2	3			1	4	7	9
3				6	8		9	
							4	1
9		8	4	2		6		

Easy # 905

	1			7			3	
4		5	2	1				
7	2				6	5	4	1
9			7	3		4		
		4				9		
		1		5	4			6
1	5	3	8				2	4
				2	3	1		9
	9			4			5	

Easy # 906

5					6	9		
8	9	6		1	4			
4		3			5	8	6	
	4					6		
1	7	2				3	9	5
		9					4	
	8	1	6			5		7
			5	2		1	8	6
		5	3					9

Easy # 907

				5		3		
	3		7		4	1		8
	2		8		1			
	1	9	4					2
2		3		7		9		4
7					9	8	3	
			6		7		8	
6		4	2		8		5	
		2		9				

Easy # 908

6	9				3			
		3		8			5	
	8		2	9	5		3	
		8	3	6			9	
	6	1				3	4	
	5			1	7	2		
	4		1	2	6		8	
	1			3		4		
			4				2	1

Easy # 909

3	6	1		7	5			4
	5				3		7	
7	2	4					5	
4			6	5				
			1		8			
				9	2			5
	7					9	1	3
	1		7				8	
9			3	1		5	2	7

Easy # 910

	4	5		2			9	6
7		1	9				3	2
			8				4	
	3	4	7	8				
	7		6		9		8	
				4	2	5	1	
	8				7			
2	1				8	6		3
3	5			6		4	7	

Easy # 911

		1	6				4	7
6		7	3		4			
	8		7		1		6	
9			1				2	
8	2						5	9
	7				2			6
	9		5		7		1	
			2		3	8		5
5	6				8	3		

Easy # 912

1			4		9	6		
		9		2	3		7	1
	4	7		8				
	2		7			3	6	
		5		6		2		
	9	6			1		4	
				1		7	9	
9	1		6	5		8		
		3	9		8			6

Easy # 913

9		6	8					5
1		8						
2	3	4				6	1	
			2	1			6	9
6			3		5			1
3	9			7	4			
	6	3				1	7	4
						5		3
8					3	9		6

Easy # 914

	5		7		3		1	
	2	4	9		6	8	3	
				5				
6			3		7	5		1
		3		6		9		
2		7	5		1			3
				2				
	3	5	6		9	4	8	
	9		8		5		6	

Easy # 915

	7					8		
		1	4	2				
6			7	8		3		4
4	5		6			2		1
9	3		8		4		5	7
7		6			2		9	8
5		7		4	1			3
				3	6	7		
		3					4	

Easy # 916

		1			8			
	7	3		6				1
			1	7	2	4	8	
	9			1	5	3	4	
	5	4				7	1	
	2	7	4	3			6	
	1	6	9	5	7			
7				2		1	9	
			8			5		

Easy # 917

		8		9	5	6	4	
9			6	3			7	
2	1							
1	2	7	5			8		4
5		9			8	7	6	1
							8	7
	9			2	3			6
	6	3	7	5		2		

Easy # 918

	7				5		2	4
2				8				
4	8	6	9	2	1	7		
				7			6	
7		8	6		9	4		5
	1			5				
		2	7	9	3	6	4	1
				1				9
9	6		5				3	

Easy # 919

	6		4		7	5		8
	8		1		6	9	7	
						6		
7		5			1	4		
6	3		5		2		9	7
		8	6			3		1
		3						
	7	2	9		4		3	
1		6	7		3		4	

Easy # 920

	1	9	2	5				
3				8				1
5	8				6	2	9	3
	7		8	3		9		
		1				8		
		8		2	4		1	
8	9	3	7				4	5
2				4				9
				9	5	3	8	

Easy # 921

		7				8		
1		3	6		7			5
6	2		9					
	4	1	5		2			8
3	5			6			1	4
2			1		9	5	3	
					1		5	9
9			8		4	3		1
		5				7		

Easy # 922

8			5		4			2
4		3	2		8	5		7
				1				
	7		6		5	9	1	
		2		8		7		
	6	5	9		7		8	
				5				
7		4	8		2	3		1
6			7		9			5

Easy # 923

7			2					6
	6	3		5	7		9	
1				3	6	4		
		6	7				1	
	4	2		1		3	8	
	1				4	2		
		1	4	7				8
	7		8	9		6	4	
8					1			3

Easy # 924

2								
	5		8	9				
8		4		5	7	9	1	
		6	3			1		9
	9	8	2		6	7	4	
5		2			9	8		
	2	5	1	6		4		7
				2	4		9	
								6

Easy # 925

3	5							8
		7		2	6		3	
	4	9		7				5
5	6		9	4				
9		3		6		5		7
				5	8		9	2
7				1		9	8	
	9		4	3		7		
6							5	1

Easy # 926

7	1			5		9		2
		6			7			4
			4	2				
3	9			6		8		
4	7	2		8		6	3	9
		1		7			4	5
				4	6			
9			7			5		
5		8		3			7	1

Easy # 927

				8			5	
5			7		2	4	6	
9			4		6			
6	3		2			9		
	5	9		7		2	3	
		7			3		4	5
			1		7			4
	2	1	9		4			8
	9			3				

Easy # 928

	8			6	2			7
				1		6	9	8
	7	6	8	4			1	5
2	1	8						
5								3
						4	8	1
8	4			3	6	1	5	
7	2	1		5				
6			2	9			7	

Easy # 929

	7				3	5		
	6	5			1	4		8
8	1	4		9	5			
2						7		
7	5	3				9	4	6
		1						2
			2	4		1	8	7
1		8	5			3	2	
		7	1				5	

Easy # 930

								9
	9		2		6	5		8
	5		3		9		6	4
8		6			3			2
	7	9	8		1	6	4	
5			9			3		7
1	6		4		2		7	
9		3	6		7		2	
7								

Easy # 931

	6			8	5		4	
3	4						2	7
			2		7			
5	8		7		9		3	2
		1		3		5		
4	3		5		2		8	9
			1		3			
8	2						1	5
	5		9	2			6	

Easy # 932

	8	1	6			2		
	6	3						
4	7	9					8	3
			9	3		1		8
		8	4		2	3		
1		4		5	7			
8	4					7	3	5
						4	2	
		6			4	8	1	

Easy # 933

				1	3	4	5	
7				8				1
4	5	1	6			8		3
	4			7	8	2		
	2						4	
		6	4	5			1	
3		4			9	1	7	5
5				4				2
	1	2	7	3				

Easy # 934

8			2	1		9		7
		1			9		5	
7			8	3			6	2
	2	7	4	8				
				2	6	4	8	
3	7			4	8			5
	9		3			7		
5		4		6	7			9

Easy # 935

1			3					5
	9	5		4	1	6		
2				9	5		7	
	5		1			2		
	3	7		2		8	9	
		2			7		3	
	2		7	1				8
		1	8	6		7	5	
8					2			9

Easy # 936

6	5				3			
		3		1			7	
	1		4	5	7		3	
		1	3	6			5	
	6	2				3	9	
	7			2	8	4		
	9		2	4	6		1	
	2			3		9		
			9				4	2

Easy # 937

		2		6	9			
								8
	1	9	7	5		3	6	
		4				2	1	
2	8	1	6		5	7	3	4
	7	6				8		
	6	7		3	8	5	4	
9								
			9	7		6		

Easy # 938

	8	1	7			5		
4	7			8	9		2	
9			3	5				
2	4		5		7		6	
				4				
	1		9		2		3	7
				7	8			5
	5		6	3			7	2
		6			5	1	8	

Easy # 939

2	8				9			
	3	1				5	8	
		7			8			4
3					7	9	5	
	7	6	9		5	4	3	
	5	8	6					2
8			7			2		
	1	2				8	9	
			5				6	1

Easy # 940

		9				3		
1			3		7	5		6
					4		2	7
9			2		1	6	8	
8	6			7			1	5
	5	1	4		6			2
4	1		6					
6		5	8		9			4
		3				1		

Easy # 941

	4			8	9		6	
9		6				3		
	2		4	7				9
		7	6				3	8
2	1			9			5	6
4	6				8	9		
8				2	7		4	
		4				2		3
	5		9	3			8	

Easy # 942

		3		6	2	7		
4		2	9				6	3
5	9			7				
	5					1	7	
	4		7		5		3	
	8	7					2	
				5			1	2
7	2				1	8		6
		8	2	4		9		

Easy # 943

		9				6	7	1
		2	7	1		3		
7			8					
	4	6	1					9
	9	8	3		2	7	6	
3					8	1	2	
					7			6
		1		5	3	4		
9	5	7				2		

Easy # 944

			2				1	8
	1	5				4	3	
6			1			7		
	5	2	7					3
	3	6	5		2	9	7	
8					9	1	5	
		8			7			1
	2	1				8	4	
4	9				5			

Easy # 945

1	4			6	7			5
		7	8			1		
		9		4	1		2	
	1		7					9
2	8			9			4	3
9					2		8	
	9		2	7		3		
		3			9	4		
7			3	5			1	2

Easy # 946

2								3
	5		6		4		8	9
					8	2	5	
	1		8		5	9		2
	9	2		7		8	4	
8		4	2		1		6	
	7	1	5					
9	8		7		3		2	
3								6

Easy # 947

	2	3		6	9			4
4							1	
		8	7		4		3	
2			1				9	8
	5	1		4		2	6	
8	4				3			7
	1		8		6	7		
	8							1
9			4	5		3	8	

Easy # 948

5				6	4			
1	9		8	3			5	6
						4		
8							6	5
6	3	9	1		5	2	8	7
2	7							9
		8						
3	5			1	6		7	4
			4	5				2

Easy # 949

8					2			
	5		9		3	6		
2		9		4			8	5
				2	6	4	5	
5	8						6	1
	6	2	7	1				
6	7			5		1		9
		8	2		9		7	
			1					8

Easy # 950

		3	6		8	4	2	
					4	3		5
	5						9	
		7	4		3		5	2
5		2		1		8		4
8	4		5		7	6		
	9						6	
7		1	3					
	2	4	1		9	5		

Easy # 951

9	6	2		4				
1			2	7		9		
3		8		5	1	4	6	
						3	8	6
4								5
2	3	6						
	1	9	3	8		6		4
		3		1	2			9
				6		7	1	3

Easy # 952

6				1	7		3	
3	4			5		6		
	8	5				9		
	3	2	4	8				
8		6		9		3		1
				7	3	8	9	
		4				1	8	
		8		6			7	3
	1		9	2				6

Easy # 953

				1				
6		8	5		3	2	7	
	3				6			5
	8	1			9	3		6
3	5		2		8		1	4
2		7	4			5	9	
9			1				6	
	2	5	6		4	9		1
				9				

Easy # 954

8		9		5				4
			6		8	9	5	1
			4	9		2		
2		3		6	9	4		7
9		4	2	1		5		3
		8		4	2			
7	9	6	3		5			
4				7		3		5

Easy # 955

	1	6		8	5			
	5		9	2				
					6		3	5
1	2				8		5	4
		3		4		8		
5	4		2				9	7
9	6		1					
				9	4		7	
			8	6		9	2	

Easy # 956

				4	8	2		5
4	5	2	6				8	9
	3			9			4	
		5		3	9			1
		1				5		
6			5	2		4		
	2			5			1	
5	8				7	3	2	4
1		4	3	8				

Easy # 957

7	2	4				6	9	
6	8							
5	9		8					3
			7	6		9		5
9			4		3			6
4		5		1	2			
8					4		5	9
							3	4
	4	9				1	6	2

Easy # 958

		6		9				
	2				8	5	6	
7	9	5	4	6	1			2
				2			7	
9		2	7		4	8		5
	1			8				
6			2	4	3	1	5	7
	7	4	8				3	
				1		4		

Easy # 959

7		6		4	8		3	
8			1	9	5		6	
								8
	6				4	2	1	
3	8		6		7		9	5
	5	2	9				8	
6								
	2		4	6	3			1
	4		7	8		9		6

Easy # 960

	3		4	9		1		2
5		1	6					7
9		2	1	5	3			
		9					6	3
3	2					7		
			8	4	1	5		6
2					9	3		1
1		8		3	2		7	

Easy # 961

			1	9	4		3	
	9					8		
		3	8		7	6	9	
9	8		7				5	3
	1		5		6		2	
3	5				9		1	6
	2	7	9		3	1		
		9					7	
	6		2	7	5			

Easy # 962

			6	8	7	9		1
6		7		5	2		3	
		2			4	7		5
	2	5						3
4						5	9	
7		1	9			3		
	5		8	4		2		7
2		4	7	1	5			

Easy # 963

	6	3	9			7		
	9	1						
2	8	4					6	1
			4	1		3		6
		6	2		7	1		
3		2		5	8			
6	2					8	1	5
						2	7	
		9			2	6	3	

Easy # 964

		4	5		3			1
	5			6		4	3	
2			8		4	9	5	
				5		3		4
			9		2			
7		2		3				
	7	1	4		6			9
	8	6		1			4	
4			7		5	6		

Easy # 965

5								
8	7		5		9		6	
3		6	7		4		5	
4			9			7		3
	8	7	1		3	5	2	
2		9			5			6
	4		2		7	9		5
	2		4		8		7	1
								2

Easy # 966

	7	1	5				2	
	3	6	7	4	9			
8			2	1		9	7	
		8					1	2
3	1					5		
	2	7		5	4			1
			1	6	7	2	5	
	8				3	7	6	

Easy # 967

6		8	3					5
		9		8	5			7
1	5						3	9
		1	6					4
	6	5	7		9	1	8	
2					8	7		
8	7						9	1
9			8	3		5		
5					2	3		8

Easy # 968

	1						7	6
	7	4	1		3			
5		3		6	8		1	4
		2			6		8	5
		5				1		
1	8		3			6		
4	5		2	7		8		1
			5		1	4	6	
8	2						9	

Easy # 969

		1	2					5
		9	5	7			6	4
7			1	3				
9				1		6	4	
2		6	9		4	3		8
	3	7		6				1
				9	1			3
1	4			8	3	9		
6					5	8		

Easy # 970

			4		6			
6	1							3
8		4	2	9	3		7	
7		5		6	8	2		4
		8				7		
1		3	7	2		9		8
	8		1	3	2	6		9
4							8	7
			8		4			

Easy # 971

6	1							3
3	9	8		2	5		7	
					4		8	1
				9	6	7		2
	6		2		8		1	
9		2	7	1				
5	4		8					
	8		3	7		2	4	5
7							3	6

Easy # 972

		6	8		9		3	
8			1	7		2	6	
		5						8
3		1			5			2
	2	7		8		4	5	
9			6			8		3
5						3		
	6	3		4	8			1
	9		7		3	5		

Easy # 973

5			2				8	
	4	9				2		
	8	2	9					1
9	2			7			1	4
		1	4		9	5		
4	3			8			2	6
2					7	1	5	
		4				3	6	
	7				6			2

Easy # 974

9			1	3		2		4
		3			2		8	
4			9	6			5	1
	1	4	7	9				
				1	5	7	9	
6	4			7	9			8
	2		6			4		
8		7		5	4			2

Easy # 975

2	1		7		4	3	9	
		4			2			1
				1				
	9	1			7		5	3
7		2	6		3	9		8
4	8		1			6	2	
				2				
9			4			8		
	3	5	8		9		6	4

Easy # 976

4			8		7			3
		1	4	9				5
9				6	1	8		
6				2	4	1		7
	4						9	
1		9	6	3				8
		4	3	7				1
3				8	2	7		
8			5		9			2

Easy # 977

1				4		9		
	9	6			1			
		8	5	6	9	1		
6			3	9		7		
4		1				5		9
		2		5	4			8
		4	7	2	6	8		
			4			2	5	
		7		8				4

Easy # 978

6	8	1			3		9	
		4	8		7			1
2			9	1		3		
		7					2	4
		3				8		
9	5					1		
		9		3	8			5
5			7		2	9		
	3		5			4	7	6

Easy # 979

		2	5				6	
9	7		6		8	5		2
				2				
1		7	8				2	9
	9	3	7		4	8	5	
5	4				2	6		3
				5				
4		6	9		3		1	7
	3				6	9		

Easy # 980

	3	1		9	7		4	
		7	2	6				
8	9		3					6
	1	4	6		3		5	
				1				
	8		7		4	3	2	
5					6		9	8
				3	9	6		
	6		5	2		4	3	

Easy # 981

5	2			3		9		
					4			8
8	3				7	2	6	
			1	5			9	3
	1		3	7	9		4	
3	9			4	8			
	5	2	9				3	6
7			4					
		3		6			1	2

Easy # 982

7		3	2					
	6						5	
1	8		7		6			4
	1	9	4		3			5
8		4		7		1		9
3			1		2	8	4	
2			5		9		8	1
	4						6	
					1	4		2

Easy # 983

		4		8			6	7
		2	9			8	1	
	7			1			2	
	8	3		2			4	
9	4		3		8		7	1
	5			4		3	9	
	2			6			8	
	6	5			2	7		
4	9			3		6		

Easy # 984

		5	8		2	7		
2		9			5		8	
			9		4		2	5
		3			8			6
6		1				3		7
5			3			2		
1	7		4		3			
	4		7			5		1
		8	2		1	6		

Easy # 985

			1		4	8		6
	2		6		5		9	
6	7				8	4		
	5				1			7
8	1						6	2
2			9				1	
		9	7				3	5
	8		5		9		7	
7		5	4		3			

Easy # 986

3		1		6			2	5
			3					4
	2	6	4				7	9
9		5	7	1				
4			8		6			3
				4	3	1		2
2	7				8	9	3	
1					4			
8	6			7		5		1

Easy # 987

9		8	2				5	
1		5			8	2	7	
	2				1			
5	7	3						8
4	8						3	9
6						1	4	5
			7				8	
	5	7	8			9		3
	6				3	5		7

Easy # 988

		8			6	1		
	4		1	7			8	5
9			8	5		3		
	3				1			8
5	2			3			9	6
6			9				3	
		2		1	9			3
8	9			4	2		1	
		5	3			2		

Easy # 989

2			3	5			9	
		7				2		8
		8		9			6	1
				1	6	8		3
	8	9		3		6	2	
6		5	7	8				
7	6			4		9		
8		4				3		
	9			2	1			6

Easy # 990

		1				4		6
	8				4	1		3
3					1		7	
8	6			2			4	1
		7	4		6	8		
1	5			3			6	9
	1		5					2
7		8	2				1	
5		9				6		

Easy # 991

	6				2	8		4
		2			5		6	
7						5		1
	7	1		9		6	5	
4			7		3			8
	3	6		2		4	7	
3		7						6
	8		6			9		
6		9	3				4	

Easy # 992

		1			5		2	
6	4		2	7			1	9
			4	1			7	
9		7	3	4				
8								2
				9	2	3		7
	8			5	6			
2	7			3	1		5	4
	5		9			7		

Easy # 993

	8					1	6	5
		5	1	6		8	4	3
	6		8				9	
				5	4	3		
			6		9			
		2	7	3				
	3				1		8	
6	7	1		8	3	2		
2	4	8					3	

Easy # 994

		3						
	2		8	9				
4		8		2	6		7	9
5			1			9		7
8	9		3		5		4	6
3		2			9			8
2	3		7	5		6		4
				3	4		9	
						5		

Easy # 995

1	7				3			
		6			7			5
	9	8				2	7	
9					6	3	2	
	6	4	3		2	5	9	
	2	7	4					1
	8	1				7	3	
7			6			1		
			2				4	8

Easy # 996

5		8	2	6		7	9	1
	6		8					
4						8		
7		6			8	1	5	
		3	4		7	2		
	8	5	9			3		4
		9						7
					3		1	
3	1	7		9	2	4		5

Easy # 997

	6		5	9			7	3
8			7	1				
		5						9
	7	4	6				8	1
	2	3	9		7	4	5	
6	5				1	2	9	
3						7		
				3	6			5
5	4			7	8		3	

Easy # 998

2	5				3		1	
9		4						3
		7	4	9	1			2
7				3			2	
	6	3				4	9	
	2			4				8
6			3	5	2	8		
5						2		6
	3		8				7	1

Easy # 999

							7	4
9				8	1		3	
6		1	2	3		5		
	6	5			2	9	4	7
1	4	9	5			3	2	
		7		2	9	8		1
	1		8	7				3
5	9							

Easy # 1000

8		9		2			4	3
6			3			5		
				8	6			
		1		5			9	7
9	7	5		1		8	3	6
2	6			3		4		
			5	6				
		2			3			9
4	3			7		1		2

Easy # 1001

	8			2	3	4		7
4	6			7				
		3	8		4		9	
5			7				3	4
	1			3			2	
3	9				6			1
	3		4		5	7		
				8			6	5
6		7	9	1			4	

Easy # 1002

			3	7		9		
	2	4		1			7	
			2		4	6	1	5
	4	2		8	3	7	5	
	1	7	5	6		4	3	
2	8	5	9		6			
	7			2		5	9	
		3		5	7			

Easy # 1003

	6				8			2
		1				3	8	
2					6	7	5	
1	3			4			2	8
		7	1		9	5		
9	2			6			7	1
	4	2	9					7
	1	9				2		
5			2				4	

Easy # 1004

3			7	1			2	
	8			2	4			7
	6		5		3		4	
7	5		4	9			1	
		2				4		
	3			6	1		7	2
	9		2		8		3	
5			9	3			6	
	7			5	6			4

Easy # 1005

5					2		9	3
9	7		5	8	6			2
4			3			5		
			7	9		6		8
1								4
8		7		2	4			
		1			7			5
7			1	6	3		4	9
3	2		9					6

Easy # 1006

9		6			2	7		
5	2	4			3			
	1		9				3	
2	5			6			8	
3				9				4
	6			8			2	3
	9				1		4	
			6			1	5	7
		5	4			2		9

Easy # 1007

	3	8		2			5	1
	7				5			
	4	9			7	8		2
				1	9		4	3
	5		2		6		7	
1	8		5	7				
4		5	6			9	8	
			7				1	
3	1			9		2	6	

Easy # 1008

5					9			
6		4			2	3	1	
2		7		4			8	5
				9	3		5	6
9			2		7			3
1	8		4	5				
3	5			7		6		8
	7	6	9			4		1
			3					9

Easy # 1009

			6					7
7		2		9			5	3
	4	8	3				9	1
1		7	4	6				
4			5		3			6
				7	9	2		8
8	9				6	5	1	
2	1			5		7		4
6					4			

Easy # 1010

	7	9	8	5	3			
			4				5	
3				1		8	9	
	3	1	2	6		7		
	2	5				9	3	
		8		9	5	2	6	
	6	3		7				9
	9				4			
			9	3	1	4	2	

Easy # 1011

	8	1		7	6			9
	2				4			7
				9	8		6	
5		6		2			8	
2	3		9		1		7	6
	9			8		1		2
	5		8	6				
8			3				4	
9			4	5		2	1	

Easy # 1012

2	3			9	1	5		
4					6		3	8
			5	8	3		2	9
5		6					9	
	4					2		5
6	8		3	1	7			
3	5		9					2
		4	2	5			7	3

Easy # 1013

		9		2	8			6
3	2			9				8
7					3	2		
		7			6	3	1	
5			3		1			9
	3	4	8			7		
		8	5					1
9				1			2	7
2			9	8		6		

Easy # 1014

		2			5			4
8	9				7			
	1	4				2	8	
2					9	4	7	
	3	6	7		1	9	5	
	7	1	5					3
	4	7				8	3	
			1				4	2
6			4			5		

Easy # 1015

3		9		5		8		
	7		9		8			3
6		8	3		2		4	
9	3			8				
			4		6			
				9			1	4
	6		5		3	1		7
5			8		1		3	
		3		7		2		5

Easy # 1016

1			9		3	6		
		9		5	1		4	3
	3	7		4				
	8		4			1	3	
		2		1		5		
	1	6			7		2	
				9		7	8	
4	7		6	2		3		
		1	3		8			4

Easy # 1017

	8		1			3		
7	6		3	9	2		4	
	3				4		1	6
			7	6		2	9	
	5						8	
	9	7		4	8			
4	1		6				2	
	7		5	2	1		6	8
		5			7		3	

Easy # 1018

7	2		4				3	
1	3		6			7		5
				3	7	2	6	
	6	8	7			3		
	7						1	
		3			8	4	9	
	8	2	9	4				
3		1			6		4	9
	4				3		2	6

Easy # 1019

9	4				1	5		7
7	8			5		2	6	
	1				6			
				2	9	8	4	
	6		5		3		1	
	7	2	6	1				
			1				2	
	2	8		9			3	5
6		4	3				7	9

Easy # 1020

	1		9	7	4		6	
			1				7	9
	9			8		1		
	3			9	5	7		
	4	9				8	1	
		6	8	4			2	
		8		6			3	
4	2				8			
	6		7	2	3		8	

Easy # 1021

6	2				1	9		
			4		9	1	2	
7			2		3			8
3					4		6	
4	1						7	2
	7		8					4
1			3		8			6
	6	3	9		5			
		8	6				3	5

Easy # 1022

3	1				7	2		
	6				2		9	
	8	5	4	9				7
7		6				4	2	
8								9
	9	4				3		6
9				6	8	7	5	
	7		3				4	
		8	5				1	2

Easy # 1023

			6	9			7	
		5			8		9	
1	9		2	4			5	3
5		4	3	8				
3								7
				1	4	5		8
8	2			5	3		1	6
	3		9			2		
	5			2	1			

Easy # 1024

9	6			2	8			
8			7	5				
					6	8		4
5		9			2	3		8
	4			3			2	
3		8	5			1		7
6		7	9					
				7	3			1
			2	6			7	5

Easy # 1025

1						3		
		2	5	3	6			
3		5	1		7			4
	7	6			1	4	2	
		4	6		2	5		
	1	8	3			6	7	
7			8		3	1		2
			4	1	9	7		
		1						8

Easy # 1026

				3			7	
	4	7	2			5		
5			9	7	8	3	4	6
		6		5				
4	2		8		6		5	3
				2		9		
6	9	4	1	8	5			7
		1			2	6	8	
	8			9				

Easy # 1027

3	1			4	7	6		8
		5	3	6				
	4							
2			5				8	6
1		3	4		6	5		2
7	5				9			4
							6	
				5	2	8		
5		7	1	8			2	3

Easy # 1028

	2		8	3		4		7
	5		2	7	6			1
7								
	9	5	4				3	
6	3		7		8		4	9
	7				2	5	1	
								3
3			1	4	9		7	
8		7		2	3		6	

Easy # 1029

9	8			1	2			
		1	3	9			4	5
					4		2	
5	1	3		4				
	9	8	2		3	1	7	
				8		3	5	6
	4		1					
2	3			6	8	7		
			7	2			8	3

Easy # 1030

	1	2		9	3		8	
		3	6	7				
4	9		1					7
	2	8	7		1		5	
				2				
	4		3		8	1	6	
5					7		9	4
				1	9	7		
	7		5	6		8	1	

Easy # 1031

		5					2	
1			2					
	2	8	3	1		9	4	7
	1	4			2		9	8
	6		5		4		3	
2	8		7			5	6	
9	4	6		7	3	8	5	
					6			9
	7					4		

Easy # 1032

						7		5
2				3	6	9		
9	6		7	4			3	
	2	4			5	1	7	9
3	7	1	4			8	5	
	5			2	4		9	8
		2	9	6				7
1		3						

Easy # 1033

					3	5		
	9							4
3	4	5		9	2		8	7
	7	1	9				3	8
	3		8		4		2	
4	6				1	7	5	
7	1		2	6		9	4	5
8							1	
		6	1					

Easy # 1034

	7			9				5
	5		7	6	4		9	
			5			3	6	
	6			3	5			9
5	1						3	2
4			8	2			7	
	4	2			1			
	9		3	4	2		1	
1				5			2	

Easy # 1035

	9	2				7		
3	4			2		1		
1				6	8		3	
	3	5	4	9				
9		1		7		3		6
				8	3	9	7	
	6		7	5				1
		9		1			8	3
		4				6	9	

Easy # 1036

9				5	2		6	
	8	2		4	9	5		
7			6					9
	3		2			6		
	7	9		6		2	3	
		6			5		8	
8					3			5
		4	5	1		8	7	
	2		8	7				6

Easy # 1037

		6		7		3	9	
	9		6		3			4
1			5		9	6	8	
				6			3	9
			8		1			
2	1			3				
	4	2	9		7			8
9			2		6		7	
	7	5		4		9		

Easy # 1038

6	1		2			8		
	4		1	8			2	7
8		7		4				
			8			6		
7		5	6	2	1	4		9
		2			3			
				1		7		8
1	3			7	5		9	
		4			8		1	3

Easy # 1039

	7		9	4	3			1
		9	1			6	5	
	4						7	3
		3		2			1	
9		7				8		2
	6			9		3		
2	8						9	
	3	4			9	5		
6			2	8	5		3	

Easy # 1040

5							4	6
	4			8	3	7		
		6	4	2		3		
	2	5			6			8
	6	1		4		9	7	
4			2			6	3	
		2		5	4	1		
		3	8	7			2	
7	5							3

Easy # 1041

		8	1	9	2	5		
2			5	7				
	6				8	2		7
	1	2	8	3				
	4	3				1	7	
				1	4	6	8	
3		7	4				2	
				5	9			1
		1	6	2	3	7		

Easy # 1042

5	4			8	3	7		
								5
2			1	5	7	6		
		8			4	9	6	
9		4	3		5	8		1
	6	2	7			5		
		5	9	4	2			8
8								
		1	8	7			5	3

Easy # 1043

				1			8	3
9					8	1	7	
	1	7		3	6	2		
5					7			
6	3		4	5	1		2	9
			8					4
		9	1	8		5	3	
	4	1	5					8
3	8			9				

Easy # 1044

		2	8		5		9	
6			2			1	3	8
	9			6	4	2		
2		9					7	
	6						4	
	8					3		5
		5	9	7			6	
4	7	1			6			9
	3		4		8	7		

Easy # 1045

	7	6		1	2		5	
		5	3	7				
8	3		5					2
	1	7	6		4		8	
				9				
	2		7		5	6	9	
5					7		3	8
				5	1	4		
	6		4	3		9	7	

Easy # 1046

		7	8		3	5		9
5						3		
			2	5	9			4
7	4		3				8	2
9			4		2			7
2	8				5		3	6
8			1	3	7			
		6						3
3		4	5		6	8		

Easy # 1047

7	4			1				6
6	3		4		8	9		
		2	7		6		4	
	7	4		6				
			9		3			
				7		5	9	
	1		6		5	4		
		3	1		4		2	5
4				2			1	8

Easy # 1048

1	2			4	3			
					7		9	
		3	2	6			1	4
6	5	1		2				
	3	7	1		4	2	8	
				9		1	7	5
5	9			8	1	7		
	4		9					
			4	7			2	8

Easy # 1049

	3		4	5	6			
1	5		7		9			3
9							5	
	8	3			7	5	9	
	2		1		8		6	
	6	1	5			3	8	
	7							5
6			3		5		2	7
			8	7	2		1	

Easy # 1050

	2		4	7	9		8	1
		4			2		3	
6	9		8				7	
	5	2		6	1			
	4						1	
			2	8		7	5	
	3				6		9	8
	1		9			3		
2	8		3	5	7		6	

Solution

Solution # 1

2	4	9	3	6	5	7	8	1
7	3	6	4	1	8	9	5	2
8	1	5	7	9	2	4	6	3
5	6	2	1	8	9	3	7	4
4	8	1	6	7	3	2	9	5
9	7	3	2	5	4	8	1	6
1	2	8	9	3	6	5	4	7
6	9	4	5	2	7	1	3	8
3	5	7	8	4	1	6	2	9

Solution # 2

2	7	1	4	8	6	3	9	5
8	5	4	2	9	3	1	7	6
3	9	6	7	1	5	2	8	4
6	3	7	9	5	2	8	4	1
9	1	8	6	7	4	5	3	2
5	4	2	1	3	8	7	6	9
1	8	5	3	4	9	6	2	7
4	2	3	5	6	7	9	1	8
7	6	9	8	2	1	4	5	3

Solution # 3

8	9	6	7	2	3	5	1	4
5	1	2	8	4	6	7	3	9
4	3	7	9	1	5	6	8	2
6	4	3	1	9	8	2	7	5
1	8	9	2	5	7	3	4	6
2	7	5	6	3	4	8	9	1
3	6	1	4	7	2	9	5	8
9	5	8	3	6	1	4	2	7
7	2	4	5	8	9	1	6	3

Solution # 4

6	9	3	2	5	4	8	7	1
5	1	2	8	7	6	4	9	3
8	4	7	9	3	1	2	6	5
3	5	6	4	2	9	7	1	8
9	2	1	3	8	7	6	5	4
4	7	8	1	6	5	9	3	2
7	6	4	5	1	8	3	2	9
1	3	9	6	4	2	5	8	7
2	8	5	7	9	3	1	4	6

Solution # 5

8	4	1	9	7	6	5	2	3
2	9	7	1	5	3	8	4	6
6	3	5	4	8	2	9	7	1
7	1	9	6	4	5	2	3	8
3	8	4	2	9	7	6	1	5
5	6	2	8	3	1	7	9	4
1	2	3	7	6	8	4	5	9
9	5	6	3	2	4	1	8	7
4	7	8	5	1	9	3	6	2

Solution # 6

3	4	1	2	5	9	6	7	8
5	7	6	8	4	1	2	3	9
8	2	9	7	6	3	1	4	5
7	8	4	3	2	6	9	5	1
6	3	2	9	1	5	7	8	4
9	1	5	4	7	8	3	6	2
2	9	3	6	8	4	5	1	7
1	6	8	5	9	7	4	2	3
4	5	7	1	3	2	8	9	6

Solution # 7

2	3	9	7	5	4	8	6	1
4	7	8	6	3	1	2	5	9
1	6	5	9	8	2	3	4	7
3	2	6	8	7	5	9	1	4
7	8	4	2	1	9	5	3	6
9	5	1	3	4	6	7	8	2
8	9	2	4	6	3	1	7	5
5	4	3	1	2	7	6	9	8
6	1	7	5	9	8	4	2	3

Solution # 8

3	8	6	5	1	7	4	9	2
9	1	5	4	2	8	3	6	7
2	7	4	6	3	9	1	5	8
4	9	7	3	5	2	6	8	1
8	2	1	7	6	4	9	3	5
5	6	3	8	9	1	7	2	4
1	5	9	2	7	3	8	4	6
7	4	2	9	8	6	5	1	3
6	3	8	1	4	5	2	7	9

Solution # 9

2	6	5	8	7	4	9	1	3
4	7	3	2	9	1	5	6	8
1	9	8	5	3	6	2	4	7
6	4	7	1	5	8	3	9	2
5	2	9	3	6	7	4	8	1
8	3	1	4	2	9	7	5	6
7	5	4	6	8	2	1	3	9
9	1	6	7	4	3	8	2	5
3	8	2	9	1	5	6	7	4

Solution # 10

4	8	9	1	6	7	3	5	2
5	2	7	4	3	8	9	6	1
1	6	3	9	5	2	4	8	7
8	9	4	5	7	1	2	3	6
6	5	2	3	9	4	1	7	8
3	7	1	2	8	6	5	4	9
7	3	5	8	2	9	6	1	4
2	1	8	6	4	3	7	9	5
9	4	6	7	1	5	8	2	3

Solution # 11

3	9	5	8	4	2	6	7	1
2	4	8	7	6	1	5	3	9
6	1	7	5	3	9	4	8	2
5	6	4	1	7	3	9	2	8
9	7	3	2	8	6	1	4	5
1	8	2	4	9	5	7	6	3
4	5	1	3	2	7	8	9	6
7	2	9	6	5	8	3	1	4
8	3	6	9	1	4	2	5	7

Solution # 12

1	2	8	7	6	3	5	9	4
5	3	6	9	8	4	1	2	7
7	9	4	2	1	5	6	3	8
6	8	3	5	9	1	7	4	2
9	5	7	6	4	2	8	1	3
2	4	1	3	7	8	9	6	5
3	7	2	1	5	6	4	8	9
8	1	5	4	3	9	2	7	6
4	6	9	8	2	7	3	5	1

Solution # 13

2	3	7	6	9	1	5	8	4
1	5	4	3	8	7	2	9	6
9	8	6	5	2	4	3	1	7
8	7	9	2	5	6	4	3	1
6	1	3	4	7	9	8	2	5
4	2	5	1	3	8	6	7	9
5	9	1	8	6	2	7	4	3
3	4	8	7	1	5	9	6	2
7	6	2	9	4	3	1	5	8

Solution # 14

9	7	3	8	2	6	4	5	1
6	4	2	1	5	3	9	7	8
5	1	8	4	7	9	6	3	2
3	2	6	9	4	8	7	1	5
1	8	9	7	3	5	2	6	4
7	5	4	2	6	1	8	9	3
4	6	7	5	1	2	3	8	9
2	9	5	3	8	7	1	4	6
8	3	1	6	9	4	5	2	7

Solution # 15

3	9	5	1	2	8	6	7	4
8	7	4	3	5	6	2	1	9
1	6	2	4	7	9	5	8	3
6	2	9	8	3	7	1	4	5
4	8	1	6	9	5	7	3	2
7	5	3	2	4	1	9	6	8
2	1	6	9	8	3	4	5	7
5	4	8	7	6	2	3	9	1
9	3	7	5	1	4	8	2	6

Solution # 16

9	7	3	2	8	5	4	1	6
1	8	4	9	7	6	5	2	3
6	2	5	4	1	3	7	8	9
8	6	7	1	5	4	9	3	2
2	4	1	3	9	8	6	5	7
5	3	9	7	6	2	1	4	8
3	1	2	6	4	9	8	7	5
7	9	8	5	3	1	2	6	4
4	5	6	8	2	7	3	9	1

Solution # 17

8	2	1	4	6	9	5	3	7
4	5	6	2	3	7	1	9	8
7	3	9	1	8	5	2	4	6
2	4	8	6	7	1	9	5	3
3	1	5	8	9	2	6	7	4
6	9	7	3	5	4	8	2	1
9	8	3	7	2	6	4	1	5
5	7	4	9	1	8	3	6	2
1	6	2	5	4	3	7	8	9

Solution # 18

9	3	4	8	5	2	6	1	7
1	6	2	4	3	7	9	5	8
7	8	5	6	1	9	3	2	4
8	4	3	7	6	5	2	9	1
6	9	7	2	4	1	5	8	3
2	5	1	3	9	8	4	7	6
5	2	6	1	7	4	8	3	9
3	1	8	9	2	6	7	4	5
4	7	9	5	8	3	1	6	2

Solution # 19

2	5	6	9	4	7	3	8	1
4	8	7	3	2	1	5	9	6
1	3	9	5	8	6	7	4	2
9	2	4	7	3	8	1	6	5
5	7	8	1	6	2	4	3	9
6	1	3	4	9	5	2	7	8
3	4	5	8	1	9	6	2	7
8	6	1	2	7	3	9	5	4
7	9	2	6	5	4	8	1	3

Solution # 20

8	5	2	1	9	4	7	6	3
3	9	6	8	7	2	1	4	5
4	7	1	6	3	5	2	9	8
2	6	7	4	5	3	8	1	9
9	1	8	2	6	7	5	3	4
5	3	4	9	8	1	6	7	2
6	4	5	7	2	9	3	8	1
1	8	3	5	4	6	9	2	7
7	2	9	3	1	8	4	5	6

Solution # 21

6	8	4	9	2	3	7	5	1
5	2	1	7	4	8	9	6	3
9	3	7	6	1	5	2	8	4
7	5	8	3	9	2	1	4	6
1	6	3	8	7	4	5	2	9
4	9	2	5	6	1	3	7	8
3	7	9	4	5	6	8	1	2
2	4	5	1	8	9	6	3	7
8	1	6	2	3	7	4	9	5

Solution # 22

7	2	8	5	3	9	6	4	1
3	9	6	4	7	1	2	8	5
4	1	5	8	2	6	3	9	7
2	5	7	1	8	4	9	6	3
1	6	4	3	9	2	5	7	8
8	3	9	6	5	7	4	1	2
6	8	3	9	1	5	7	2	4
9	7	1	2	4	3	8	5	6
5	4	2	7	6	8	1	3	9

Solution # 23

1	4	8	2	3	9	6	5	7
9	5	3	6	4	7	1	2	8
2	7	6	1	8	5	3	4	9
3	1	4	9	2	8	5	7	6
6	9	2	7	5	4	8	1	3
5	8	7	3	6	1	2	9	4
7	3	9	5	1	6	4	8	2
8	2	5	4	7	3	9	6	1
4	6	1	8	9	2	7	3	5

Solution # 24

7	6	5	4	3	2	1	8	9
4	3	1	8	5	9	2	7	6
8	2	9	7	6	1	4	3	5
5	9	3	1	2	8	7	6	4
2	1	7	5	4	6	3	9	8
6	4	8	9	7	3	5	1	2
3	5	6	2	8	7	9	4	1
9	7	4	6	1	5	8	2	3
1	8	2	3	9	4	6	5	7

Solution # 25

7	6	3	4	9	8	2	5	1
5	2	4	3	6	1	9	8	7
1	9	8	7	2	5	4	6	3
6	3	7	5	4	9	1	2	8
4	1	2	8	3	6	5	7	9
9	8	5	1	7	2	3	4	6
2	7	6	9	5	3	8	1	4
8	4	9	2	1	7	6	3	5
3	5	1	6	8	4	7	9	2

Solution # 26

5	6	7	4	9	8	1	2	3
9	2	3	5	1	7	4	8	6
1	4	8	6	3	2	7	5	9
6	3	5	7	8	9	2	1	4
7	1	4	2	6	5	3	9	8
8	9	2	1	4	3	5	6	7
3	7	1	9	2	6	8	4	5
2	5	6	8	7	4	9	3	1
4	8	9	3	5	1	6	7	2

Solution # 27

4	8	5	3	1	7	2	6	9
9	1	2	6	4	5	7	8	3
7	3	6	9	2	8	5	1	4
3	9	1	4	7	6	8	5	2
5	2	8	1	9	3	6	4	7
6	7	4	8	5	2	9	3	1
2	6	7	5	3	1	4	9	8
1	5	9	2	8	4	3	7	6
8	4	3	7	6	9	1	2	5

Solution # 28

4	6	1	5	3	2	7	8	9
9	8	2	4	1	7	6	5	3
7	3	5	9	6	8	2	1	4
6	4	3	1	8	5	9	2	7
5	7	9	6	2	3	8	4	1
1	2	8	7	9	4	5	3	6
3	5	4	2	7	6	1	9	8
2	9	6	8	4	1	3	7	5
8	1	7	3	5	9	4	6	2

Solution # 29

1	3	4	6	2	7	5	9	8
5	7	8	1	4	9	2	3	6
2	9	6	5	3	8	1	7	4
7	8	5	9	1	3	6	4	2
9	4	2	8	7	6	3	5	1
6	1	3	2	5	4	7	8	9
3	2	1	4	8	5	9	6	7
4	5	9	7	6	1	8	2	3
8	6	7	3	9	2	4	1	5

Solution # 30

5	4	3	6	1	9	8	7	2
9	8	7	3	5	2	6	4	1
2	1	6	4	8	7	9	5	3
4	3	9	5	2	6	7	1	8
8	6	2	1	7	3	4	9	5
1	7	5	8	9	4	3	2	6
3	2	1	9	4	8	5	6	7
7	9	8	2	6	5	1	3	4
6	5	4	7	3	1	2	8	9

Solution # 31

1	5	9	2	3	6	8	4	7
6	2	4	8	7	1	9	5	3
3	7	8	4	9	5	6	1	2
5	6	1	9	4	2	7	3	8
8	9	2	7	5	3	4	6	1
4	3	7	1	6	8	5	2	9
2	4	6	3	8	9	1	7	5
7	8	3	5	1	4	2	9	6
9	1	5	6	2	7	3	8	4

Solution # 32

9	4	7	5	6	3	1	8	2
3	1	2	4	8	9	7	6	5
8	5	6	7	2	1	3	9	4
7	3	5	8	1	2	9	4	6
2	9	1	6	4	5	8	3	7
6	8	4	9	3	7	2	5	1
4	6	9	2	7	8	5	1	3
1	2	8	3	5	4	6	7	9
5	7	3	1	9	6	4	2	8

Solution # 33

7	9	3	4	2	5	1	6	8
2	1	5	9	6	8	7	4	3
6	8	4	3	1	7	2	9	5
8	5	6	1	7	9	3	2	4
3	4	1	8	5	2	6	7	9
9	2	7	6	3	4	8	5	1
4	7	2	5	8	1	9	3	6
1	6	9	2	4	3	5	8	7
5	3	8	7	9	6	4	1	2

Solution # 34

1	6	9	7	4	3	5	2	8
3	4	5	8	2	1	9	6	7
8	7	2	6	5	9	1	3	4
5	8	6	1	7	2	3	4	9
7	2	3	5	9	4	6	8	1
4	9	1	3	8	6	7	5	2
9	1	4	2	3	5	8	7	6
2	3	8	9	6	7	4	1	5
6	5	7	4	1	8	2	9	3

Solution # 35

5	8	6	7	1	9	4	2	3
3	1	9	5	4	2	8	6	7
2	4	7	6	3	8	9	1	5
9	6	4	1	2	5	3	7	8
8	2	3	4	7	6	1	5	9
7	5	1	8	9	3	6	4	2
6	7	2	9	8	1	5	3	4
4	9	5	3	6	7	2	8	1
1	3	8	2	5	4	7	9	6

Solution # 36

2	3	5	8	4	1	9	6	7
8	1	6	7	9	2	5	4	3
7	4	9	6	5	3	8	2	1
6	9	7	3	2	5	1	8	4
1	5	8	9	7	4	6	3	2
4	2	3	1	8	6	7	5	9
9	8	2	4	6	7	3	1	5
5	7	1	2	3	8	4	9	6
3	6	4	5	1	9	2	7	8

Solution # 37

3	1	6	9	8	5	4	7	2
4	7	8	2	1	6	5	3	9
2	9	5	3	4	7	6	8	1
7	3	2	1	9	4	8	6	5
8	5	1	6	2	3	9	4	7
9	6	4	7	5	8	2	1	3
1	2	3	4	6	9	7	5	8
5	4	9	8	7	1	3	2	6
6	8	7	5	3	2	1	9	4

Solution # 38

5	4	8	7	6	9	3	1	2
1	7	2	3	4	5	9	8	6
3	6	9	8	2	1	7	4	5
2	1	7	9	5	3	4	6	8
8	3	4	2	1	6	5	9	7
9	5	6	4	8	7	2	3	1
4	2	1	5	3	8	6	7	9
7	8	3	6	9	2	1	5	4
6	9	5	1	7	4	8	2	3

Solution # 39

2	5	3	6	8	7	4	9	1
6	1	7	2	4	9	3	5	8
9	4	8	5	3	1	6	2	7
4	2	6	1	5	8	9	7	3
8	9	1	3	7	2	5	4	6
3	7	5	9	6	4	8	1	2
7	6	4	8	2	5	1	3	9
1	8	2	4	9	3	7	6	5
5	3	9	7	1	6	2	8	4

Solution # 40

8	5	4	1	9	2	7	3	6
3	1	9	4	7	6	2	8	5
7	6	2	3	5	8	1	4	9
2	7	6	9	3	1	8	5	4
4	8	3	5	2	7	9	6	1
5	9	1	6	8	4	3	7	2
1	3	8	2	6	5	4	9	7
9	2	5	7	4	3	6	1	8
6	4	7	8	1	9	5	2	3

Solution # 41

8	5	3	2	6	4	7	9	1
2	4	7	3	9	1	8	5	6
1	6	9	7	8	5	2	4	3
3	7	1	8	2	9	4	6	5
4	8	6	5	7	3	1	2	9
5	9	2	1	4	6	3	7	8
9	2	4	6	1	8	5	3	7
7	3	8	9	5	2	6	1	4
6	1	5	4	3	7	9	8	2

Solution # 42

7	8	1	9	5	4	3	2	6
4	5	9	6	3	2	1	7	8
2	3	6	7	8	1	5	9	4
6	1	3	8	2	5	7	4	9
5	9	7	4	1	6	8	3	2
8	4	2	3	7	9	6	1	5
9	7	4	5	6	3	2	8	1
1	6	8	2	9	7	4	5	3
3	2	5	1	4	8	9	6	7

Solution # 43

7	3	1	5	4	6	9	8	2
4	5	9	8	7	2	1	6	3
6	2	8	1	9	3	4	5	7
8	1	2	7	3	5	6	4	9
3	4	6	9	8	1	2	7	5
5	9	7	2	6	4	8	3	1
2	6	4	3	5	9	7	1	8
9	8	5	6	1	7	3	2	4
1	7	3	4	2	8	5	9	6

Solution # 44

2	4	3	6	8	1	5	7	9
1	6	8	7	9	5	3	2	4
5	7	9	4	2	3	1	8	6
7	5	4	2	1	6	9	3	8
9	3	6	8	5	7	2	4	1
8	2	1	3	4	9	7	6	5
4	8	5	1	7	2	6	9	3
3	1	7	9	6	4	8	5	2
6	9	2	5	3	8	4	1	7

Solution # 45

1	9	6	7	3	2	5	4	8
8	7	3	5	9	4	2	6	1
4	5	2	8	1	6	3	9	7
7	6	9	2	4	1	8	3	5
3	2	8	6	5	9	7	1	4
5	1	4	3	8	7	6	2	9
9	3	7	1	2	8	4	5	6
6	4	5	9	7	3	1	8	2
2	8	1	4	6	5	9	7	3

Solution # 46

3	6	2	4	8	9	5	1	7
8	5	7	1	2	3	4	9	6
9	4	1	7	6	5	8	2	3
2	8	9	6	1	7	3	5	4
4	7	3	5	9	8	2	6	1
6	1	5	3	4	2	9	7	8
7	3	8	9	5	1	6	4	2
5	2	4	8	7	6	1	3	9
1	9	6	2	3	4	7	8	5

Solution # 47

3	1	4	6	9	2	5	7	8
8	6	7	3	4	5	9	2	1
9	5	2	7	8	1	3	6	4
4	9	3	5	6	7	8	1	2
7	2	6	8	1	3	4	9	5
5	8	1	9	2	4	7	3	6
2	7	5	4	3	6	1	8	9
1	4	9	2	7	8	6	5	3
6	3	8	1	5	9	2	4	7

Solution # 48

5	6	7	2	8	9	1	4	3
4	8	2	5	3	1	9	7	6
1	9	3	4	6	7	2	5	8
6	1	5	9	2	4	8	3	7
7	4	8	3	1	5	6	2	9
3	2	9	6	7	8	4	1	5
2	7	1	8	9	3	5	6	4
8	5	6	7	4	2	3	9	1
9	3	4	1	5	6	7	8	2

Solution # 49

4	6	8	2	1	5	7	9	3
9	5	1	3	7	4	2	6	8
7	3	2	6	8	9	5	1	4
2	9	3	7	4	6	8	5	1
1	4	6	8	5	3	9	2	7
8	7	5	1	9	2	4	3	6
6	2	7	5	3	8	1	4	9
5	1	9	4	6	7	3	8	2
3	8	4	9	2	1	6	7	5

Solution # 50

3	9	7	5	1	4	8	2	6
4	6	1	9	2	8	3	5	7
5	8	2	3	6	7	4	9	1
2	4	8	6	7	9	5	1	3
6	5	3	8	4	1	9	7	2
7	1	9	2	3	5	6	4	8
1	7	5	4	8	6	2	3	9
9	2	6	1	5	3	7	8	4
8	3	4	7	9	2	1	6	5

Solution # 51

1	8	9	4	2	5	3	6	7
3	7	2	6	8	9	1	4	5
4	6	5	1	7	3	9	2	8
6	9	3	5	4	2	8	7	1
7	5	1	9	6	8	2	3	4
8	2	4	3	1	7	6	5	9
5	4	6	8	3	1	7	9	2
2	3	8	7	9	4	5	1	6
9	1	7	2	5	6	4	8	3

Solution # 52

3	8	7	4	9	6	5	1	2
1	4	9	5	7	2	6	8	3
6	2	5	8	1	3	9	7	4
9	6	4	7	5	1	2	3	8
2	5	8	3	6	4	1	9	7
7	3	1	9	2	8	4	5	6
4	7	2	1	3	5	8	6	9
8	1	3	6	4	9	7	2	5
5	9	6	2	8	7	3	4	1

Solution # 53

4	1	8	2	9	3	7	6	5
2	6	7	8	4	5	3	9	1
3	5	9	1	7	6	8	2	4
1	2	4	6	8	9	5	7	3
8	9	3	5	2	7	4	1	6
5	7	6	3	1	4	9	8	2
6	3	1	9	5	8	2	4	7
9	4	5	7	6	2	1	3	8
7	8	2	4	3	1	6	5	9

Solution # 54

2	8	7	5	3	6	1	4	9
5	3	1	4	2	9	7	6	8
4	6	9	1	7	8	5	3	2
8	2	5	6	9	7	4	1	3
6	1	3	8	4	2	9	5	7
7	9	4	3	1	5	2	8	6
3	5	2	7	6	1	8	9	4
9	4	8	2	5	3	6	7	1
1	7	6	9	8	4	3	2	5

Solution # 55

4	8	6	1	5	7	2	9	3
7	5	2	9	6	3	8	4	1
3	9	1	2	4	8	5	6	7
1	2	3	8	9	5	4	7	6
5	7	9	4	2	6	3	1	8
6	4	8	7	3	1	9	2	5
2	6	7	3	8	4	1	5	9
9	3	5	6	1	2	7	8	4
8	1	4	5	7	9	6	3	2

Solution # 56

6	5	7	3	1	8	4	9	2
8	3	9	5	2	4	6	7	1
2	1	4	7	6	9	5	8	3
7	2	3	1	4	5	9	6	8
4	8	6	9	3	2	7	1	5
1	9	5	6	8	7	3	2	4
3	4	8	2	7	6	1	5	9
5	7	2	4	9	1	8	3	6
9	6	1	8	5	3	2	4	7

Solution # 57

6	5	1	3	7	9	8	4	2
3	2	9	8	4	1	6	5	7
8	4	7	5	2	6	9	1	3
1	8	4	2	5	7	3	6	9
5	3	6	9	1	8	2	7	4
9	7	2	4	6	3	1	8	5
7	9	3	1	8	4	5	2	6
4	1	5	6	3	2	7	9	8
2	6	8	7	9	5	4	3	1

Solution # 58

4	1	2	6	3	7	5	8	9
8	9	3	5	2	1	7	6	4
6	7	5	9	8	4	3	2	1
9	3	1	7	6	5	2	4	8
5	2	4	3	9	8	6	1	7
7	6	8	1	4	2	9	5	3
3	8	9	2	1	6	4	7	5
2	4	7	8	5	3	1	9	6
1	5	6	4	7	9	8	3	2

Solution # 59

8	7	3	6	1	2	4	9	5
1	6	5	8	4	9	2	7	3
4	9	2	3	5	7	8	6	1
2	3	9	4	7	8	1	5	6
5	4	6	9	3	1	7	8	2
7	1	8	5	2	6	9	3	4
9	2	1	7	6	3	5	4	8
3	5	7	1	8	4	6	2	9
6	8	4	2	9	5	3	1	7

Solution # 60

5	7	3	9	4	6	8	2	1
2	8	6	7	1	5	4	9	3
4	1	9	3	8	2	6	5	7
8	5	7	4	9	1	2	3	6
6	3	1	2	5	8	9	7	4
9	2	4	6	3	7	5	1	8
1	4	5	8	7	9	3	6	2
7	6	8	5	2	3	1	4	9
3	9	2	1	6	4	7	8	5

Solution # 61

3	5	8	4	1	9	2	6	7
1	7	6	3	8	2	9	5	4
2	9	4	6	5	7	8	3	1
7	8	5	1	6	4	3	9	2
4	2	1	9	3	5	6	7	8
6	3	9	7	2	8	1	4	5
5	6	3	8	7	1	4	2	9
8	4	7	2	9	3	5	1	6
9	1	2	5	4	6	7	8	3

Solution # 62

5	1	3	9	6	4	2	7	8
4	8	7	2	1	3	5	9	6
9	6	2	8	7	5	1	4	3
6	5	9	3	4	1	8	2	7
8	3	1	7	2	6	4	5	9
7	2	4	5	9	8	3	6	1
3	9	8	4	5	7	6	1	2
2	4	6	1	3	9	7	8	5
1	7	5	6	8	2	9	3	4

Solution # 63

7	4	1	3	2	6	9	5	8
3	9	5	8	1	4	6	2	7
2	8	6	7	5	9	4	3	1
6	2	7	4	9	5	1	8	3
9	3	4	1	8	2	5	7	6
5	1	8	6	7	3	2	9	4
8	7	9	2	4	1	3	6	5
1	6	2	5	3	8	7	4	9
4	5	3	9	6	7	8	1	2

Solution # 64

9	7	5	4	3	1	8	2	6
4	8	3	6	9	2	1	7	5
6	2	1	8	5	7	3	9	4
7	6	8	1	4	3	2	5	9
1	9	4	2	6	5	7	8	3
5	3	2	7	8	9	6	4	1
8	1	6	9	2	4	5	3	7
3	4	7	5	1	8	9	6	2
2	5	9	3	7	6	4	1	8

Solution # 65

8	9	1	4	7	3	5	6	2
4	5	6	2	9	1	7	3	8
3	7	2	6	8	5	1	9	4
1	6	7	9	3	4	8	2	5
5	3	8	7	2	6	9	4	1
2	4	9	5	1	8	6	7	3
6	2	3	1	5	9	4	8	7
9	8	5	3	4	7	2	1	6
7	1	4	8	6	2	3	5	9

Solution # 66

6	7	3	1	5	2	8	9	4
4	2	1	3	8	9	5	6	7
5	8	9	4	7	6	2	1	3
9	6	7	5	2	4	1	3	8
3	1	5	6	9	8	7	4	2
8	4	2	7	1	3	6	5	9
7	5	4	8	3	1	9	2	6
1	9	6	2	4	7	3	8	5
2	3	8	9	6	5	4	7	1

Solution # 67

3	2	1	6	9	7	4	5	8
7	5	6	3	8	4	9	2	1
8	4	9	1	2	5	7	6	3
1	9	4	5	7	8	2	3	6
2	3	5	4	6	9	8	1	7
6	7	8	2	3	1	5	4	9
5	8	3	7	4	6	1	9	2
9	1	2	8	5	3	6	7	4
4	6	7	9	1	2	3	8	5

Solution # 68

4	6	5	3	7	2	1	8	9
2	7	9	4	8	1	6	3	5
3	1	8	6	5	9	4	2	7
7	5	4	1	9	3	2	6	8
1	8	3	2	6	7	5	9	4
6	9	2	8	4	5	7	1	3
5	2	7	9	3	6	8	4	1
9	4	1	5	2	8	3	7	6
8	3	6	7	1	4	9	5	2

Solution # 69

9	4	1	5	8	3	2	7	6
8	7	2	6	1	4	5	3	9
5	6	3	2	7	9	1	4	8
2	8	5	1	3	6	7	9	4
7	9	4	8	2	5	3	6	1
3	1	6	9	4	7	8	2	5
6	3	7	4	5	8	9	1	2
4	2	8	7	9	1	6	5	3
1	5	9	3	6	2	4	8	7

Solution # 70

7	8	2	9	6	4	1	5	3
1	3	4	2	5	8	7	9	6
9	6	5	3	1	7	2	4	8
4	5	6	7	8	9	3	1	2
2	9	1	6	4	3	8	7	5
3	7	8	5	2	1	9	6	4
6	2	7	8	9	5	4	3	1
8	1	9	4	3	6	5	2	7
5	4	3	1	7	2	6	8	9

Solution # 71

6	2	5	9	3	1	4	8	7
4	9	7	8	5	6	3	2	1
1	3	8	7	2	4	9	5	6
7	5	9	6	1	3	8	4	2
2	4	1	5	9	8	6	7	3
3	8	6	2	4	7	1	9	5
8	7	3	4	6	2	5	1	9
9	6	2	1	8	5	7	3	4
5	1	4	3	7	9	2	6	8

Solution # 72

8	3	5	6	4	1	9	7	2
7	4	2	9	8	5	3	1	6
9	1	6	7	2	3	8	4	5
1	8	3	4	5	2	7	6	9
6	5	4	8	9	7	1	2	3
2	9	7	1	3	6	4	5	8
3	6	1	5	7	8	2	9	4
4	7	8	2	6	9	5	3	1
5	2	9	3	1	4	6	8	7

Solution # 73

8	7	2	1	4	9	6	5	3
6	5	4	3	7	8	1	9	2
1	3	9	5	6	2	4	7	8
9	2	8	7	3	1	5	6	4
7	1	5	6	8	4	2	3	9
4	6	3	2	9	5	7	8	1
5	4	6	8	2	3	9	1	7
3	9	1	4	5	7	8	2	6
2	8	7	9	1	6	3	4	5

Solution # 74

5	9	1	7	8	3	2	4	6
6	4	3	2	1	5	8	7	9
8	7	2	4	9	6	1	3	5
9	8	4	1	5	2	3	6	7
1	3	7	8	6	4	5	9	2
2	5	6	9	3	7	4	1	8
4	6	9	3	2	8	7	5	1
3	2	5	6	7	1	9	8	4
7	1	8	5	4	9	6	2	3

Solution # 75

6	1	9	4	3	7	2	5	8
2	4	8	9	1	5	6	7	3
5	3	7	8	6	2	1	4	9
4	9	5	6	7	3	8	2	1
3	7	2	1	8	9	5	6	4
1	8	6	2	5	4	9	3	7
7	6	3	5	9	8	4	1	2
9	5	4	7	2	1	3	8	6
8	2	1	3	4	6	7	9	5

Solution # 76

8	9	4	7	2	5	3	1	6
5	2	1	4	3	6	8	7	9
6	7	3	8	9	1	2	4	5
4	6	8	3	5	9	7	2	1
2	5	7	6	1	4	9	8	3
1	3	9	2	8	7	5	6	4
9	8	6	1	7	3	4	5	2
7	4	5	9	6	2	1	3	8
3	1	2	5	4	8	6	9	7

Solution # 77

9	2	6	3	5	8	4	7	1
8	4	3	7	1	2	5	9	6
7	1	5	4	9	6	3	8	2
2	6	8	9	7	5	1	4	3
4	3	9	6	2	1	7	5	8
1	5	7	8	4	3	2	6	9
5	9	2	1	6	7	8	3	4
3	7	4	2	8	9	6	1	5
6	8	1	5	3	4	9	2	7

Solution # 78

1	8	7	4	6	3	9	5	2
5	9	4	1	7	2	3	8	6
2	6	3	8	5	9	1	4	7
6	7	1	3	8	5	4	2	9
4	5	2	9	1	6	8	7	3
8	3	9	2	4	7	5	6	1
9	2	6	5	3	8	7	1	4
3	1	5	7	2	4	6	9	8
7	4	8	6	9	1	2	3	5

Solution # 79

8	6	3	2	1	9	7	5	4
9	5	4	7	3	6	1	8	2
7	2	1	5	4	8	9	3	6
4	8	5	1	6	7	3	2	9
3	9	7	8	2	4	5	6	1
2	1	6	3	9	5	4	7	8
1	7	8	4	5	2	6	9	3
6	4	2	9	7	3	8	1	5
5	3	9	6	8	1	2	4	7

Solution # 80

1	4	2	8	9	6	7	5	3
7	6	3	5	1	2	8	9	4
8	5	9	3	4	7	2	1	6
6	8	5	9	2	4	3	7	1
3	2	1	6	7	8	5	4	9
9	7	4	1	5	3	6	2	8
5	9	8	7	3	1	4	6	2
4	3	7	2	6	9	1	8	5
2	1	6	4	8	5	9	3	7

Solution # 81

2	4	5	1	3	7	6	9	8
9	3	8	2	4	6	1	5	7
7	6	1	9	8	5	4	2	3
4	7	6	3	2	8	9	1	5
5	8	2	7	1	9	3	4	6
3	1	9	6	5	4	7	8	2
6	5	4	8	9	3	2	7	1
8	2	7	4	6	1	5	3	9
1	9	3	5	7	2	8	6	4

Solution # 82

1	6	2	7	8	4	9	5	3
5	9	4	6	2	3	1	7	8
8	7	3	9	5	1	6	2	4
7	3	1	5	6	8	4	9	2
4	2	8	3	9	7	5	1	6
6	5	9	4	1	2	8	3	7
2	1	6	8	3	5	7	4	9
9	4	5	2	7	6	3	8	1
3	8	7	1	4	9	2	6	5

Solution # 83

7	4	9	6	1	3	8	5	2
1	8	3	9	5	2	7	6	4
6	5	2	8	7	4	9	1	3
9	6	7	2	8	5	4	3	1
8	1	4	7	3	6	2	9	5
3	2	5	1	4	9	6	7	8
4	7	6	3	2	1	5	8	9
2	3	8	5	9	7	1	4	6
5	9	1	4	6	8	3	2	7

Solution # 84

1	4	7	3	6	5	9	2	8
6	9	5	7	8	2	1	3	4
3	8	2	9	1	4	7	6	5
9	3	1	6	4	7	5	8	2
5	6	8	1	2	3	4	7	9
7	2	4	5	9	8	3	1	6
8	5	9	2	7	1	6	4	3
2	7	6	4	3	9	8	5	1
4	1	3	8	5	6	2	9	7

Solution # 85

8	1	3	2	5	9	6	7	4
6	4	2	1	8	7	3	5	9
9	5	7	4	3	6	2	1	8
5	2	1	7	4	8	9	3	6
4	7	8	6	9	3	5	2	1
3	9	6	5	1	2	4	8	7
1	3	5	9	7	4	8	6	2
7	6	4	8	2	5	1	9	3
2	8	9	3	6	1	7	4	5

Solution # 86

6	1	7	8	2	3	4	9	5
2	9	4	5	1	7	3	6	8
5	8	3	9	4	6	1	7	2
8	3	5	1	6	9	2	4	7
7	4	1	2	5	8	6	3	9
9	2	6	7	3	4	8	5	1
4	5	8	6	9	1	7	2	3
1	6	2	3	7	5	9	8	4
3	7	9	4	8	2	5	1	6

Solution # 87

5	8	1	7	3	2	9	4	6
9	2	3	4	6	5	7	8	1
7	4	6	9	1	8	3	5	2
1	6	8	3	4	7	2	9	5
4	9	5	6	2	1	8	3	7
3	7	2	5	8	9	6	1	4
6	5	9	8	7	4	1	2	3
2	3	4	1	9	6	5	7	8
8	1	7	2	5	3	4	6	9

Solution # 88

4	9	7	5	1	6	2	3	8
3	5	2	8	9	7	4	6	1
8	6	1	4	3	2	9	5	7
7	3	9	6	2	1	5	8	4
6	8	4	7	5	9	1	2	3
2	1	5	3	8	4	7	9	6
5	2	3	1	7	8	6	4	9
9	7	6	2	4	3	8	1	5
1	4	8	9	6	5	3	7	2

Solution # 89

2	4	6	1	7	5	9	3	8
3	5	8	4	9	6	2	7	1
7	9	1	2	8	3	6	4	5
1	2	5	9	4	7	8	6	3
9	8	3	6	2	1	7	5	4
6	7	4	3	5	8	1	2	9
8	6	7	5	3	9	4	1	2
4	3	9	7	1	2	5	8	6
5	1	2	8	6	4	3	9	7

Solution # 90

3	7	6	8	5	2	1	4	9
1	5	4	9	7	3	8	2	6
8	9	2	1	6	4	3	5	7
9	2	1	3	4	6	7	8	5
4	6	5	7	8	1	2	9	3
7	8	3	5	2	9	4	6	1
2	1	8	6	3	5	9	7	4
5	4	9	2	1	7	6	3	8
6	3	7	4	9	8	5	1	2

Solution # 91

6	1	5	4	2	7	8	3	9
7	8	2	3	6	9	5	4	1
9	4	3	8	1	5	6	2	7
4	5	8	2	3	1	7	9	6
2	6	9	7	4	8	1	5	3
1	3	7	5	9	6	2	8	4
8	2	4	1	7	3	9	6	5
5	7	6	9	8	4	3	1	2
3	9	1	6	5	2	4	7	8

Solution # 92

6	2	3	9	7	8	1	5	4
1	8	7	3	5	4	2	6	9
4	9	5	2	1	6	3	7	8
8	3	4	6	2	1	7	9	5
9	7	6	4	3	5	8	2	1
2	5	1	8	9	7	6	4	3
7	4	9	1	6	3	5	8	2
5	1	8	7	4	2	9	3	6
3	6	2	5	8	9	4	1	7

Solution # 93

3	9	7	2	5	1	4	8	6
6	4	8	7	3	9	5	2	1
5	1	2	4	8	6	3	9	7
4	8	5	6	7	3	2	1	9
7	6	9	1	4	2	8	5	3
1	2	3	8	9	5	7	6	4
8	3	6	5	1	4	9	7	2
2	5	4	9	6	7	1	3	8
9	7	1	3	2	8	6	4	5

Solution # 94

1	3	8	6	9	2	5	7	4
4	2	6	1	7	5	8	9	3
5	9	7	3	4	8	6	1	2
9	1	2	8	5	7	3	4	6
7	5	3	9	6	4	1	2	8
6	8	4	2	1	3	7	5	9
8	4	1	5	2	6	9	3	7
3	7	9	4	8	1	2	6	5
2	6	5	7	3	9	4	8	1

Solution # 95

5	2	1	7	6	9	3	4	8
9	4	7	3	1	8	6	2	5
8	3	6	2	5	4	1	7	9
4	7	3	9	8	6	2	5	1
6	5	8	1	4	2	7	9	3
2	1	9	5	7	3	8	6	4
1	6	2	4	3	5	9	8	7
3	8	5	6	9	7	4	1	2
7	9	4	8	2	1	5	3	6

Solution # 96

9	5	8	2	1	6	4	3	7
6	7	4	8	3	5	1	9	2
2	3	1	9	7	4	6	8	5
3	4	7	5	6	1	9	2	8
8	9	6	3	2	7	5	4	1
5	1	2	4	8	9	7	6	3
1	8	5	6	9	2	3	7	4
4	6	3	7	5	8	2	1	9
7	2	9	1	4	3	8	5	6

Solution # 97

8	1	7	5	9	4	2	6	3
3	9	5	2	6	8	1	4	7
6	2	4	7	1	3	5	8	9
1	4	2	9	8	6	7	3	5
7	5	8	4	3	2	6	9	1
9	6	3	1	5	7	4	2	8
5	8	6	3	2	1	9	7	4
4	3	1	6	7	9	8	5	2
2	7	9	8	4	5	3	1	6

Solution # 98

8	1	7	3	5	4	2	9	6
9	2	4	8	6	1	3	7	5
5	6	3	9	7	2	8	1	4
1	9	5	4	8	6	7	2	3
6	7	8	2	9	3	5	4	1
3	4	2	7	1	5	9	6	8
4	3	6	5	2	7	1	8	9
7	8	1	6	3	9	4	5	2
2	5	9	1	4	8	6	3	7

Solution # 99

1	6	3	9	2	4	8	7	5
7	8	9	3	1	5	4	6	2
2	4	5	8	6	7	1	9	3
9	5	8	6	7	3	2	1	4
6	2	1	4	5	8	7	3	9
3	7	4	1	9	2	5	8	6
4	1	2	7	3	6	9	5	8
5	9	6	2	8	1	3	4	7
8	3	7	5	4	9	6	2	1

Solution # 100

7	1	5	8	2	3	9	6	4
2	9	6	4	7	1	5	8	3
4	3	8	9	6	5	1	7	2
1	2	9	3	8	4	6	5	7
5	7	4	2	9	6	3	1	8
6	8	3	1	5	7	2	4	9
9	6	2	7	1	8	4	3	5
3	5	7	6	4	2	8	9	1
8	4	1	5	3	9	7	2	6

Solution # 101

3	4	9	1	7	2	8	5	6
1	8	5	9	6	3	4	7	2
7	2	6	5	8	4	3	9	1
8	9	2	7	5	1	6	4	3
6	5	1	3	4	9	7	2	8
4	7	3	8	2	6	5	1	9
5	1	7	2	3	8	9	6	4
9	3	4	6	1	7	2	8	5
2	6	8	4	9	5	1	3	7

Solution # 102

1	2	5	7	6	9	8	4	3
4	7	8	5	2	3	1	6	9
6	9	3	8	4	1	2	7	5
3	4	6	2	9	8	7	5	1
9	8	1	3	7	5	4	2	6
2	5	7	4	1	6	9	3	8
7	1	9	6	5	2	3	8	4
5	3	2	9	8	4	6	1	7
8	6	4	1	3	7	5	9	2

Solution # 103

4	8	5	3	9	7	2	6	1
7	6	3	2	1	8	5	9	4
1	9	2	6	5	4	8	7	3
5	2	6	8	3	1	9	4	7
3	1	8	7	4	9	6	5	2
9	7	4	5	2	6	1	3	8
6	4	1	9	7	2	3	8	5
8	3	7	1	6	5	4	2	9
2	5	9	4	8	3	7	1	6

Solution # 104

5	4	1	3	9	7	8	6	2
7	6	3	2	5	8	9	1	4
8	2	9	6	4	1	5	7	3
9	8	6	7	2	3	4	5	1
4	1	2	9	8	5	6	3	7
3	7	5	4	1	6	2	9	8
2	3	4	1	6	9	7	8	5
1	9	8	5	7	2	3	4	6
6	5	7	8	3	4	1	2	9

Solution # 105

5	8	6	3	2	4	1	9	7
1	3	2	7	5	9	4	6	8
9	7	4	1	6	8	5	3	2
2	9	7	5	3	6	8	4	1
6	4	5	8	7	1	3	2	9
8	1	3	9	4	2	7	5	6
7	6	9	4	8	5	2	1	3
3	5	1	2	9	7	6	8	4
4	2	8	6	1	3	9	7	5

Solution # 106

3	6	1	4	2	8	7	9	5
4	9	2	6	7	5	3	8	1
7	5	8	1	9	3	6	4	2
2	7	3	9	4	6	1	5	8
5	4	6	8	1	7	2	3	9
1	8	9	5	3	2	4	7	6
8	1	5	3	6	4	9	2	7
6	3	7	2	8	9	5	1	4
9	2	4	7	5	1	8	6	3

Solution # 107

9	3	2	6	5	7	1	8	4
1	4	6	9	3	8	5	7	2
5	8	7	2	1	4	3	9	6
2	9	4	8	7	1	6	5	3
8	7	5	3	4	6	2	1	9
6	1	3	5	9	2	8	4	7
3	5	1	4	6	9	7	2	8
7	2	9	1	8	3	4	6	5
4	6	8	7	2	5	9	3	1

Solution # 108

1	6	8	2	3	7	4	9	5
5	4	2	9	6	1	3	8	7
3	7	9	5	8	4	2	1	6
9	3	6	8	4	5	1	7	2
7	8	5	3	1	2	6	4	9
2	1	4	6	7	9	8	5	3
6	5	7	1	2	8	9	3	4
8	9	3	4	5	6	7	2	1
4	2	1	7	9	3	5	6	8

Solution # 109

5	8	9	6	4	1	3	2	7
4	2	6	3	5	7	1	9	8
3	7	1	8	2	9	6	5	4
2	6	5	7	8	3	4	1	9
9	1	7	5	6	4	8	3	2
8	3	4	1	9	2	5	7	6
6	9	8	2	3	5	7	4	1
7	4	3	9	1	8	2	6	5
1	5	2	4	7	6	9	8	3

Solution # 110

3	8	5	6	4	9	2	7	1
2	1	9	8	7	3	6	4	5
6	4	7	1	5	2	9	3	8
4	6	3	2	8	7	1	5	9
9	7	8	4	1	5	3	6	2
1	5	2	3	9	6	7	8	4
5	2	6	9	3	8	4	1	7
7	3	4	5	2	1	8	9	6
8	9	1	7	6	4	5	2	3

Solution # 111

2	7	9	6	4	5	8	1	3
3	6	1	8	7	2	5	4	9
4	8	5	9	3	1	6	2	7
8	4	6	7	5	3	2	9	1
9	1	3	2	8	4	7	6	5
5	2	7	1	6	9	4	3	8
6	5	2	3	1	8	9	7	4
1	9	8	4	2	7	3	5	6
7	3	4	5	9	6	1	8	2

Solution # 112

9	3	6	2	4	5	1	7	8
5	7	8	1	6	3	9	2	4
4	1	2	7	8	9	5	3	6
2	9	1	5	3	4	8	6	7
7	4	5	8	2	6	3	9	1
6	8	3	9	7	1	4	5	2
8	2	4	3	5	7	6	1	9
1	5	7	6	9	8	2	4	3
3	6	9	4	1	2	7	8	5

Solution # 113

3	5	9	1	8	4	2	6	7
1	7	6	5	9	2	4	3	8
4	8	2	3	7	6	5	1	9
6	3	5	2	4	7	8	9	1
8	2	4	9	6	1	7	5	3
9	1	7	8	5	3	6	4	2
7	9	8	6	1	5	3	2	4
5	4	3	7	2	9	1	8	6
2	6	1	4	3	8	9	7	5

Solution # 114

6	9	8	4	7	5	2	3	1
2	4	7	1	3	8	9	6	5
5	3	1	9	2	6	7	8	4
8	6	3	7	1	4	5	9	2
1	2	5	3	8	9	6	4	7
4	7	9	5	6	2	8	1	3
9	8	4	2	5	3	1	7	6
7	5	6	8	4	1	3	2	9
3	1	2	6	9	7	4	5	8

Solution # 115

5	3	8	7	2	4	1	6	9
4	6	9	8	1	5	3	7	2
1	7	2	9	3	6	8	4	5
9	4	3	2	6	1	5	8	7
6	8	7	5	4	9	2	3	1
2	5	1	3	8	7	6	9	4
7	1	4	6	5	3	9	2	8
8	9	6	1	7	2	4	5	3
3	2	5	4	9	8	7	1	6

Solution # 116

7	1	8	5	2	4	6	3	9
2	4	9	3	6	8	7	5	1
6	5	3	1	7	9	4	2	8
9	7	2	6	1	5	8	4	3
8	6	4	9	3	2	5	1	7
5	3	1	8	4	7	9	6	2
4	2	6	7	8	1	3	9	5
1	8	5	4	9	3	2	7	6
3	9	7	2	5	6	1	8	4

Solution # 117

5	6	7	4	8	3	2	9	1
3	4	8	9	2	1	5	7	6
1	9	2	7	6	5	8	3	4
8	2	5	6	9	7	4	1	3
6	1	3	5	4	8	7	2	9
9	7	4	1	3	2	6	8	5
2	8	1	3	5	4	9	6	7
7	5	6	8	1	9	3	4	2
4	3	9	2	7	6	1	5	8

Solution # 118

7	9	5	3	6	8	1	2	4
2	8	4	7	9	1	6	5	3
3	1	6	2	4	5	8	7	9
1	5	3	4	7	6	9	8	2
9	6	8	1	2	3	7	4	5
4	2	7	8	5	9	3	6	1
5	7	9	6	1	4	2	3	8
8	4	2	9	3	7	5	1	6
6	3	1	5	8	2	4	9	7

Solution # 119

2	3	7	6	9	5	1	4	8
4	8	5	1	3	7	9	2	6
1	9	6	4	2	8	7	3	5
3	6	4	2	7	1	5	8	9
8	7	9	5	4	3	6	1	2
5	2	1	9	8	6	4	7	3
9	5	2	8	1	4	3	6	7
6	1	3	7	5	2	8	9	4
7	4	8	3	6	9	2	5	1

Solution # 120

9	5	7	6	1	8	4	3	2
1	3	6	4	2	5	9	7	8
4	8	2	3	9	7	5	6	1
2	9	4	5	3	6	1	8	7
6	7	3	1	8	4	2	9	5
5	1	8	9	7	2	6	4	3
3	4	1	8	5	9	7	2	6
8	2	9	7	6	1	3	5	4
7	6	5	2	4	3	8	1	9

Solution # 121

2	1	4	3	9	5	7	8	6
6	8	7	1	4	2	5	9	3
9	3	5	7	6	8	4	1	2
1	9	3	8	2	4	6	5	7
4	2	8	6	5	7	9	3	1
5	7	6	9	3	1	8	2	4
7	4	9	5	1	3	2	6	8
3	6	2	4	8	9	1	7	5
8	5	1	2	7	6	3	4	9

Solution # 122

8	9	6	2	5	4	7	3	1
4	7	3	6	9	1	2	5	8
5	1	2	7	8	3	6	4	9
7	4	8	9	6	2	3	1	5
3	2	9	4	1	5	8	6	7
1	6	5	3	7	8	9	2	4
6	5	1	8	2	7	4	9	3
2	8	4	1	3	9	5	7	6
9	3	7	5	4	6	1	8	2

Solution # 123

4	7	9	8	5	6	2	3	1
3	5	6	1	4	2	8	7	9
2	1	8	3	7	9	4	5	6
9	6	4	7	8	3	1	2	5
8	2	7	5	6	1	3	9	4
1	3	5	2	9	4	6	8	7
6	4	2	9	3	5	7	1	8
7	9	3	4	1	8	5	6	2
5	8	1	6	2	7	9	4	3

Solution # 124

3	2	9	6	5	7	1	8	4
7	4	8	2	3	1	5	6	9
1	6	5	4	8	9	7	3	2
9	1	6	8	2	5	3	4	7
5	3	7	1	4	6	2	9	8
4	8	2	7	9	3	6	1	5
2	9	3	5	1	8	4	7	6
8	7	4	3	6	2	9	5	1
6	5	1	9	7	4	8	2	3

Solution # 125

4	3	8	6	9	5	7	2	1
1	5	7	8	4	2	9	6	3
2	9	6	7	1	3	4	8	5
7	2	5	4	6	1	8	3	9
3	1	9	5	8	7	2	4	6
6	8	4	3	2	9	5	1	7
5	7	1	2	3	8	6	9	4
9	6	2	1	5	4	3	7	8
8	4	3	9	7	6	1	5	2

Solution # 126

2	7	8	3	6	1	4	5	9
1	6	4	5	9	2	7	3	8
9	3	5	8	4	7	1	6	2
4	5	1	2	7	6	9	8	3
3	9	2	1	8	4	6	7	5
7	8	6	9	3	5	2	4	1
8	1	7	4	5	9	3	2	6
6	2	3	7	1	8	5	9	4
5	4	9	6	2	3	8	1	7

Solution # 127

1	9	6	3	8	5	7	4	2
2	4	8	7	1	9	6	5	3
5	7	3	2	4	6	8	9	1
3	8	2	1	5	7	9	6	4
9	6	4	8	2	3	1	7	5
7	1	5	6	9	4	3	2	8
4	5	7	9	3	8	2	1	6
8	2	9	4	6	1	5	3	7
6	3	1	5	7	2	4	8	9

Solution # 128

3	2	7	6	8	4	1	5	9
8	6	4	1	5	9	3	2	7
5	1	9	7	3	2	4	8	6
2	5	1	3	9	8	6	7	4
6	7	8	4	2	1	9	3	5
4	9	3	5	7	6	8	1	2
7	4	6	8	1	5	2	9	3
1	3	2	9	6	7	5	4	8
9	8	5	2	4	3	7	6	1

Solution # 129

6	1	2	7	3	9	5	8	4
9	8	3	5	1	4	6	7	2
7	4	5	6	2	8	3	1	9
3	6	1	9	7	2	8	4	5
5	9	7	4	8	1	2	6	3
8	2	4	3	5	6	7	9	1
4	3	9	8	6	5	1	2	7
2	7	8	1	4	3	9	5	6
1	5	6	2	9	7	4	3	8

Solution # 130

9	8	3	1	7	2	5	6	4
6	2	4	5	9	3	1	7	8
5	1	7	8	4	6	3	9	2
7	6	8	2	3	1	9	4	5
4	9	2	7	6	5	8	3	1
3	5	1	9	8	4	6	2	7
2	4	9	6	1	8	7	5	3
1	3	6	4	5	7	2	8	9
8	7	5	3	2	9	4	1	6

Solution # 131

9	1	3	7	6	5	4	8	2
8	5	6	4	2	3	9	1	7
7	2	4	8	9	1	3	5	6
4	3	9	1	5	2	6	7	8
2	8	5	9	7	6	1	4	3
1	6	7	3	8	4	5	2	9
3	4	2	6	1	8	7	9	5
6	7	8	5	4	9	2	3	1
5	9	1	2	3	7	8	6	4

Solution # 132

9	8	4	1	6	5	3	2	7
7	2	5	8	9	3	1	6	4
3	1	6	2	7	4	8	9	5
2	5	3	9	1	7	4	8	6
4	6	7	5	8	2	9	3	1
1	9	8	4	3	6	7	5	2
5	7	2	3	4	8	6	1	9
6	3	1	7	5	9	2	4	8
8	4	9	6	2	1	5	7	3

Solution # 133

4	6	2	1	3	8	9	5	7
7	9	8	5	6	2	3	1	4
1	3	5	9	7	4	6	8	2
5	4	3	2	8	6	7	9	1
6	1	7	3	5	9	4	2	8
8	2	9	7	4	1	5	3	6
2	7	1	4	9	3	8	6	5
9	5	6	8	2	7	1	4	3
3	8	4	6	1	5	2	7	9

Solution # 134

8	9	4	2	6	3	7	5	1
3	1	7	5	8	9	2	6	4
6	5	2	7	1	4	8	9	3
4	6	3	8	2	7	5	1	9
5	7	9	6	3	1	4	2	8
1	2	8	9	4	5	6	3	7
2	3	1	4	7	6	9	8	5
7	8	5	1	9	2	3	4	6
9	4	6	3	5	8	1	7	2

Solution # 135

1	8	6	5	2	3	7	4	9
2	5	4	9	6	7	3	1	8
9	3	7	1	8	4	6	2	5
5	7	9	2	1	8	4	6	3
4	1	3	6	7	9	5	8	2
6	2	8	3	4	5	1	9	7
8	4	5	7	9	1	2	3	6
7	6	1	8	3	2	9	5	4
3	9	2	4	5	6	8	7	1

Solution # 136

1	3	4	2	9	6	8	7	5
2	7	9	1	8	5	6	4	3
6	5	8	3	7	4	2	1	9
3	6	1	7	4	2	5	9	8
9	8	7	5	6	3	1	2	4
5	4	2	9	1	8	7	3	6
8	9	5	4	2	7	3	6	1
4	2	3	6	5	1	9	8	7
7	1	6	8	3	9	4	5	2

Solution # 137

4	1	9	5	3	7	6	2	8
7	5	6	4	8	2	9	3	1
3	8	2	1	6	9	4	7	5
2	3	8	6	7	1	5	9	4
5	7	1	9	2	4	3	8	6
9	6	4	8	5	3	7	1	2
1	2	7	3	4	5	8	6	9
8	4	3	2	9	6	1	5	7
6	9	5	7	1	8	2	4	3

Solution # 138

5	7	3	8	6	4	2	1	9
1	6	2	7	9	5	8	3	4
9	8	4	3	1	2	7	6	5
3	9	1	2	7	6	5	4	8
8	2	6	5	4	9	1	7	3
4	5	7	1	3	8	9	2	6
2	3	9	4	8	1	6	5	7
6	4	5	9	2	7	3	8	1
7	1	8	6	5	3	4	9	2

Solution # 139

7	6	8	9	5	4	2	1	3
2	4	5	3	1	7	8	6	9
9	3	1	8	2	6	4	7	5
1	9	6	2	3	8	5	4	7
5	7	2	4	6	9	3	8	1
4	8	3	1	7	5	9	2	6
6	1	9	5	4	2	7	3	8
8	2	7	6	9	3	1	5	4
3	5	4	7	8	1	6	9	2

Solution # 140

6	8	1	9	2	4	5	3	7
9	3	4	6	5	7	8	1	2
5	2	7	8	1	3	6	9	4
2	9	6	3	7	1	4	5	8
1	5	8	4	6	9	7	2	3
7	4	3	2	8	5	1	6	9
3	1	9	7	4	6	2	8	5
4	6	2	5	9	8	3	7	1
8	7	5	1	3	2	9	4	6

Solution # 141

6	7	5	4	3	1	8	2	9
1	9	3	2	7	8	5	6	4
2	8	4	5	6	9	3	7	1
8	1	9	3	4	2	6	5	7
3	5	2	6	1	7	4	9	8
7	4	6	9	8	5	1	3	2
5	6	7	1	9	4	2	8	3
9	3	1	8	2	6	7	4	5
4	2	8	7	5	3	9	1	6

Solution # 142

8	1	7	6	4	2	9	5	3
9	4	2	7	5	3	8	6	1
5	3	6	1	8	9	4	2	7
7	8	4	3	9	6	5	1	2
6	5	3	2	1	4	7	9	8
2	9	1	5	7	8	6	3	4
4	2	8	9	6	1	3	7	5
3	7	9	8	2	5	1	4	6
1	6	5	4	3	7	2	8	9

Solution # 143

5	9	7	6	4	3	1	8	2
6	1	8	2	7	9	4	3	5
4	2	3	8	1	5	6	9	7
3	7	4	9	5	1	8	2	6
2	6	9	3	8	7	5	1	4
1	8	5	4	6	2	3	7	9
7	5	6	1	2	8	9	4	3
9	4	1	7	3	6	2	5	8
8	3	2	5	9	4	7	6	1

Solution # 144

9	1	8	5	7	2	6	4	3
3	4	7	9	6	8	2	5	1
5	6	2	4	3	1	9	8	7
2	3	9	1	5	7	8	6	4
7	5	6	8	9	4	1	3	2
4	8	1	6	2	3	7	9	5
8	9	3	2	1	5	4	7	6
1	7	4	3	8	6	5	2	9
6	2	5	7	4	9	3	1	8

Solution # 145

3	4	5	7	1	2	8	9	6
9	7	8	5	4	6	3	1	2
1	2	6	8	9	3	4	7	5
6	9	1	4	2	8	7	5	3
2	8	3	6	7	5	9	4	1
4	5	7	9	3	1	2	6	8
7	3	2	1	5	4	6	8	9
5	6	4	2	8	9	1	3	7
8	1	9	3	6	7	5	2	4

Solution # 146

6	8	1	7	4	3	2	9	5
2	9	3	5	8	1	7	6	4
7	5	4	2	9	6	8	1	3
9	1	8	3	6	2	5	4	7
4	7	6	8	5	9	1	3	2
3	2	5	1	7	4	9	8	6
5	3	2	4	1	8	6	7	9
1	6	7	9	3	5	4	2	8
8	4	9	6	2	7	3	5	1

Solution # 147

9	5	6	1	8	7	2	3	4
3	4	7	2	9	6	5	1	8
2	8	1	5	3	4	7	6	9
1	9	2	6	7	5	8	4	3
5	3	8	4	2	9	1	7	6
7	6	4	3	1	8	9	2	5
4	7	5	9	6	1	3	8	2
8	2	9	7	4	3	6	5	1
6	1	3	8	5	2	4	9	7

Solution # 148

6	5	4	1	7	2	3	8	9
3	9	8	6	5	4	1	2	7
1	2	7	3	9	8	5	4	6
7	6	9	8	3	5	2	1	4
5	4	3	2	1	9	7	6	8
2	8	1	7	4	6	9	5	3
4	3	2	9	6	1	8	7	5
8	7	6	5	2	3	4	9	1
9	1	5	4	8	7	6	3	2

Solution # 149

7	6	4	5	1	9	3	8	2
1	8	5	3	2	6	7	4	9
3	9	2	8	7	4	6	5	1
2	7	3	6	8	5	1	9	4
5	4	8	1	9	3	2	7	6
6	1	9	7	4	2	5	3	8
8	3	1	9	6	7	4	2	5
9	2	7	4	5	1	8	6	3
4	5	6	2	3	8	9	1	7

Solution # 150

1	9	8	2	3	4	5	6	7
2	5	6	7	8	9	4	1	3
3	7	4	1	6	5	9	2	8
6	2	9	3	5	8	1	7	4
5	8	7	4	9	1	6	3	2
4	3	1	6	7	2	8	9	5
7	1	3	5	4	6	2	8	9
9	4	2	8	1	3	7	5	6
8	6	5	9	2	7	3	4	1

Solution # 151

6	5	2	1	8	9	7	4	3
3	4	9	7	2	6	1	8	5
8	1	7	4	5	3	9	2	6
4	7	3	2	1	5	6	9	8
5	2	8	9	6	7	4	3	1
1	9	6	3	4	8	5	7	2
7	6	4	8	3	1	2	5	9
2	3	5	6	9	4	8	1	7
9	8	1	5	7	2	3	6	4

Solution # 152

1	9	3	8	5	6	2	4	7
8	4	6	2	7	3	1	5	9
2	5	7	4	9	1	6	3	8
7	8	1	9	3	4	5	2	6
3	2	5	6	1	8	9	7	4
4	6	9	5	2	7	8	1	3
6	3	8	1	4	2	7	9	5
9	7	2	3	8	5	4	6	1
5	1	4	7	6	9	3	8	2

Solution # 153

9	7	4	5	8	3	6	1	2
2	8	5	6	4	1	9	7	3
3	1	6	2	9	7	8	4	5
1	6	2	9	5	4	3	8	7
8	4	9	3	7	2	1	5	6
7	5	3	1	6	8	2	9	4
4	9	1	7	3	6	5	2	8
6	2	7	8	1	5	4	3	9
5	3	8	4	2	9	7	6	1

Solution # 154

2	8	6	4	1	3	9	7	5
5	3	1	2	7	9	8	6	4
7	4	9	5	6	8	3	2	1
9	7	5	3	8	6	1	4	2
8	2	3	1	4	7	6	5	9
6	1	4	9	5	2	7	8	3
4	5	8	7	3	1	2	9	6
3	6	2	8	9	5	4	1	7
1	9	7	6	2	4	5	3	8

Solution # 155

6	9	3	2	4	5	1	8	7
4	2	8	1	3	7	5	6	9
5	7	1	6	9	8	4	3	2
8	3	9	5	7	6	2	4	1
1	6	4	9	8	2	7	5	3
2	5	7	3	1	4	6	9	8
3	8	2	4	5	1	9	7	6
7	4	6	8	2	9	3	1	5
9	1	5	7	6	3	8	2	4

Solution # 156

7	2	1	5	4	6	9	3	8
9	4	5	8	1	3	7	6	2
8	3	6	7	9	2	1	4	5
4	6	8	3	2	7	5	9	1
1	7	3	9	5	4	8	2	6
5	9	2	1	6	8	4	7	3
3	1	4	2	8	9	6	5	7
6	8	7	4	3	5	2	1	9
2	5	9	6	7	1	3	8	4

Solution # 157

8	3	7	2	4	6	1	5	9
5	4	1	3	9	7	2	8	6
2	6	9	8	5	1	4	7	3
7	9	5	4	8	2	3	6	1
1	2	3	6	7	9	8	4	5
6	8	4	1	3	5	9	2	7
3	5	6	9	2	4	7	1	8
4	7	8	5	1	3	6	9	2
9	1	2	7	6	8	5	3	4

Solution # 158

8	6	3	7	9	5	4	2	1
9	5	2	6	4	1	8	7	3
1	4	7	3	8	2	5	9	6
7	8	5	2	1	3	9	6	4
2	9	4	8	5	6	1	3	7
3	1	6	4	7	9	2	8	5
6	7	8	1	2	4	3	5	9
5	3	1	9	6	8	7	4	2
4	2	9	5	3	7	6	1	8

Solution # 159

9	4	2	5	1	3	8	6	7
5	8	7	6	9	2	3	4	1
6	1	3	4	7	8	5	9	2
4	3	8	9	6	1	2	7	5
1	7	9	2	4	5	6	8	3
2	6	5	8	3	7	4	1	9
8	5	1	7	2	4	9	3	6
3	2	6	1	8	9	7	5	4
7	9	4	3	5	6	1	2	8

Solution # 160

5	9	2	6	1	4	8	7	3
8	4	6	3	7	9	1	2	5
3	1	7	2	8	5	6	4	9
9	2	8	7	5	1	4	3	6
6	5	3	4	9	2	7	1	8
4	7	1	8	6	3	5	9	2
1	3	9	5	4	8	2	6	7
7	8	4	9	2	6	3	5	1
2	6	5	1	3	7	9	8	4

Solution # 161

1	6	8	2	3	4	9	5	7
5	7	2	1	6	9	3	8	4
3	9	4	8	7	5	6	1	2
6	8	3	5	9	2	7	4	1
7	2	5	4	1	3	8	6	9
9	4	1	7	8	6	2	3	5
8	5	9	6	4	7	1	2	3
4	3	6	9	2	1	5	7	8
2	1	7	3	5	8	4	9	6

Solution # 162

5	1	7	9	8	3	4	2	6
8	9	2	4	1	6	7	3	5
4	6	3	5	2	7	8	9	1
1	8	5	6	9	2	3	4	7
6	3	9	8	7	4	5	1	2
2	7	4	1	3	5	6	8	9
3	2	6	7	4	1	9	5	8
9	5	1	3	6	8	2	7	4
7	4	8	2	5	9	1	6	3

Solution # 163

7	5	6	3	1	8	9	2	4
9	8	4	6	7	2	3	5	1
2	1	3	9	5	4	6	7	8
6	4	2	5	8	3	1	9	7
1	3	5	4	9	7	8	6	2
8	7	9	2	6	1	5	4	3
5	2	8	1	4	6	7	3	9
3	9	7	8	2	5	4	1	6
4	6	1	7	3	9	2	8	5

Solution # 164

6	4	8	7	2	3	1	5	9
3	5	7	4	1	9	6	2	8
1	2	9	5	8	6	3	4	7
9	8	1	2	6	5	4	7	3
2	3	6	9	4	7	5	8	1
4	7	5	8	3	1	9	6	2
7	1	2	6	9	4	8	3	5
8	6	3	1	5	2	7	9	4
5	9	4	3	7	8	2	1	6

Solution # 165

3	2	4	6	9	8	5	1	7
9	7	6	4	5	1	2	3	8
8	1	5	2	3	7	9	4	6
7	5	1	8	2	3	6	9	4
6	9	8	5	1	4	7	2	3
2	4	3	9	7	6	1	8	5
4	8	2	1	6	5	3	7	9
5	3	9	7	8	2	4	6	1
1	6	7	3	4	9	8	5	2

Solution # 166

1	7	2	9	8	3	6	5	4
6	4	8	7	1	5	2	9	3
5	3	9	2	4	6	8	1	7
3	6	5	8	2	7	9	4	1
2	8	1	5	9	4	7	3	6
7	9	4	3	6	1	5	2	8
9	1	3	6	5	8	4	7	2
8	2	7	4	3	9	1	6	5
4	5	6	1	7	2	3	8	9

Solution # 167

1	9	2	5	6	4	7	3	8
5	7	6	1	8	3	4	2	9
3	4	8	9	2	7	5	6	1
6	8	4	2	5	1	9	7	3
2	3	7	4	9	8	6	1	5
9	5	1	7	3	6	8	4	2
7	1	5	3	4	9	2	8	6
4	6	9	8	1	2	3	5	7
8	2	3	6	7	5	1	9	4

Solution # 168

9	5	1	4	8	6	2	7	3
7	8	2	1	3	5	9	6	4
6	3	4	2	9	7	5	8	1
1	2	8	5	7	9	4	3	6
3	4	6	8	2	1	7	9	5
5	9	7	6	4	3	8	1	2
4	7	5	3	1	8	6	2	9
8	6	3	9	5	2	1	4	7
2	1	9	7	6	4	3	5	8

Solution # 169

6	5	9	1	2	4	7	3	8
3	2	7	6	8	5	1	9	4
4	8	1	9	7	3	6	2	5
2	3	8	5	6	1	9	4	7
9	4	6	7	3	8	5	1	2
1	7	5	4	9	2	8	6	3
7	1	4	2	5	9	3	8	6
5	9	3	8	4	6	2	7	1
8	6	2	3	1	7	4	5	9

Solution # 170

3	7	4	8	5	1	9	2	6
1	6	9	2	3	4	7	5	8
8	2	5	7	6	9	4	3	1
6	8	1	9	4	5	2	7	3
4	5	7	6	2	3	8	1	9
2	9	3	1	7	8	5	6	4
5	1	8	3	9	7	6	4	2
9	4	6	5	1	2	3	8	7
7	3	2	4	8	6	1	9	5

Solution # 171

8	7	6	2	3	5	9	4	1
4	9	2	1	6	8	7	3	5
3	1	5	7	9	4	2	6	8
5	2	8	9	7	6	3	1	4
9	6	3	4	2	1	5	8	7
1	4	7	5	8	3	6	9	2
7	8	4	6	5	9	1	2	3
2	3	9	8	1	7	4	5	6
6	5	1	3	4	2	8	7	9

Solution # 172

4	6	8	5	3	7	2	1	9
3	5	1	8	2	9	6	4	7
7	2	9	1	6	4	3	8	5
5	1	4	7	9	6	8	3	2
9	7	2	3	4	8	1	5	6
6	8	3	2	5	1	7	9	4
2	3	6	4	8	5	9	7	1
8	4	7	9	1	2	5	6	3
1	9	5	6	7	3	4	2	8

Solution # 173

6	5	1	9	7	3	8	2	4
7	2	3	6	8	4	5	9	1
8	4	9	2	1	5	6	7	3
9	7	6	8	3	2	1	4	5
4	1	8	7	5	6	2	3	9
2	3	5	4	9	1	7	6	8
5	8	2	3	6	9	4	1	7
3	6	7	1	4	8	9	5	2
1	9	4	5	2	7	3	8	6

Solution # 174

1	8	6	4	2	3	9	5	7
5	7	3	8	9	1	4	6	2
4	9	2	5	7	6	3	1	8
6	1	7	3	8	9	5	2	4
3	2	8	7	5	4	6	9	1
9	4	5	6	1	2	7	8	3
7	6	9	2	3	8	1	4	5
2	3	1	9	4	5	8	7	6
8	5	4	1	6	7	2	3	9

Solution # 175

6	9	8	2	5	7	1	4	3
4	2	3	1	6	9	5	8	7
1	7	5	8	4	3	6	2	9
2	4	1	7	8	5	9	3	6
8	3	6	4	9	1	2	7	5
9	5	7	6	3	2	4	1	8
3	1	4	9	7	6	8	5	2
5	6	2	3	1	8	7	9	4
7	8	9	5	2	4	3	6	1

Solution # 176

1	2	3	7	9	5	8	4	6
4	6	9	2	3	8	1	5	7
8	5	7	1	6	4	3	2	9
6	8	1	5	4	3	7	9	2
7	4	5	8	2	9	6	1	3
9	3	2	6	1	7	4	8	5
5	1	6	4	7	2	9	3	8
3	7	8	9	5	1	2	6	4
2	9	4	3	8	6	5	7	1

Solution # 177

4	5	6	2	8	1	3	9	7
9	7	2	6	5	3	8	4	1
8	1	3	9	4	7	5	2	6
7	8	9	5	6	4	1	3	2
2	3	4	1	7	8	6	5	9
1	6	5	3	2	9	7	8	4
3	9	8	7	1	2	4	6	5
6	2	1	4	3	5	9	7	8
5	4	7	8	9	6	2	1	3

Solution # 178

4	8	7	2	6	5	9	1	3
2	1	5	3	7	9	6	4	8
9	6	3	8	1	4	5	7	2
3	4	2	7	9	1	8	5	6
5	7	6	4	2	8	1	3	9
8	9	1	5	3	6	7	2	4
7	2	9	6	5	3	4	8	1
6	3	8	1	4	7	2	9	5
1	5	4	9	8	2	3	6	7

Solution # 179

6	2	8	7	5	4	3	1	9
1	9	3	8	6	2	5	7	4
5	7	4	3	9	1	6	8	2
2	8	7	5	4	3	1	9	6
9	6	5	2	1	8	4	3	7
4	3	1	6	7	9	8	2	5
7	4	2	1	8	6	9	5	3
3	1	6	9	2	5	7	4	8
8	5	9	4	3	7	2	6	1

Solution # 180

6	4	1	8	2	5	7	3	9
2	3	9	4	6	7	5	8	1
7	5	8	3	9	1	6	4	2
5	1	3	2	7	4	9	6	8
4	2	6	5	8	9	1	7	3
9	8	7	1	3	6	4	2	5
8	9	4	7	1	2	3	5	6
3	6	5	9	4	8	2	1	7
1	7	2	6	5	3	8	9	4

Solution # 181

3	6	2	1	7	5	9	8	4
1	8	4	9	3	6	2	5	7
7	5	9	2	8	4	3	6	1
6	4	7	3	9	2	5	1	8
2	3	1	5	4	8	7	9	6
5	9	8	6	1	7	4	3	2
9	2	6	7	5	1	8	4	3
4	1	5	8	2	3	6	7	9
8	7	3	4	6	9	1	2	5

Solution # 182

2	6	3	9	1	4	5	8	7
5	4	8	6	7	3	2	1	9
7	9	1	5	2	8	6	4	3
1	2	9	3	8	6	7	5	4
4	3	6	2	5	7	1	9	8
8	7	5	4	9	1	3	6	2
3	1	4	8	6	2	9	7	5
9	8	7	1	3	5	4	2	6
6	5	2	7	4	9	8	3	1

Solution # 183

1	8	7	4	3	5	2	9	6
9	6	3	2	1	8	7	4	5
4	2	5	7	6	9	3	8	1
6	1	8	5	2	3	4	7	9
5	9	2	8	4	7	6	1	3
3	7	4	1	9	6	5	2	8
7	4	9	3	5	1	8	6	2
2	3	6	9	8	4	1	5	7
8	5	1	6	7	2	9	3	4

Solution # 184

6	4	3	7	5	1	8	2	9
1	7	9	2	6	8	3	4	5
5	2	8	9	3	4	1	7	6
2	3	7	4	9	5	6	8	1
4	1	5	6	8	7	9	3	2
9	8	6	1	2	3	4	5	7
3	9	1	5	4	2	7	6	8
7	5	4	8	1	6	2	9	3
8	6	2	3	7	9	5	1	4

Solution # 185

9	2	1	3	7	8	6	5	4
3	6	7	5	9	4	8	1	2
8	5	4	6	2	1	3	7	9
7	8	2	1	5	6	9	4	3
6	9	5	4	3	7	1	2	8
4	1	3	2	8	9	7	6	5
2	3	6	9	1	5	4	8	7
1	7	9	8	4	2	5	3	6
5	4	8	7	6	3	2	9	1

Solution # 186

6	3	1	9	5	2	8	4	7
4	2	5	8	3	7	1	9	6
9	8	7	6	4	1	3	5	2
3	7	6	5	1	4	2	8	9
5	9	4	3	2	8	6	7	1
8	1	2	7	6	9	5	3	4
7	4	3	1	8	6	9	2	5
1	5	9	2	7	3	4	6	8
2	6	8	4	9	5	7	1	3

Solution # 187

8	7	3	5	2	6	4	1	9
1	2	5	7	4	9	6	8	3
9	6	4	8	3	1	7	5	2
7	9	6	1	5	3	2	4	8
5	4	2	6	8	7	3	9	1
3	1	8	4	9	2	5	7	6
4	5	9	3	6	8	1	2	7
2	3	7	9	1	4	8	6	5
6	8	1	2	7	5	9	3	4

Solution # 188

7	5	9	6	1	4	2	8	3
6	3	4	8	2	5	9	1	7
8	1	2	3	9	7	5	6	4
1	4	6	2	3	8	7	9	5
3	8	5	1	7	9	4	2	6
9	2	7	5	4	6	8	3	1
5	9	1	4	8	3	6	7	2
2	6	8	7	5	1	3	4	9
4	7	3	9	6	2	1	5	8

Solution # 189

4	8	1	9	7	6	5	3	2
2	3	7	8	4	5	6	1	9
6	5	9	1	2	3	8	4	7
8	9	2	7	3	1	4	5	6
3	4	6	5	9	2	7	8	1
7	1	5	4	6	8	2	9	3
1	2	8	3	5	7	9	6	4
9	6	3	2	8	4	1	7	5
5	7	4	6	1	9	3	2	8

Solution # 190

1	3	8	6	9	5	7	4	2
9	5	6	4	7	2	1	8	3
2	7	4	8	1	3	6	9	5
6	1	7	3	8	9	5	2	4
5	8	9	2	4	1	3	7	6
3	4	2	7	5	6	8	1	9
8	2	5	1	6	4	9	3	7
4	9	1	5	3	7	2	6	8
7	6	3	9	2	8	4	5	1

Solution # 191

3	1	7	9	5	2	4	6	8
5	4	8	6	1	7	3	2	9
9	2	6	4	8	3	1	5	7
6	3	4	5	9	8	7	1	2
1	7	5	2	6	4	9	8	3
8	9	2	7	3	1	5	4	6
4	6	3	1	2	9	8	7	5
2	8	1	3	7	5	6	9	4
7	5	9	8	4	6	2	3	1

Solution # 192

7	1	3	9	8	5	6	4	2
9	2	8	7	6	4	5	1	3
5	6	4	2	3	1	8	9	7
3	9	2	8	1	7	4	5	6
8	5	7	6	4	2	1	3	9
1	4	6	5	9	3	2	7	8
4	8	1	3	7	6	9	2	5
2	7	9	4	5	8	3	6	1
6	3	5	1	2	9	7	8	4

Solution # 193

3	8	9	6	4	1	7	2	5
7	6	2	9	5	8	3	4	1
5	4	1	7	2	3	6	9	8
1	2	7	4	9	6	5	8	3
4	9	6	8	3	5	1	7	2
8	5	3	1	7	2	9	6	4
9	1	4	5	8	7	2	3	6
2	7	5	3	6	4	8	1	9
6	3	8	2	1	9	4	5	7

Solution # 194

5	9	7	6	1	2	3	4	8
3	1	6	4	8	7	9	2	5
4	8	2	3	9	5	1	7	6
9	6	8	2	5	3	4	1	7
1	3	5	7	4	8	2	6	9
2	7	4	9	6	1	5	8	3
6	5	9	1	7	4	8	3	2
8	4	3	5	2	6	7	9	1
7	2	1	8	3	9	6	5	4

Solution # 195

2	9	1	7	6	4	8	3	5
7	3	4	5	8	2	9	6	1
5	8	6	9	1	3	4	2	7
6	1	5	8	4	9	3	7	2
9	7	2	6	3	1	5	4	8
8	4	3	2	7	5	1	9	6
4	2	8	1	9	7	6	5	3
3	6	7	4	5	8	2	1	9
1	5	9	3	2	6	7	8	4

Solution # 196

5	3	9	2	7	8	1	6	4
2	7	6	9	1	4	8	3	5
4	8	1	3	5	6	2	9	7
9	5	2	7	4	3	6	8	1
1	6	3	5	8	9	4	7	2
7	4	8	1	6	2	9	5	3
8	1	5	4	9	7	3	2	6
6	2	4	8	3	5	7	1	9
3	9	7	6	2	1	5	4	8

Solution # 197

1	5	6	7	3	4	8	2	9
4	3	9	2	8	5	7	6	1
2	8	7	1	6	9	5	4	3
9	4	5	3	7	1	2	8	6
3	7	8	6	4	2	9	1	5
6	2	1	9	5	8	3	7	4
7	9	4	5	2	6	1	3	8
5	6	3	8	1	7	4	9	2
8	1	2	4	9	3	6	5	7

Solution # 198

5	4	9	7	6	3	8	2	1
8	2	7	9	1	4	3	5	6
1	6	3	8	5	2	4	9	7
7	8	6	2	4	1	5	3	9
2	9	5	3	7	8	1	6	4
4	3	1	5	9	6	7	8	2
3	7	8	4	2	9	6	1	5
9	1	4	6	8	5	2	7	3
6	5	2	1	3	7	9	4	8

Solution # 199

5	6	3	2	1	7	4	8	9
7	2	9	6	8	4	1	5	3
4	8	1	3	9	5	2	7	6
9	5	7	8	2	1	3	6	4
1	4	2	5	6	3	8	9	7
8	3	6	4	7	9	5	2	1
3	9	8	7	4	2	6	1	5
2	1	5	9	3	6	7	4	8
6	7	4	1	5	8	9	3	2

Solution # 200

3	4	8	1	7	2	6	9	5
2	5	1	3	9	6	4	7	8
9	6	7	5	8	4	1	3	2
4	2	5	8	6	7	9	1	3
7	9	3	4	2	1	5	8	6
1	8	6	9	5	3	7	2	4
6	1	2	7	4	8	3	5	9
5	7	4	2	3	9	8	6	1
8	3	9	6	1	5	2	4	7

Solution # 201

9	8	6	4	5	1	7	3	2
2	1	3	8	6	7	5	9	4
7	4	5	3	2	9	6	8	1
5	7	9	1	8	2	4	6	3
1	3	4	7	9	6	2	5	8
6	2	8	5	3	4	9	1	7
4	5	7	9	1	8	3	2	6
3	6	1	2	7	5	8	4	9
8	9	2	6	4	3	1	7	5

Solution # 202

5	3	2	1	8	6	7	9	4
7	9	4	2	3	5	8	6	1
1	6	8	9	7	4	2	3	5
9	8	6	3	2	1	5	4	7
4	7	5	6	9	8	3	1	2
2	1	3	5	4	7	9	8	6
3	4	7	8	6	2	1	5	9
6	5	9	7	1	3	4	2	8
8	2	1	4	5	9	6	7	3

Solution # 203

9	7	1	3	5	4	2	8	6
5	3	2	8	9	6	7	4	1
8	4	6	1	2	7	9	3	5
2	9	3	6	1	8	4	5	7
7	6	4	9	3	5	8	1	2
1	5	8	4	7	2	6	9	3
3	1	7	2	8	9	5	6	4
6	8	5	7	4	1	3	2	9
4	2	9	5	6	3	1	7	8

Solution # 204

8	9	3	1	2	4	5	7	6
7	5	2	8	9	6	3	4	1
4	1	6	5	3	7	2	9	8
3	2	7	9	5	1	8	6	4
6	8	1	2	4	3	9	5	7
9	4	5	6	7	8	1	3	2
2	6	4	3	8	5	7	1	9
1	3	9	7	6	2	4	8	5
5	7	8	4	1	9	6	2	3

Solution # 205

5	8	9	6	4	3	2	1	7
1	4	3	8	7	2	5	9	6
2	6	7	1	9	5	3	4	8
6	7	5	4	2	9	8	3	1
9	2	1	3	5	8	7	6	4
4	3	8	7	1	6	9	2	5
3	5	6	9	8	4	1	7	2
8	1	4	2	3	7	6	5	9
7	9	2	5	6	1	4	8	3

Solution # 206

8	2	3	6	5	9	7	1	4
9	5	4	1	7	2	6	8	3
6	1	7	3	8	4	9	5	2
4	9	1	5	2	6	8	3	7
5	7	8	4	1	3	2	9	6
2	3	6	7	9	8	1	4	5
1	4	5	8	6	7	3	2	9
7	8	2	9	3	5	4	6	1
3	6	9	2	4	1	5	7	8

Solution # 207

6	2	4	7	9	5	1	8	3
1	9	7	3	8	6	2	4	5
8	3	5	1	4	2	6	9	7
9	7	3	2	6	8	4	5	1
5	6	1	9	7	4	3	2	8
2	4	8	5	1	3	9	7	6
3	8	2	6	5	9	7	1	4
4	1	6	8	2	7	5	3	9
7	5	9	4	3	1	8	6	2

Solution # 208

6	8	4	9	3	1	5	2	7
5	9	2	8	4	7	1	3	6
7	1	3	6	2	5	4	9	8
4	6	8	5	9	3	2	7	1
9	2	7	1	8	4	6	5	3
3	5	1	7	6	2	9	8	4
2	4	6	3	5	8	7	1	9
1	3	9	2	7	6	8	4	5
8	7	5	4	1	9	3	6	2

Solution # 209

3	9	7	6	4	2	8	5	1
2	6	1	8	7	5	4	9	3
5	8	4	3	9	1	7	2	6
7	4	5	9	8	6	3	1	2
1	3	6	4	2	7	9	8	5
9	2	8	1	5	3	6	7	4
6	7	9	5	1	4	2	3	8
4	1	2	7	3	8	5	6	9
8	5	3	2	6	9	1	4	7

Solution # 210

2	6	7	4	5	1	9	8	3
3	5	8	7	2	9	4	6	1
4	9	1	6	8	3	5	2	7
6	7	9	2	3	5	8	1	4
8	3	4	9	1	7	2	5	6
5	1	2	8	6	4	7	3	9
1	4	5	3	9	2	6	7	8
9	8	3	5	7	6	1	4	2
7	2	6	1	4	8	3	9	5

Solution # 211

2	3	4	6	5	1	9	8	7
5	9	1	8	7	2	6	3	4
6	7	8	9	4	3	2	1	5
7	1	6	2	8	5	4	9	3
3	8	5	4	1	9	7	6	2
4	2	9	3	6	7	1	5	8
8	6	7	1	3	4	5	2	9
9	5	3	7	2	6	8	4	1
1	4	2	5	9	8	3	7	6

Solution # 212

4	6	3	2	8	7	9	1	5
2	5	7	1	9	6	3	8	4
1	8	9	5	3	4	6	2	7
8	7	2	9	5	1	4	3	6
5	1	6	8	4	3	7	9	2
3	9	4	6	7	2	1	5	8
6	3	8	7	1	5	2	4	9
9	2	1	4	6	8	5	7	3
7	4	5	3	2	9	8	6	1

Solution # 213

8	9	5	1	2	7	6	4	3
3	4	1	9	5	6	2	8	7
6	2	7	3	8	4	9	1	5
1	6	9	8	3	5	7	2	4
2	7	8	4	1	9	3	5	6
5	3	4	7	6	2	8	9	1
7	8	3	5	9	1	4	6	2
9	1	6	2	4	3	5	7	8
4	5	2	6	7	8	1	3	9

Solution # 214

7	5	6	8	1	4	9	2	3
8	3	9	2	7	6	5	1	4
2	1	4	9	3	5	7	8	6
5	8	3	4	6	2	1	7	9
1	6	2	5	9	7	4	3	8
4	9	7	3	8	1	6	5	2
3	7	8	6	5	9	2	4	1
6	2	1	7	4	8	3	9	5
9	4	5	1	2	3	8	6	7

Solution # 215

4	6	1	8	2	3	5	7	9
7	3	9	6	5	4	8	2	1
8	2	5	9	1	7	3	6	4
1	4	2	5	7	9	6	3	8
3	5	8	4	6	2	9	1	7
6	9	7	1	3	8	4	5	2
2	8	6	3	9	1	7	4	5
9	7	3	2	4	5	1	8	6
5	1	4	7	8	6	2	9	3

Solution # 216

6	5	7	1	2	8	3	9	4
4	1	9	7	5	3	6	2	8
2	8	3	6	4	9	1	7	5
7	4	6	9	3	1	5	8	2
1	2	5	8	6	4	7	3	9
3	9	8	2	7	5	4	6	1
5	6	4	3	8	2	9	1	7
9	7	2	5	1	6	8	4	3
8	3	1	4	9	7	2	5	6

Solution # 217

4	2	1	3	7	9	6	8	5
6	5	7	8	1	4	9	2	3
8	3	9	5	2	6	7	1	4
1	7	8	4	3	2	5	6	9
2	4	5	9	6	7	8	3	1
3	9	6	1	8	5	2	4	7
5	1	3	2	9	8	4	7	6
7	8	4	6	5	1	3	9	2
9	6	2	7	4	3	1	5	8

Solution # 218

3	5	4	1	6	9	2	8	7
7	9	1	8	2	3	5	4	6
6	2	8	5	4	7	9	1	3
9	1	7	3	8	6	4	5	2
2	4	3	9	7	5	8	6	1
8	6	5	4	1	2	3	7	9
5	7	9	6	3	4	1	2	8
1	3	2	7	5	8	6	9	4
4	8	6	2	9	1	7	3	5

Solution # 219

9	6	4	1	7	5	2	3	8
2	3	5	6	8	4	1	9	7
7	1	8	2	3	9	5	4	6
3	4	9	7	2	1	8	6	5
5	8	2	3	9	6	7	1	4
6	7	1	5	4	8	3	2	9
4	5	7	9	1	2	6	8	3
1	9	6	8	5	3	4	7	2
8	2	3	4	6	7	9	5	1

Solution # 220

5	3	9	6	2	1	7	8	4
6	7	4	9	8	5	2	3	1
2	8	1	7	4	3	6	5	9
4	5	2	1	3	8	9	6	7
9	1	7	4	5	6	3	2	8
3	6	8	2	7	9	1	4	5
1	4	3	8	6	7	5	9	2
8	9	5	3	1	2	4	7	6
7	2	6	5	9	4	8	1	3

Solution # 221

2	3	5	1	4	6	9	8	7
8	1	7	3	2	9	5	4	6
4	6	9	5	8	7	3	1	2
3	5	2	7	1	4	6	9	8
9	8	1	6	3	2	4	7	5
6	7	4	9	5	8	2	3	1
5	2	8	4	7	3	1	6	9
1	4	6	8	9	5	7	2	3
7	9	3	2	6	1	8	5	4

Solution # 222

2	5	9	4	1	7	8	6	3
4	7	8	9	3	6	1	2	5
6	3	1	8	2	5	9	4	7
3	6	2	1	5	8	4	7	9
5	1	7	6	9	4	2	3	8
8	9	4	2	7	3	5	1	6
1	4	3	5	6	9	7	8	2
7	2	5	3	8	1	6	9	4
9	8	6	7	4	2	3	5	1

Solution # 223

5	4	2	8	7	3	1	9	6
9	1	6	5	2	4	7	8	3
3	7	8	6	1	9	5	4	2
2	5	1	4	8	7	6	3	9
4	8	3	9	5	6	2	7	1
6	9	7	2	3	1	8	5	4
8	3	9	7	6	2	4	1	5
7	2	4	1	9	5	3	6	8
1	6	5	3	4	8	9	2	7

Solution # 224

1	5	4	3	9	7	6	8	2
9	6	2	8	4	1	3	7	5
8	3	7	2	6	5	1	4	9
2	7	8	5	1	9	4	3	6
3	4	9	6	7	8	2	5	1
6	1	5	4	3	2	7	9	8
7	2	1	9	8	4	5	6	3
5	8	3	7	2	6	9	1	4
4	9	6	1	5	3	8	2	7

Solution # 225

6	8	9	3	5	2	7	1	4
1	4	2	9	7	6	3	5	8
7	5	3	4	8	1	2	6	9
4	3	8	7	6	9	5	2	1
2	9	7	5	1	8	4	3	6
5	1	6	2	4	3	9	8	7
8	2	5	1	9	7	6	4	3
9	6	4	8	3	5	1	7	2
3	7	1	6	2	4	8	9	5

Solution # 226

1	9	6	5	4	8	7	2	3
8	2	3	1	7	6	9	4	5
4	5	7	3	2	9	8	1	6
6	3	1	9	8	7	4	5	2
5	7	9	2	1	4	6	3	8
2	8	4	6	5	3	1	7	9
9	1	5	4	6	2	3	8	7
7	6	2	8	3	1	5	9	4
3	4	8	7	9	5	2	6	1

Solution # 227

4	3	7	8	1	6	5	9	2
5	6	8	7	2	9	1	4	3
2	1	9	4	5	3	7	8	6
1	7	6	5	3	8	4	2	9
9	8	4	1	6	2	3	7	5
3	2	5	9	7	4	8	6	1
7	9	3	6	8	1	2	5	4
8	4	1	2	9	5	6	3	7
6	5	2	3	4	7	9	1	8

Solution # 228

3	6	5	2	8	7	1	9	4
2	9	4	5	1	6	8	7	3
7	8	1	4	9	3	5	2	6
8	2	3	6	4	1	7	5	9
9	1	7	8	3	5	4	6	2
4	5	6	7	2	9	3	1	8
1	4	2	9	7	8	6	3	5
6	3	8	1	5	2	9	4	7
5	7	9	3	6	4	2	8	1

Solution # 229

4	3	2	1	5	8	7	6	9
6	8	9	4	7	3	2	1	5
1	5	7	2	9	6	8	3	4
5	6	4	9	8	2	1	7	3
8	7	1	3	6	4	5	9	2
2	9	3	7	1	5	6	4	8
3	4	8	6	2	1	9	5	7
7	2	6	5	3	9	4	8	1
9	1	5	8	4	7	3	2	6

Solution # 230

8	1	6	2	5	4	9	3	7
2	5	9	8	7	3	1	4	6
7	3	4	1	9	6	5	8	2
4	6	3	7	8	1	2	9	5
1	7	2	5	3	9	4	6	8
5	9	8	4	6	2	7	1	3
3	4	7	6	1	5	8	2	9
6	8	1	9	2	7	3	5	4
9	2	5	3	4	8	6	7	1

Solution # 231

2	9	1	5	8	6	7	4	3
7	8	3	4	2	1	9	5	6
6	5	4	9	3	7	2	8	1
9	4	8	2	1	3	5	6	7
3	7	2	6	5	8	1	9	4
5	1	6	7	4	9	3	2	8
1	2	5	8	7	4	6	3	9
8	3	9	1	6	5	4	7	2
4	6	7	3	9	2	8	1	5

Solution # 232

7	3	2	8	1	4	6	5	9
5	8	4	6	3	9	7	1	2
1	6	9	7	5	2	4	8	3
6	7	3	5	2	8	9	4	1
2	5	8	4	9	1	3	7	6
9	4	1	3	7	6	5	2	8
3	1	6	2	4	5	8	9	7
4	9	7	1	8	3	2	6	5
8	2	5	9	6	7	1	3	4

Solution # 233

6	8	3	2	9	5	4	1	7
7	1	9	3	8	4	5	6	2
4	2	5	1	6	7	3	9	8
2	9	4	8	5	6	1	7	3
5	3	1	7	4	2	9	8	6
8	6	7	9	1	3	2	5	4
1	5	2	6	3	8	7	4	9
9	7	8	4	2	1	6	3	5
3	4	6	5	7	9	8	2	1

Solution # 234

2	3	4	1	7	9	8	6	5
7	6	8	2	5	4	9	3	1
1	9	5	8	6	3	4	2	7
5	1	6	9	4	8	2	7	3
3	8	9	7	2	5	1	4	6
4	7	2	3	1	6	5	8	9
8	5	3	4	9	7	6	1	2
6	2	7	5	8	1	3	9	4
9	4	1	6	3	2	7	5	8

Solution # 235

5	7	9	4	6	1	2	3	8
3	4	6	9	2	8	1	7	5
1	2	8	7	5	3	4	9	6
2	9	4	1	8	5	7	6	3
6	8	5	3	7	2	9	4	1
7	1	3	6	4	9	8	5	2
4	5	7	2	1	6	3	8	9
9	6	1	8	3	4	5	2	7
8	3	2	5	9	7	6	1	4

Solution # 236

6	3	4	7	5	1	9	8	2
9	1	2	8	6	3	4	7	5
7	8	5	9	2	4	3	1	6
5	9	8	2	4	7	1	6	3
1	2	7	6	3	9	5	4	8
4	6	3	5	1	8	7	2	9
2	5	1	4	9	6	8	3	7
3	7	6	1	8	5	2	9	4
8	4	9	3	7	2	6	5	1

Solution # 237

1	2	9	3	8	7	5	4	6
5	7	6	2	9	4	1	3	8
3	8	4	6	5	1	7	9	2
9	6	7	5	2	3	8	1	4
4	5	1	9	7	8	6	2	3
2	3	8	4	1	6	9	7	5
6	1	3	8	4	9	2	5	7
8	9	2	7	3	5	4	6	1
7	4	5	1	6	2	3	8	9

Solution # 238

3	4	7	5	1	2	6	9	8
2	9	5	6	3	8	4	7	1
6	1	8	4	9	7	5	3	2
5	7	6	3	8	9	1	2	4
1	3	2	7	5	4	8	6	9
9	8	4	2	6	1	7	5	3
8	5	3	1	2	6	9	4	7
4	6	1	9	7	3	2	8	5
7	2	9	8	4	5	3	1	6

Solution # 239

8	5	7	6	2	1	3	4	9
3	1	4	9	5	8	6	7	2
9	2	6	4	3	7	1	8	5
2	7	5	3	8	6	9	1	4
6	8	1	7	9	4	2	5	3
4	3	9	2	1	5	8	6	7
7	4	3	8	6	2	5	9	1
5	6	2	1	7	9	4	3	8
1	9	8	5	4	3	7	2	6

Solution # 240

4	5	9	1	8	7	2	3	6
2	8	1	4	6	3	5	7	9
3	7	6	5	9	2	4	8	1
9	4	7	3	2	6	1	5	8
8	6	5	9	4	1	3	2	7
1	2	3	8	7	5	9	6	4
7	9	2	6	5	4	8	1	3
6	1	4	2	3	8	7	9	5
5	3	8	7	1	9	6	4	2

Solution # 241

9	2	1	7	5	8	6	4	3
6	3	8	9	1	4	7	5	2
7	5	4	6	3	2	8	9	1
3	6	9	5	4	7	1	2	8
5	8	7	2	6	1	9	3	4
4	1	2	8	9	3	5	7	6
1	7	6	4	2	9	3	8	5
8	4	5	3	7	6	2	1	9
2	9	3	1	8	5	4	6	7

Solution # 242

3	1	5	8	6	4	9	2	7
4	9	2	5	1	7	6	3	8
6	8	7	3	9	2	5	4	1
7	6	1	2	5	3	4	8	9
2	5	4	7	8	9	3	1	6
9	3	8	6	4	1	2	7	5
8	2	6	1	3	5	7	9	4
1	4	3	9	7	6	8	5	2
5	7	9	4	2	8	1	6	3

Solution # 243

6	2	4	9	1	5	7	3	8
7	1	8	2	3	6	5	9	4
5	3	9	7	4	8	2	1	6
1	9	5	8	2	4	6	7	3
4	7	2	1	6	3	8	5	9
3	8	6	5	7	9	4	2	1
9	6	3	4	5	2	1	8	7
8	5	7	6	9	1	3	4	2
2	4	1	3	8	7	9	6	5

Solution # 244

4	8	2	3	7	9	1	5	6
3	7	1	5	6	8	9	4	2
9	5	6	2	4	1	7	8	3
8	1	9	6	3	5	2	7	4
6	3	7	4	8	2	5	1	9
2	4	5	9	1	7	3	6	8
1	6	8	7	9	3	4	2	5
7	2	3	8	5	4	6	9	1
5	9	4	1	2	6	8	3	7

Solution # 245

7	9	3	4	5	8	6	2	1
1	4	6	3	2	9	5	7	8
5	2	8	7	6	1	9	4	3
6	1	5	9	7	2	8	3	4
8	7	9	1	3	4	2	5	6
4	3	2	6	8	5	1	9	7
2	5	7	8	1	3	4	6	9
9	6	1	5	4	7	3	8	2
3	8	4	2	9	6	7	1	5

Solution # 246

1	4	7	2	9	8	6	5	3
5	9	3	4	6	1	2	7	8
8	6	2	3	7	5	9	1	4
7	3	4	6	1	2	8	9	5
2	1	6	8	5	9	4	3	7
9	5	8	7	4	3	1	2	6
4	8	5	9	2	7	3	6	1
3	7	9	1	8	6	5	4	2
6	2	1	5	3	4	7	8	9

Solution # 247

8	6	1	2	5	4	7	9	3
7	5	4	3	6	9	2	8	1
9	3	2	1	7	8	6	5	4
2	7	8	6	9	1	3	4	5
1	4	3	7	8	5	9	6	2
5	9	6	4	2	3	1	7	8
6	8	5	9	3	2	4	1	7
4	2	9	8	1	7	5	3	6
3	1	7	5	4	6	8	2	9

Solution # 248

4	8	3	9	5	2	7	1	6
7	1	5	3	4	6	2	8	9
2	9	6	1	7	8	3	5	4
5	2	7	4	9	1	6	3	8
8	6	1	2	3	5	4	9	7
3	4	9	6	8	7	5	2	1
6	3	4	5	1	9	8	7	2
9	7	2	8	6	3	1	4	5
1	5	8	7	2	4	9	6	3

Solution # 249

6	8	4	7	3	5	1	9	2
9	3	7	4	2	1	6	8	5
1	2	5	9	6	8	3	7	4
3	4	1	5	8	7	2	6	9
2	5	6	3	1	9	8	4	7
8	7	9	6	4	2	5	1	3
7	1	8	2	9	3	4	5	6
5	6	2	1	7	4	9	3	8
4	9	3	8	5	6	7	2	1

Solution # 250

4	1	6	8	3	7	2	9	5
7	5	3	6	2	9	1	8	4
2	9	8	5	4	1	7	6	3
8	6	2	3	9	4	5	7	1
5	4	1	7	8	2	6	3	9
9	3	7	1	6	5	4	2	8
6	7	9	4	1	8	3	5	2
3	2	4	9	5	6	8	1	7
1	8	5	2	7	3	9	4	6

Solution # 251

7	1	5	2	4	9	3	8	6
9	3	6	7	8	5	4	1	2
4	8	2	1	3	6	7	5	9
2	6	8	5	7	1	9	3	4
1	9	3	8	2	4	6	7	5
5	4	7	6	9	3	1	2	8
3	5	9	4	1	8	2	6	7
6	7	1	9	5	2	8	4	3
8	2	4	3	6	7	5	9	1

Solution # 252

7	5	8	1	2	9	4	3	6
6	1	4	8	3	7	2	9	5
2	9	3	6	5	4	7	8	1
1	7	6	5	9	8	3	2	4
4	3	5	7	6	2	8	1	9
8	2	9	4	1	3	5	6	7
9	4	2	3	7	6	1	5	8
5	6	7	2	8	1	9	4	3
3	8	1	9	4	5	6	7	2

Solution # 253

7	3	6	5	2	9	4	8	1
4	5	9	1	7	8	6	3	2
2	1	8	6	3	4	5	9	7
8	2	7	3	1	5	9	4	6
5	6	4	2	9	7	8	1	3
1	9	3	4	8	6	2	7	5
3	7	5	8	4	2	1	6	9
9	4	2	7	6	1	3	5	8
6	8	1	9	5	3	7	2	4

Solution # 254

7	5	4	6	8	1	3	2	9
3	9	1	7	2	4	8	6	5
8	6	2	3	5	9	4	1	7
5	8	7	9	6	2	1	3	4
9	1	6	8	4	3	7	5	2
2	4	3	5	1	7	9	8	6
6	7	5	1	9	8	2	4	3
1	2	9	4	3	5	6	7	8
4	3	8	2	7	6	5	9	1

Solution # 255

5	8	1	3	4	7	6	2	9
3	9	2	5	6	8	1	4	7
7	4	6	1	9	2	3	5	8
9	5	7	4	2	1	8	3	6
6	2	8	9	3	5	7	1	4
4	1	3	8	7	6	2	9	5
1	7	9	6	5	3	4	8	2
2	3	5	7	8	4	9	6	1
8	6	4	2	1	9	5	7	3

Solution # 256

5	4	9	8	1	6	7	2	3
2	3	8	9	5	7	1	4	6
6	1	7	3	4	2	5	8	9
8	5	4	2	9	3	6	1	7
7	2	3	1	6	5	8	9	4
1	9	6	7	8	4	3	5	2
4	6	2	5	3	1	9	7	8
9	7	1	6	2	8	4	3	5
3	8	5	4	7	9	2	6	1

Solution # 257

4	1	3	9	2	6	5	8	7
7	5	8	4	1	3	9	6	2
9	2	6	8	5	7	3	4	1
2	7	9	5	6	1	4	3	8
8	4	1	3	7	9	6	2	5
6	3	5	2	8	4	1	7	9
3	8	4	1	9	2	7	5	6
1	6	2	7	3	5	8	9	4
5	9	7	6	4	8	2	1	3

Solution # 258

4	2	7	5	9	1	8	6	3
6	8	1	7	2	3	4	5	9
3	9	5	8	4	6	7	1	2
9	1	2	4	6	8	5	3	7
5	6	4	2	3	7	1	9	8
8	7	3	9	1	5	6	2	4
2	5	6	3	8	4	9	7	1
1	3	8	6	7	9	2	4	5
7	4	9	1	5	2	3	8	6

Solution # 259

8	7	9	1	3	5	6	4	2
1	4	5	2	6	9	7	8	3
3	6	2	8	4	7	9	1	5
7	9	4	3	5	1	8	2	6
6	5	3	4	8	2	1	7	9
2	1	8	9	7	6	3	5	4
5	8	7	6	9	4	2	3	1
4	2	6	7	1	3	5	9	8
9	3	1	5	2	8	4	6	7

Solution # 260

9	3	8	4	7	6	2	1	5
4	1	2	9	5	8	3	7	6
6	5	7	2	3	1	4	8	9
1	2	6	3	9	7	8	5	4
5	7	3	8	4	2	9	6	1
8	9	4	6	1	5	7	3	2
2	4	5	7	6	3	1	9	8
7	8	1	5	2	9	6	4	3
3	6	9	1	8	4	5	2	7

Solution # 261

6	7	3	5	1	4	2	9	8
5	4	1	9	8	2	7	6	3
9	8	2	6	3	7	4	1	5
4	5	9	8	7	6	1	3	2
2	1	6	3	4	5	8	7	9
8	3	7	2	9	1	5	4	6
1	6	8	4	2	3	9	5	7
3	9	4	7	5	8	6	2	1
7	2	5	1	6	9	3	8	4

Solution # 262

9	6	7	1	5	8	4	2	3
2	3	1	7	6	4	5	9	8
4	8	5	9	2	3	6	7	1
3	4	2	8	1	9	7	6	5
8	7	9	5	4	6	3	1	2
1	5	6	2	3	7	8	4	9
7	9	4	3	8	1	2	5	6
6	2	8	4	9	5	1	3	7
5	1	3	6	7	2	9	8	4

Solution # 263

7	8	2	3	4	6	5	9	1
9	3	5	8	2	1	4	6	7
1	6	4	5	7	9	8	2	3
8	4	7	6	3	2	1	5	9
6	5	3	9	1	7	2	4	8
2	1	9	4	5	8	3	7	6
3	2	1	7	9	4	6	8	5
4	9	6	1	8	5	7	3	2
5	7	8	2	6	3	9	1	4

Solution # 264

9	7	3	4	6	2	1	5	8
8	6	2	1	5	3	4	9	7
5	1	4	7	9	8	3	6	2
4	5	7	9	8	1	6	2	3
1	3	9	6	2	7	8	4	5
2	8	6	5	3	4	9	7	1
3	9	5	8	7	6	2	1	4
7	4	8	2	1	9	5	3	6
6	2	1	3	4	5	7	8	9

Solution # 265

2	4	3	5	1	6	9	8	7
9	6	5	7	4	8	1	3	2
7	1	8	9	3	2	5	6	4
6	5	2	1	7	3	8	4	9
4	3	1	8	9	5	7	2	6
8	7	9	2	6	4	3	1	5
1	2	7	6	8	9	4	5	3
3	8	6	4	5	7	2	9	1
5	9	4	3	2	1	6	7	8

Solution # 266

8	2	7	5	6	4	3	9	1
1	5	9	8	2	3	4	6	7
3	4	6	1	9	7	8	5	2
6	9	3	2	4	1	7	8	5
7	1	4	3	5	8	6	2	9
5	8	2	9	7	6	1	3	4
4	7	5	6	8	2	9	1	3
2	3	8	4	1	9	5	7	6
9	6	1	7	3	5	2	4	8

Solution # 267

5	9	2	3	7	1	6	8	4
4	8	7	6	9	2	1	5	3
1	3	6	5	4	8	2	7	9
6	5	8	2	1	4	3	9	7
9	4	1	7	8	3	5	2	6
2	7	3	9	6	5	4	1	8
7	2	4	8	5	6	9	3	1
3	6	9	1	2	7	8	4	5
8	1	5	4	3	9	7	6	2

Solution # 268

9	3	1	5	4	6	2	7	8
7	5	6	1	8	2	4	9	3
2	4	8	3	9	7	1	6	5
5	2	9	8	3	1	6	4	7
3	1	4	6	7	5	9	8	2
8	6	7	9	2	4	5	3	1
1	8	3	2	6	9	7	5	4
4	9	2	7	5	3	8	1	6
6	7	5	4	1	8	3	2	9

Solution # 269

2	4	7	6	5	1	8	3	9
9	8	6	3	7	4	5	1	2
5	1	3	9	8	2	6	4	7
7	2	9	4	1	8	3	6	5
1	3	8	7	6	5	2	9	4
6	5	4	2	3	9	7	8	1
3	6	5	1	9	7	4	2	8
8	9	2	5	4	3	1	7	6
4	7	1	8	2	6	9	5	3

Solution # 270

3	2	4	8	5	9	6	7	1
1	6	9	3	7	4	8	5	2
8	5	7	6	1	2	9	3	4
4	1	2	5	6	8	7	9	3
7	9	6	1	2	3	5	4	8
5	3	8	9	4	7	1	2	6
9	4	5	2	8	6	3	1	7
2	8	3	7	9	1	4	6	5
6	7	1	4	3	5	2	8	9

Solution # 271

3	8	7	4	2	6	9	1	5
1	5	6	9	7	3	8	2	4
4	9	2	8	1	5	3	7	6
8	7	5	6	4	9	1	3	2
2	3	9	1	5	8	4	6	7
6	1	4	7	3	2	5	8	9
7	2	8	3	9	4	6	5	1
5	4	3	2	6	1	7	9	8
9	6	1	5	8	7	2	4	3

Solution # 272

3	6	2	7	8	1	9	5	4
7	4	9	3	2	5	1	8	6
5	1	8	6	9	4	3	7	2
2	5	6	8	7	3	4	9	1
1	8	3	2	4	9	7	6	5
9	7	4	1	5	6	2	3	8
8	3	7	5	1	2	6	4	9
4	2	5	9	6	7	8	1	3
6	9	1	4	3	8	5	2	7

Solution # 273

5	7	4	8	3	2	6	1	9
9	3	2	4	6	1	8	7	5
6	1	8	7	9	5	2	4	3
4	2	5	1	7	6	9	3	8
7	8	3	9	2	4	5	6	1
1	6	9	5	8	3	7	2	4
2	4	6	3	5	9	1	8	7
3	9	7	2	1	8	4	5	6
8	5	1	6	4	7	3	9	2

Solution # 274

3	6	5	2	7	4	8	9	1
1	8	7	3	9	6	5	2	4
2	9	4	8	5	1	7	6	3
9	7	3	6	8	2	1	4	5
5	1	8	7	4	9	6	3	2
6	4	2	5	1	3	9	8	7
4	5	6	1	3	8	2	7	9
8	3	1	9	2	7	4	5	6
7	2	9	4	6	5	3	1	8

Solution # 275

6	7	5	4	8	3	1	2	9
4	2	1	9	7	6	5	3	8
9	3	8	5	1	2	6	7	4
1	6	3	8	4	9	7	5	2
5	8	9	2	6	7	3	4	1
7	4	2	3	5	1	9	8	6
8	1	7	6	3	4	2	9	5
3	9	4	1	2	5	8	6	7
2	5	6	7	9	8	4	1	3

Solution # 276

6	4	5	1	3	8	2	9	7
7	9	3	5	2	4	1	8	6
8	2	1	9	7	6	3	4	5
1	7	2	8	4	9	6	5	3
9	8	6	3	5	2	7	1	4
3	5	4	6	1	7	8	2	9
2	6	9	4	8	3	5	7	1
5	3	7	2	9	1	4	6	8
4	1	8	7	6	5	9	3	2

Solution # 277

7	3	6	2	8	4	5	9	1
1	4	8	9	3	5	6	7	2
9	2	5	6	1	7	4	3	8
2	1	7	5	4	8	9	6	3
4	8	9	1	6	3	2	5	7
5	6	3	7	2	9	8	1	4
6	7	4	8	9	1	3	2	5
3	5	2	4	7	6	1	8	9
8	9	1	3	5	2	7	4	6

Solution # 278

5	8	6	2	4	7	9	1	3
2	4	1	9	5	3	6	7	8
7	9	3	8	1	6	5	4	2
4	6	9	7	8	1	3	2	5
8	1	5	3	9	2	7	6	4
3	2	7	5	6	4	1	8	9
6	5	8	1	2	9	4	3	7
1	7	2	4	3	5	8	9	6
9	3	4	6	7	8	2	5	1

Solution # 279

5	9	3	4	7	2	1	6	8
8	4	6	5	9	1	3	2	7
1	2	7	8	6	3	9	4	5
3	6	1	7	8	5	2	9	4
2	5	4	6	3	9	7	8	1
7	8	9	2	1	4	5	3	6
6	1	2	3	4	7	8	5	9
9	3	8	1	5	6	4	7	2
4	7	5	9	2	8	6	1	3

Solution # 280

3	4	2	1	7	8	9	5	6
9	6	8	4	5	3	1	7	2
1	7	5	6	2	9	8	3	4
7	1	3	9	4	5	6	2	8
8	5	4	7	6	2	3	1	9
6	2	9	3	8	1	5	4	7
2	9	1	8	3	4	7	6	5
5	3	7	2	9	6	4	8	1
4	8	6	5	1	7	2	9	3

Solution # 281

4	8	9	3	6	7	1	2	5
5	7	2	9	8	1	6	4	3
6	1	3	4	2	5	7	8	9
8	6	5	7	3	9	4	1	2
9	2	1	6	5	4	8	3	7
3	4	7	8	1	2	9	5	6
7	9	8	5	4	3	2	6	1
1	3	6	2	7	8	5	9	4
2	5	4	1	9	6	3	7	8

Solution # 282

3	1	7	8	9	2	4	6	5
6	5	8	4	1	3	7	9	2
2	4	9	5	7	6	1	8	3
5	8	1	3	2	9	6	4	7
7	2	6	1	8	4	5	3	9
9	3	4	7	6	5	8	2	1
4	9	5	6	3	7	2	1	8
8	7	3	2	4	1	9	5	6
1	6	2	9	5	8	3	7	4

Solution # 283

7	1	4	9	3	6	8	2	5
6	2	9	7	8	5	3	1	4
5	8	3	4	2	1	7	6	9
8	7	2	5	4	3	6	9	1
3	9	5	1	6	2	4	7	8
4	6	1	8	7	9	5	3	2
1	3	6	2	5	4	9	8	7
2	5	7	6	9	8	1	4	3
9	4	8	3	1	7	2	5	6

Solution # 284

5	9	8	4	1	7	6	2	3
1	2	3	9	5	6	8	7	4
6	7	4	2	8	3	9	5	1
8	3	7	5	9	2	1	4	6
4	6	2	8	7	1	5	3	9
9	5	1	3	6	4	7	8	2
3	1	5	7	4	9	2	6	8
7	4	9	6	2	8	3	1	5
2	8	6	1	3	5	4	9	7

Solution # 285

9	1	6	5	3	2	4	7	8
4	5	3	8	7	6	1	2	9
7	8	2	4	9	1	6	5	3
8	4	7	6	5	9	3	1	2
5	3	1	2	8	7	9	4	6
2	6	9	3	1	4	5	8	7
6	9	8	1	2	5	7	3	4
3	7	5	9	4	8	2	6	1
1	2	4	7	6	3	8	9	5

Solution # 286

5	1	2	7	4	9	8	6	3
6	8	7	3	1	5	2	9	4
3	9	4	2	6	8	1	7	5
4	6	8	5	9	1	7	3	2
2	7	5	8	3	6	4	1	9
1	3	9	4	2	7	5	8	6
8	5	6	9	7	4	3	2	1
7	2	1	6	5	3	9	4	8
9	4	3	1	8	2	6	5	7

Solution # 287

9	3	2	4	1	6	5	8	7
1	7	4	5	9	8	3	6	2
6	8	5	2	3	7	1	9	4
7	9	3	6	8	2	4	1	5
2	4	6	1	5	9	7	3	8
5	1	8	3	7	4	9	2	6
8	6	7	9	4	1	2	5	3
3	2	9	7	6	5	8	4	1
4	5	1	8	2	3	6	7	9

Solution # 288

5	8	7	4	2	6	9	3	1
1	9	6	8	7	3	5	2	4
2	4	3	5	1	9	6	8	7
4	5	2	6	8	1	7	9	3
8	7	9	3	4	2	1	5	6
3	6	1	7	9	5	8	4	2
6	1	4	9	3	8	2	7	5
9	3	5	2	6	7	4	1	8
7	2	8	1	5	4	3	6	9

Solution # 289

6	3	5	8	9	2	4	7	1
2	7	9	3	4	1	6	8	5
8	4	1	5	6	7	3	9	2
9	1	4	2	7	6	5	3	8
3	8	2	1	5	9	7	4	6
5	6	7	4	3	8	1	2	9
7	9	3	6	8	5	2	1	4
1	5	8	7	2	4	9	6	3
4	2	6	9	1	3	8	5	7

Solution # 290

1	2	3	7	6	4	9	5	8
6	4	5	8	9	3	7	1	2
9	7	8	5	1	2	4	3	6
7	3	9	1	5	8	2	6	4
8	1	2	9	4	6	3	7	5
4	5	6	2	3	7	1	8	9
2	6	7	4	8	1	5	9	3
3	9	4	6	7	5	8	2	1
5	8	1	3	2	9	6	4	7

Solution # 291

4	9	5	3	8	7	2	1	6
6	1	8	9	2	5	3	7	4
2	3	7	4	1	6	9	8	5
1	4	3	5	9	2	7	6	8
7	6	9	8	3	1	4	5	2
5	8	2	6	7	4	1	3	9
8	5	1	7	4	9	6	2	3
3	2	4	1	6	8	5	9	7
9	7	6	2	5	3	8	4	1

Solution # 292

4	1	9	6	5	2	8	7	3
2	3	6	9	7	8	4	5	1
8	5	7	1	4	3	6	2	9
7	4	3	8	6	9	2	1	5
6	9	5	2	1	4	7	3	8
1	2	8	7	3	5	9	6	4
3	6	2	4	9	1	5	8	7
9	7	1	5	8	6	3	4	2
5	8	4	3	2	7	1	9	6

Solution # 293

1	7	8	2	3	5	6	9	4
9	2	4	7	6	8	5	3	1
5	3	6	4	1	9	2	8	7
7	5	2	8	9	6	4	1	3
4	9	3	1	5	7	8	6	2
8	6	1	3	2	4	9	7	5
2	8	7	6	4	3	1	5	9
3	4	5	9	8	1	7	2	6
6	1	9	5	7	2	3	4	8

Solution # 294

9	7	3	8	1	2	5	4	6
4	6	2	3	7	5	9	8	1
8	5	1	6	4	9	7	3	2
3	1	9	7	8	6	4	2	5
7	2	5	1	9	4	3	6	8
6	8	4	2	5	3	1	7	9
2	9	7	5	3	8	6	1	4
5	3	6	4	2	1	8	9	7
1	4	8	9	6	7	2	5	3

Solution # 295

4	6	8	3	2	9	5	1	7
9	3	5	8	1	7	4	6	2
1	2	7	6	4	5	8	3	9
8	1	6	2	7	4	3	9	5
3	7	9	5	6	8	1	2	4
5	4	2	1	9	3	6	7	8
7	5	1	9	8	6	2	4	3
6	9	3	4	5	2	7	8	1
2	8	4	7	3	1	9	5	6

Solution # 296

2	5	6	3	1	4	7	9	8
9	1	3	8	7	2	6	5	4
7	4	8	6	5	9	1	2	3
4	6	2	1	8	5	3	7	9
5	9	7	2	3	6	4	8	1
3	8	1	9	4	7	2	6	5
6	2	5	4	9	3	8	1	7
1	3	9	7	6	8	5	4	2
8	7	4	5	2	1	9	3	6

Solution # 297

2	3	4	5	8	7	9	1	6
6	7	9	2	3	1	4	5	8
8	1	5	6	9	4	3	7	2
4	6	8	7	1	3	2	9	5
1	9	2	8	5	6	7	4	3
3	5	7	9	4	2	6	8	1
7	4	6	1	2	5	8	3	9
9	2	1	3	7	8	5	6	4
5	8	3	4	6	9	1	2	7

Solution # 298

7	9	2	8	1	6	4	3	5
6	1	5	4	2	3	8	7	9
8	3	4	5	9	7	1	6	2
3	7	9	6	4	5	2	1	8
4	8	1	2	3	9	7	5	6
5	2	6	7	8	1	9	4	3
2	5	8	1	6	4	3	9	7
9	4	7	3	5	8	6	2	1
1	6	3	9	7	2	5	8	4

Solution # 299

2	5	8	4	1	7	6	9	3
3	7	4	9	6	2	5	8	1
1	9	6	8	3	5	7	2	4
7	1	3	6	4	8	2	5	9
8	2	9	3	5	1	4	7	6
4	6	5	2	7	9	1	3	8
9	4	7	5	8	6	3	1	2
6	8	1	7	2	3	9	4	5
5	3	2	1	9	4	8	6	7

Solution # 300

3	8	7	1	2	6	4	9	5
2	5	4	7	9	3	6	8	1
6	1	9	8	4	5	3	2	7
8	6	2	9	3	1	5	7	4
1	7	5	2	6	4	9	3	8
4	9	3	5	8	7	2	1	6
7	2	1	4	5	9	8	6	3
5	3	8	6	1	2	7	4	9
9	4	6	3	7	8	1	5	2

Solution # 301

9	5	1	4	6	8	7	2	3
4	7	8	3	2	1	6	9	5
2	3	6	9	7	5	1	8	4
1	6	9	7	3	2	5	4	8
7	8	3	5	4	9	2	1	6
5	2	4	8	1	6	9	3	7
6	9	7	2	8	3	4	5	1
3	1	5	6	9	4	8	7	2
8	4	2	1	5	7	3	6	9

Solution # 302

3	5	2	7	8	9	1	6	4
1	7	4	2	5	6	3	8	9
6	9	8	1	4	3	5	7	2
4	1	6	9	3	8	7	2	5
2	8	9	5	7	4	6	3	1
5	3	7	6	1	2	4	9	8
8	6	5	4	2	7	9	1	3
9	2	1	3	6	5	8	4	7
7	4	3	8	9	1	2	5	6

Solution # 303

4	9	2	1	6	8	3	7	5
8	6	5	7	3	2	9	4	1
1	3	7	9	4	5	8	2	6
7	5	3	8	2	6	4	1	9
6	8	9	4	1	3	7	5	2
2	1	4	5	7	9	6	3	8
9	7	8	2	5	4	1	6	3
5	4	6	3	9	1	2	8	7
3	2	1	6	8	7	5	9	4

Solution # 304

3	9	5	1	2	8	6	4	7
4	2	1	7	3	6	9	8	5
6	8	7	4	9	5	1	3	2
2	7	6	3	1	9	8	5	4
9	3	8	5	7	4	2	1	6
5	1	4	6	8	2	3	7	9
7	6	9	8	5	3	4	2	1
1	4	3	2	6	7	5	9	8
8	5	2	9	4	1	7	6	3

Solution # 305

9	4	7	6	8	1	5	3	2
2	6	5	9	4	3	1	8	7
8	1	3	2	5	7	4	9	6
6	9	8	3	1	5	7	2	4
1	5	4	7	2	8	3	6	9
3	7	2	4	9	6	8	1	5
5	2	9	8	3	4	6	7	1
7	8	1	5	6	2	9	4	3
4	3	6	1	7	9	2	5	8

Solution # 306

9	5	2	3	1	8	7	6	4
7	6	3	9	5	4	1	2	8
8	1	4	2	6	7	5	3	9
5	8	1	6	9	2	4	7	3
3	4	9	5	7	1	2	8	6
2	7	6	4	8	3	9	5	1
1	2	8	7	4	6	3	9	5
6	9	7	1	3	5	8	4	2
4	3	5	8	2	9	6	1	7

Solution # 307

3	9	5	4	7	8	1	6	2
8	6	4	2	1	9	3	7	5
2	7	1	5	6	3	8	4	9
7	8	6	1	4	5	9	2	3
9	5	2	6	3	7	4	1	8
1	4	3	9	8	2	7	5	6
4	2	8	3	5	1	6	9	7
5	1	7	8	9	6	2	3	4
6	3	9	7	2	4	5	8	1

Solution # 308

7	3	2	4	6	8	1	5	9
6	5	8	1	9	2	4	3	7
4	9	1	7	5	3	6	2	8
1	7	5	2	4	9	3	8	6
9	8	4	5	3	6	2	7	1
3	2	6	8	1	7	9	4	5
2	6	3	9	8	5	7	1	4
8	1	9	3	7	4	5	6	2
5	4	7	6	2	1	8	9	3

Solution # 309

5	7	9	4	3	8	1	2	6
4	3	2	1	5	6	7	8	9
8	6	1	2	9	7	3	4	5
3	9	8	7	1	2	6	5	4
2	1	5	6	4	3	9	7	8
6	4	7	9	8	5	2	3	1
9	2	4	5	7	1	8	6	3
7	5	3	8	6	9	4	1	2
1	8	6	3	2	4	5	9	7

Solution # 310

9	8	1	6	5	7	4	2	3
3	5	2	9	4	1	6	8	7
6	7	4	2	8	3	1	5	9
4	9	5	1	2	6	7	3	8
8	2	3	5	7	4	9	1	6
1	6	7	8	3	9	2	4	5
7	1	8	3	9	2	5	6	4
2	3	9	4	6	5	8	7	1
5	4	6	7	1	8	3	9	2

Solution # 311

5	9	1	7	4	2	3	8	6
8	4	2	5	3	6	7	9	1
3	6	7	8	1	9	5	2	4
7	2	9	3	5	1	4	6	8
1	3	4	6	7	8	2	5	9
6	8	5	2	9	4	1	7	3
4	7	3	9	8	5	6	1	2
9	5	6	1	2	3	8	4	7
2	1	8	4	6	7	9	3	5

Solution # 312

2	7	8	3	6	5	4	1	9
5	4	3	8	1	9	7	2	6
6	1	9	7	2	4	8	3	5
7	5	1	6	9	3	2	4	8
9	3	4	5	8	2	1	6	7
8	6	2	1	4	7	9	5	3
4	2	6	9	3	8	5	7	1
3	8	5	2	7	1	6	9	4
1	9	7	4	5	6	3	8	2

Solution # 313

6	5	3	8	9	7	1	4	2
1	8	9	2	5	4	7	3	6
4	7	2	3	1	6	9	8	5
2	9	6	7	3	1	8	5	4
5	3	1	4	2	8	6	9	7
7	4	8	5	6	9	3	2	1
3	6	5	1	8	2	4	7	9
8	1	7	9	4	5	2	6	3
9	2	4	6	7	3	5	1	8

Solution # 314

6	3	4	7	5	9	1	8	2
8	5	7	2	1	6	4	3	9
1	9	2	4	8	3	5	6	7
9	8	5	6	3	7	2	1	4
7	2	6	1	4	8	9	5	3
4	1	3	5	9	2	8	7	6
5	7	8	9	6	4	3	2	1
2	4	1	3	7	5	6	9	8
3	6	9	8	2	1	7	4	5

Solution # 315

5	4	6	9	7	2	1	8	3
1	2	8	4	6	3	9	7	5
9	7	3	8	5	1	4	6	2
8	9	1	2	3	5	7	4	6
2	5	4	7	8	6	3	9	1
6	3	7	1	4	9	2	5	8
7	1	5	3	9	8	6	2	4
3	6	9	5	2	4	8	1	7
4	8	2	6	1	7	5	3	9

Solution # 316

1	2	6	4	5	8	7	9	3
8	9	4	7	6	3	5	1	2
3	5	7	2	9	1	8	6	4
6	7	1	9	4	5	3	2	8
4	8	9	1	3	2	6	5	7
2	3	5	6	8	7	9	4	1
9	1	2	8	7	6	4	3	5
5	6	8	3	1	4	2	7	9
7	4	3	5	2	9	1	8	6

Solution # 317

9	3	8	1	4	5	6	2	7
5	6	4	2	3	7	8	9	1
2	1	7	9	8	6	4	3	5
3	7	2	8	1	9	5	4	6
1	9	6	3	5	4	7	8	2
4	8	5	6	7	2	9	1	3
8	2	9	5	6	3	1	7	4
7	5	1	4	2	8	3	6	9
6	4	3	7	9	1	2	5	8

Solution # 318

6	9	4	5	3	7	8	2	1
7	8	3	1	2	4	5	6	9
1	5	2	8	6	9	7	3	4
8	3	9	4	1	5	6	7	2
2	7	5	6	9	8	1	4	3
4	6	1	3	7	2	9	8	5
3	2	8	7	5	1	4	9	6
5	4	6	9	8	3	2	1	7
9	1	7	2	4	6	3	5	8

Solution # 319

6	7	5	1	3	9	2	8	4
3	1	9	2	8	4	7	5	6
4	8	2	6	5	7	9	1	3
8	4	6	3	1	2	5	9	7
5	9	7	4	6	8	1	3	2
2	3	1	9	7	5	4	6	8
7	5	3	8	4	1	6	2	9
1	2	8	7	9	6	3	4	5
9	6	4	5	2	3	8	7	1

Solution # 320

4	8	9	6	1	3	2	5	7
1	5	2	8	7	9	4	3	6
7	6	3	2	5	4	8	9	1
3	7	1	5	6	8	9	4	2
8	2	4	7	9	1	3	6	5
6	9	5	4	3	2	7	1	8
5	1	8	3	4	7	6	2	9
2	3	6	9	8	5	1	7	4
9	4	7	1	2	6	5	8	3

Solution # 321

3	1	7	5	4	8	2	9	6
5	8	6	7	2	9	1	4	3
4	9	2	3	1	6	7	8	5
1	4	5	2	9	7	6	3	8
6	2	3	1	8	5	9	7	4
9	7	8	4	6	3	5	1	2
2	3	1	9	5	4	8	6	7
7	6	9	8	3	2	4	5	1
8	5	4	6	7	1	3	2	9

Solution # 322

1	9	4	7	2	3	8	6	5
8	5	3	4	1	6	7	2	9
6	7	2	8	5	9	1	3	4
7	6	5	2	8	4	9	1	3
2	3	9	5	6	1	4	7	8
4	1	8	9	3	7	2	5	6
5	8	1	6	4	2	3	9	7
9	2	6	3	7	8	5	4	1
3	4	7	1	9	5	6	8	2

Solution # 323

5	3	6	1	8	4	7	2	9
7	1	9	2	5	6	4	3	8
4	8	2	7	9	3	6	5	1
8	4	5	6	3	9	2	1	7
6	2	1	4	7	5	9	8	3
3	9	7	8	1	2	5	6	4
1	5	3	9	2	7	8	4	6
9	6	8	5	4	1	3	7	2
2	7	4	3	6	8	1	9	5

Solution # 324

3	6	8	1	9	2	7	4	5
1	2	4	7	5	3	6	9	8
9	7	5	4	6	8	3	2	1
4	3	7	9	1	5	8	6	2
6	8	9	2	4	7	1	5	3
5	1	2	8	3	6	9	7	4
2	5	6	3	8	9	4	1	7
7	4	3	6	2	1	5	8	9
8	9	1	5	7	4	2	3	6

Solution # 325

5	1	2	8	7	6	9	4	3
3	6	7	2	9	4	8	5	1
9	8	4	5	1	3	2	6	7
7	5	1	6	2	9	3	8	4
8	3	9	4	5	7	1	2	6
2	4	6	3	8	1	7	9	5
6	9	5	1	3	8	4	7	2
4	7	3	9	6	2	5	1	8
1	2	8	7	4	5	6	3	9

Solution # 326

5	6	7	9	2	4	3	1	8
9	3	8	7	1	5	2	6	4
4	1	2	8	6	3	5	9	7
7	4	9	5	3	2	1	8	6
8	2	3	6	7	1	4	5	9
1	5	6	4	8	9	7	3	2
2	8	5	3	4	6	9	7	1
6	9	1	2	5	7	8	4	3
3	7	4	1	9	8	6	2	5

Solution # 327

4	8	1	7	5	6	2	9	3
3	7	2	1	4	9	6	5	8
5	6	9	8	2	3	7	1	4
6	2	5	9	8	4	3	7	1
1	9	8	6	3	7	5	4	2
7	4	3	5	1	2	9	8	6
9	3	7	4	6	1	8	2	5
2	5	4	3	7	8	1	6	9
8	1	6	2	9	5	4	3	7

Solution # 328

3	9	5	4	1	7	6	2	8
7	6	8	9	2	5	4	1	3
1	2	4	6	3	8	7	9	5
6	8	9	3	5	4	1	7	2
5	7	1	2	6	9	3	8	4
4	3	2	8	7	1	5	6	9
2	4	6	7	9	3	8	5	1
8	1	7	5	4	2	9	3	6
9	5	3	1	8	6	2	4	7

Solution # 329

8	6	7	1	9	2	4	5	3
4	5	1	6	7	3	9	8	2
9	3	2	5	4	8	1	7	6
3	9	4	7	2	5	6	1	8
2	8	5	3	1	6	7	4	9
7	1	6	4	8	9	3	2	5
1	2	8	9	6	7	5	3	4
6	7	3	2	5	4	8	9	1
5	4	9	8	3	1	2	6	7

Solution # 330

7	5	1	3	2	4	6	8	9
9	8	6	1	5	7	2	4	3
4	3	2	8	6	9	7	1	5
2	6	7	5	8	1	3	9	4
5	9	4	7	3	2	1	6	8
8	1	3	9	4	6	5	2	7
1	7	5	2	9	8	4	3	6
6	2	9	4	7	3	8	5	1
3	4	8	6	1	5	9	7	2

Solution # 331

4	8	2	6	3	1	7	5	9
3	9	1	5	7	8	2	6	4
7	6	5	4	9	2	8	3	1
2	7	6	9	5	4	1	8	3
9	3	8	1	2	6	4	7	5
5	1	4	7	8	3	9	2	6
6	2	7	3	1	9	5	4	8
8	4	9	2	6	5	3	1	7
1	5	3	8	4	7	6	9	2

Solution # 332

5	7	4	2	3	9	1	6	8
1	3	9	6	7	8	5	4	2
8	6	2	1	5	4	7	9	3
9	2	8	3	4	7	6	5	1
7	1	5	9	8	6	3	2	4
6	4	3	5	2	1	8	7	9
2	9	7	8	1	5	4	3	6
3	5	1	4	6	2	9	8	7
4	8	6	7	9	3	2	1	5

Solution # 333

3	9	1	5	7	4	6	8	2
5	7	2	8	1	6	4	9	3
4	6	8	2	9	3	5	1	7
8	5	9	1	3	2	7	6	4
7	3	6	4	8	9	1	2	5
2	1	4	7	6	5	9	3	8
9	2	5	3	4	1	8	7	6
1	4	7	6	2	8	3	5	9
6	8	3	9	5	7	2	4	1

Solution # 334

1	7	8	9	5	2	6	4	3
4	2	3	6	8	1	7	9	5
6	9	5	4	7	3	1	8	2
3	4	7	8	1	5	9	2	6
9	8	2	7	3	6	4	5	1
5	1	6	2	4	9	3	7	8
8	6	9	1	2	4	5	3	7
7	3	1	5	9	8	2	6	4
2	5	4	3	6	7	8	1	9

Solution # 335

8	2	6	3	1	9	4	7	5
3	1	4	5	2	7	8	6	9
9	5	7	6	4	8	2	1	3
2	4	3	7	6	5	9	8	1
7	8	9	2	3	1	5	4	6
1	6	5	9	8	4	7	3	2
6	3	8	4	5	2	1	9	7
5	7	1	8	9	6	3	2	4
4	9	2	1	7	3	6	5	8

Solution # 336

2	5	4	7	1	3	8	9	6
9	3	7	2	6	8	1	4	5
8	6	1	5	4	9	2	3	7
1	7	8	9	3	5	6	2	4
5	4	2	6	8	7	9	1	3
6	9	3	1	2	4	7	5	8
7	2	6	3	5	1	4	8	9
4	1	5	8	9	6	3	7	2
3	8	9	4	7	2	5	6	1

Solution # 337

7	9	1	4	6	2	8	3	5
3	2	8	5	7	1	9	4	6
5	6	4	8	3	9	7	2	1
4	5	9	6	2	8	1	7	3
8	7	2	3	1	4	5	6	9
1	3	6	7	9	5	4	8	2
6	8	5	1	4	3	2	9	7
2	4	3	9	5	7	6	1	8
9	1	7	2	8	6	3	5	4

Solution # 338

3	7	1	4	9	2	5	6	8
4	8	5	6	1	3	2	9	7
6	9	2	8	5	7	4	3	1
9	1	3	7	4	5	8	2	6
7	4	8	3	2	6	9	1	5
5	2	6	1	8	9	3	7	4
1	5	9	2	6	4	7	8	3
8	3	4	9	7	1	6	5	2
2	6	7	5	3	8	1	4	9

Solution # 339

6	9	4	5	1	2	8	3	7
8	5	1	3	7	6	4	9	2
7	2	3	9	4	8	1	6	5
1	8	6	2	9	7	5	4	3
2	3	5	8	6	4	7	1	9
4	7	9	1	3	5	6	2	8
3	4	2	7	8	1	9	5	6
5	1	8	6	2	9	3	7	4
9	6	7	4	5	3	2	8	1

Solution # 340

1	5	6	8	3	7	4	9	2
3	4	9	1	2	6	7	5	8
8	2	7	4	5	9	1	3	6
9	8	3	5	1	2	6	4	7
5	6	1	9	7	4	8	2	3
2	7	4	3	6	8	9	1	5
4	9	5	6	8	3	2	7	1
7	1	8	2	4	5	3	6	9
6	3	2	7	9	1	5	8	4

Solution # 341

2	1	9	6	8	4	5	7	3
3	8	6	5	2	7	9	4	1
7	5	4	9	1	3	2	6	8
8	4	1	2	7	5	6	3	9
6	7	2	1	3	9	4	8	5
5	9	3	8	4	6	7	1	2
4	3	5	7	9	8	1	2	6
1	6	7	3	5	2	8	9	4
9	2	8	4	6	1	3	5	7

Solution # 342

6	3	8	5	9	2	4	7	1
7	5	4	1	8	3	6	2	9
9	2	1	7	4	6	3	5	8
1	7	3	4	5	8	2	9	6
5	4	9	2	6	1	7	8	3
2	8	6	3	7	9	1	4	5
4	6	7	9	3	5	8	1	2
8	9	2	6	1	4	5	3	7
3	1	5	8	2	7	9	6	4

Solution # 343

1	2	5	6	8	7	4	9	3
9	7	4	2	3	1	8	6	5
3	6	8	5	4	9	2	1	7
8	5	6	9	7	4	1	3	2
4	1	2	8	6	3	5	7	9
7	3	9	1	2	5	6	8	4
2	8	7	3	5	6	9	4	1
5	9	3	4	1	8	7	2	6
6	4	1	7	9	2	3	5	8

Solution # 344

8	9	5	3	1	2	4	6	7
4	2	1	6	7	8	3	9	5
3	6	7	4	5	9	1	8	2
5	7	9	1	6	3	2	4	8
6	4	8	7	2	5	9	1	3
1	3	2	8	9	4	7	5	6
7	8	4	9	3	6	5	2	1
2	1	6	5	4	7	8	3	9
9	5	3	2	8	1	6	7	4

Solution # 345

7	2	3	9	5	6	8	1	4
4	9	8	7	3	1	2	5	6
1	6	5	2	4	8	3	7	9
5	4	6	1	2	3	7	9	8
2	1	7	6	8	9	4	3	5
8	3	9	5	7	4	6	2	1
9	8	1	3	6	7	5	4	2
3	5	4	8	9	2	1	6	7
6	7	2	4	1	5	9	8	3

Solution # 346

6	1	5	9	3	7	4	2	8
4	3	8	1	2	6	5	9	7
7	9	2	8	5	4	3	6	1
3	4	9	5	1	2	8	7	6
1	5	6	4	7	8	2	3	9
8	2	7	3	6	9	1	5	4
5	8	4	6	9	3	7	1	2
9	7	1	2	8	5	6	4	3
2	6	3	7	4	1	9	8	5

Solution # 347

2	8	1	4	6	7	5	3	9
6	4	5	9	1	3	2	7	8
3	7	9	8	5	2	4	6	1
4	1	6	2	3	8	7	9	5
8	2	7	1	9	5	3	4	6
9	5	3	7	4	6	1	8	2
5	3	4	6	2	9	8	1	7
7	6	2	3	8	1	9	5	4
1	9	8	5	7	4	6	2	3

Solution # 348

3	6	9	2	5	7	4	1	8
4	2	5	8	1	3	9	6	7
1	8	7	6	4	9	3	5	2
7	5	4	3	9	2	1	8	6
2	3	8	1	6	4	5	7	9
6	9	1	5	7	8	2	4	3
5	4	2	7	3	6	8	9	1
8	1	6	9	2	5	7	3	4
9	7	3	4	8	1	6	2	5

Solution # 349

8	5	9	4	7	3	1	2	6
2	3	7	5	6	1	9	8	4
6	4	1	8	9	2	5	7	3
7	6	3	2	5	9	8	4	1
5	2	8	3	1	4	6	9	7
1	9	4	7	8	6	3	5	2
9	7	6	1	4	5	2	3	8
4	1	2	9	3	8	7	6	5
3	8	5	6	2	7	4	1	9

Solution # 350

9	3	8	6	4	1	7	2	5
5	1	7	2	8	3	9	4	6
6	2	4	5	9	7	3	1	8
8	6	2	9	3	4	1	5	7
3	7	5	1	6	2	4	8	9
4	9	1	8	7	5	2	6	3
2	4	9	3	5	8	6	7	1
7	5	6	4	1	9	8	3	2
1	8	3	7	2	6	5	9	4

Solution # 351

8	3	9	1	5	4	2	6	7
1	6	7	9	2	3	5	4	8
4	5	2	7	6	8	9	1	3
5	1	8	3	7	2	4	9	6
6	2	4	5	8	9	7	3	1
7	9	3	4	1	6	8	2	5
2	7	1	6	4	5	3	8	9
3	8	5	2	9	1	6	7	4
9	4	6	8	3	7	1	5	2

Solution # 352

2	8	9	3	6	5	4	7	1
7	5	1	2	8	4	6	3	9
3	4	6	7	1	9	2	5	8
1	2	7	6	4	3	9	8	5
6	3	5	9	7	8	1	2	4
4	9	8	5	2	1	3	6	7
8	7	3	4	9	2	5	1	6
5	6	4	1	3	7	8	9	2
9	1	2	8	5	6	7	4	3

Solution # 353

8	5	3	6	4	7	1	9	2
6	1	4	5	2	9	3	8	7
9	7	2	8	3	1	5	4	6
7	4	6	3	9	5	2	1	8
1	2	5	7	8	4	6	3	9
3	8	9	2	1	6	4	7	5
2	6	1	9	7	3	8	5	4
4	9	8	1	5	2	7	6	3
5	3	7	4	6	8	9	2	1

Solution # 354

3	2	7	9	1	6	8	4	5
4	9	8	7	3	5	2	6	1
6	5	1	4	8	2	3	9	7
7	1	3	5	9	4	6	2	8
9	6	4	8	2	7	5	1	3
2	8	5	1	6	3	9	7	4
5	4	6	3	7	9	1	8	2
1	3	2	6	4	8	7	5	9
8	7	9	2	5	1	4	3	6

Solution # 355

6	4	3	9	2	7	8	1	5
9	1	2	8	5	3	4	7	6
8	7	5	1	6	4	3	9	2
2	8	1	5	3	9	7	6	4
7	5	9	6	4	8	2	3	1
3	6	4	2	7	1	9	5	8
4	9	6	7	1	2	5	8	3
5	2	7	3	8	6	1	4	9
1	3	8	4	9	5	6	2	7

Solution # 356

8	2	5	4	9	7	6	1	3
3	1	6	8	2	5	4	7	9
9	7	4	1	6	3	2	5	8
6	8	3	9	4	1	7	2	5
2	5	7	3	8	6	9	4	1
1	4	9	7	5	2	8	3	6
4	6	1	5	7	8	3	9	2
5	9	2	6	3	4	1	8	7
7	3	8	2	1	9	5	6	4

Solution # 357

4	7	9	5	6	8	3	1	2
8	3	2	7	1	9	4	6	5
6	1	5	3	4	2	7	8	9
1	9	6	4	3	5	2	7	8
5	8	7	2	9	6	1	4	3
2	4	3	8	7	1	5	9	6
3	6	8	1	5	4	9	2	7
7	2	1	9	8	3	6	5	4
9	5	4	6	2	7	8	3	1

Solution # 358

5	7	1	6	2	8	4	9	3
9	6	3	7	1	4	8	2	5
4	2	8	3	5	9	6	7	1
8	4	9	2	6	3	5	1	7
7	3	2	1	8	5	9	6	4
6	1	5	4	9	7	3	8	2
2	8	4	9	3	1	7	5	6
1	5	7	8	4	6	2	3	9
3	9	6	5	7	2	1	4	8

Solution # 359

7	4	6	3	8	2	5	1	9
3	5	9	4	1	6	7	8	2
2	8	1	5	7	9	4	6	3
8	3	4	9	5	7	6	2	1
6	2	5	1	4	8	3	9	7
9	1	7	2	6	3	8	4	5
4	7	3	8	2	1	9	5	6
5	6	2	7	9	4	1	3	8
1	9	8	6	3	5	2	7	4

Solution # 360

1	5	3	2	6	8	9	4	7
6	7	9	4	3	5	1	8	2
4	2	8	1	9	7	5	3	6
3	9	5	6	4	1	7	2	8
2	4	7	3	8	9	6	5	1
8	1	6	7	5	2	4	9	3
7	6	4	9	2	3	8	1	5
9	8	2	5	1	6	3	7	4
5	3	1	8	7	4	2	6	9

Solution # 361

3	4	8	6	9	7	2	1	5
9	7	5	2	1	3	8	6	4
6	2	1	5	4	8	7	9	3
7	9	4	8	3	1	5	2	6
5	8	6	7	2	4	1	3	9
1	3	2	9	6	5	4	7	8
8	1	9	4	7	6	3	5	2
4	6	3	1	5	2	9	8	7
2	5	7	3	8	9	6	4	1

Solution # 362

8	5	6	3	4	7	2	9	1
9	7	3	2	5	1	8	6	4
4	1	2	6	9	8	5	3	7
2	6	4	8	1	5	3	7	9
3	9	7	4	2	6	1	8	5
5	8	1	7	3	9	4	2	6
7	4	9	5	8	2	6	1	3
1	3	8	9	6	4	7	5	2
6	2	5	1	7	3	9	4	8

Solution # 363

7	9	8	2	3	1	6	5	4
5	4	1	9	6	7	3	8	2
3	2	6	8	5	4	1	9	7
9	6	4	1	8	3	7	2	5
1	3	5	7	2	9	4	6	8
8	7	2	5	4	6	9	3	1
2	1	9	6	7	8	5	4	3
4	8	7	3	9	5	2	1	6
6	5	3	4	1	2	8	7	9

Solution # 364

8	1	4	9	5	2	6	3	7
7	5	2	8	6	3	9	4	1
6	3	9	7	4	1	2	8	5
1	4	6	2	3	5	7	9	8
2	8	7	1	9	4	3	5	6
3	9	5	6	7	8	4	1	2
4	2	1	3	8	6	5	7	9
9	6	3	5	1	7	8	2	4
5	7	8	4	2	9	1	6	3

Solution # 365

3	5	1	4	2	7	6	9	8
6	7	4	1	8	9	5	3	2
9	8	2	6	5	3	4	7	1
7	9	8	3	4	1	2	5	6
4	6	5	2	7	8	9	1	3
1	2	3	9	6	5	8	4	7
5	3	7	8	9	2	1	6	4
2	1	6	5	3	4	7	8	9
8	4	9	7	1	6	3	2	5

Solution # 366

7	6	5	9	3	4	1	2	8
2	9	3	1	7	8	6	5	4
8	4	1	5	6	2	9	3	7
1	8	9	3	5	6	4	7	2
4	5	2	7	8	1	3	6	9
6	3	7	4	2	9	8	1	5
9	2	6	8	1	5	7	4	3
5	7	8	6	4	3	2	9	1
3	1	4	2	9	7	5	8	6

Solution # 367

7	9	2	8	5	1	3	4	6
4	8	1	6	3	2	9	5	7
6	3	5	9	4	7	1	8	2
2	4	9	3	7	5	6	1	8
8	6	7	1	2	9	5	3	4
1	5	3	4	8	6	2	7	9
5	2	6	7	1	4	8	9	3
3	1	4	2	9	8	7	6	5
9	7	8	5	6	3	4	2	1

Solution # 368

2	6	4	8	9	7	1	5	3
1	5	9	3	6	2	4	8	7
3	8	7	5	4	1	9	6	2
6	9	2	4	8	5	7	3	1
5	7	1	2	3	9	8	4	6
4	3	8	7	1	6	5	2	9
7	4	6	9	2	8	3	1	5
8	1	5	6	7	3	2	9	4
9	2	3	1	5	4	6	7	8

Solution # 369

6	7	9	2	3	4	8	1	5
1	3	2	6	5	8	9	7	4
5	4	8	7	9	1	3	6	2
2	9	6	3	8	7	5	4	1
7	1	5	9	4	6	2	3	8
4	8	3	5	1	2	7	9	6
9	5	4	8	6	3	1	2	7
8	2	1	4	7	9	6	5	3
3	6	7	1	2	5	4	8	9

Solution # 370

7	9	5	3	6	2	8	1	4
1	3	4	8	9	7	2	5	6
8	6	2	1	5	4	7	3	9
3	2	7	4	1	5	6	9	8
9	4	8	2	3	6	5	7	1
6	5	1	7	8	9	3	4	2
2	8	9	5	4	3	1	6	7
4	1	3	6	7	8	9	2	5
5	7	6	9	2	1	4	8	3

Solution # 371

6	1	7	8	3	4	2	9	5
5	3	9	2	1	6	4	7	8
4	2	8	7	9	5	1	3	6
8	9	2	5	6	1	7	4	3
3	4	6	9	2	7	5	8	1
1	7	5	3	4	8	6	2	9
2	5	3	6	7	9	8	1	4
9	6	4	1	8	2	3	5	7
7	8	1	4	5	3	9	6	2

Solution # 372

6	5	8	1	4	3	9	2	7
9	3	1	7	2	5	6	8	4
7	4	2	8	9	6	1	3	5
2	1	9	6	7	4	3	5	8
5	7	6	2	3	8	4	1	9
3	8	4	9	5	1	7	6	2
1	9	7	5	6	2	8	4	3
8	2	3	4	1	7	5	9	6
4	6	5	3	8	9	2	7	1

Solution # 373

7	4	1	5	9	6	2	8	3
8	3	2	1	4	7	9	5	6
9	6	5	2	8	3	4	1	7
2	9	4	6	5	8	3	7	1
3	5	8	7	1	9	6	4	2
6	1	7	4	3	2	5	9	8
4	2	6	8	7	5	1	3	9
1	8	9	3	6	4	7	2	5
5	7	3	9	2	1	8	6	4

Solution # 374

2	6	4	3	1	8	7	5	9
7	9	3	6	5	4	8	1	2
5	1	8	7	2	9	3	4	6
3	7	1	4	9	6	5	2	8
4	2	9	1	8	5	6	7	3
6	8	5	2	7	3	1	9	4
1	3	6	5	4	2	9	8	7
8	5	2	9	3	7	4	6	1
9	4	7	8	6	1	2	3	5

Solution # 375

6	2	3	5	9	7	1	4	8
4	1	9	2	3	8	6	7	5
5	7	8	1	4	6	2	9	3
8	3	7	9	2	4	5	6	1
2	6	4	7	5	1	3	8	9
1	9	5	8	6	3	4	2	7
3	8	6	4	1	9	7	5	2
7	4	2	3	8	5	9	1	6
9	5	1	6	7	2	8	3	4

Solution # 376

6	9	3	2	5	1	4	7	8
7	5	1	9	4	8	2	6	3
4	8	2	7	6	3	9	5	1
2	7	4	1	9	5	8	3	6
9	6	8	4	3	7	1	2	5
3	1	5	6	8	2	7	4	9
5	2	6	8	7	9	3	1	4
8	4	7	3	1	6	5	9	2
1	3	9	5	2	4	6	8	7

Solution # 377

8	6	1	3	7	4	2	9	5
4	9	7	5	2	1	8	6	3
3	2	5	8	9	6	7	4	1
7	5	6	9	4	8	1	3	2
1	4	8	6	3	2	9	5	7
2	3	9	7	1	5	6	8	4
5	7	4	1	6	9	3	2	8
9	1	2	4	8	3	5	7	6
6	8	3	2	5	7	4	1	9

Solution # 378

3	8	1	4	5	9	7	2	6
6	4	9	7	8	2	5	3	1
7	5	2	3	1	6	4	8	9
5	1	3	2	9	7	8	6	4
2	6	7	5	4	8	1	9	3
8	9	4	6	3	1	2	5	7
4	7	5	8	6	3	9	1	2
9	3	8	1	2	4	6	7	5
1	2	6	9	7	5	3	4	8

Solution # 379

7	2	3	8	9	6	5	1	4
5	9	4	1	3	7	6	2	8
1	6	8	2	5	4	3	9	7
3	7	5	6	1	9	4	8	2
4	1	6	7	8	2	9	5	3
2	8	9	3	4	5	7	6	1
8	4	2	9	6	3	1	7	5
9	5	1	4	7	8	2	3	6
6	3	7	5	2	1	8	4	9

Solution # 380

2	1	5	8	7	4	3	6	9
4	3	7	9	6	1	2	5	8
8	9	6	2	5	3	1	4	7
5	7	3	6	2	8	9	1	4
6	4	1	3	9	7	5	8	2
9	2	8	1	4	5	7	3	6
7	6	4	5	1	2	8	9	3
1	8	2	4	3	9	6	7	5
3	5	9	7	8	6	4	2	1

Solution # 381

4	2	6	9	3	5	7	8	1
8	9	3	1	2	7	5	4	6
1	5	7	4	8	6	3	2	9
3	1	5	7	6	4	8	9	2
2	7	4	8	9	3	6	1	5
9	6	8	2	5	1	4	3	7
7	3	2	6	1	8	9	5	4
5	4	1	3	7	9	2	6	8
6	8	9	5	4	2	1	7	3

Solution # 382

3	5	6	7	9	8	2	1	4
9	4	1	5	2	3	8	7	6
2	8	7	1	6	4	9	5	3
5	1	2	3	7	9	4	6	8
8	6	4	2	1	5	7	3	9
7	9	3	4	8	6	1	2	5
4	7	9	6	3	2	5	8	1
6	2	5	8	4	1	3	9	7
1	3	8	9	5	7	6	4	2

Solution # 383

4	8	2	3	1	5	9	7	6
5	7	1	6	2	9	3	4	8
9	3	6	8	4	7	2	1	5
3	2	4	9	8	6	1	5	7
7	6	9	1	5	4	8	3	2
1	5	8	7	3	2	4	6	9
8	1	7	5	9	3	6	2	4
2	9	5	4	6	1	7	8	3
6	4	3	2	7	8	5	9	1

Solution # 384

9	5	4	8	7	1	2	3	6
2	3	8	4	9	6	5	1	7
1	7	6	5	2	3	8	4	9
5	4	2	9	3	8	7	6	1
8	6	9	7	1	5	4	2	3
7	1	3	6	4	2	9	8	5
3	9	1	2	8	7	6	5	4
6	2	7	3	5	4	1	9	8
4	8	5	1	6	9	3	7	2

Solution # 385

5	7	8	4	6	1	9	2	3
2	1	9	5	7	3	4	8	6
3	4	6	9	2	8	7	1	5
1	8	3	7	9	5	6	4	2
9	2	7	6	3	4	8	5	1
6	5	4	8	1	2	3	9	7
8	6	5	1	4	7	2	3	9
4	9	2	3	5	6	1	7	8
7	3	1	2	8	9	5	6	4

Solution # 386

3	1	8	7	5	9	6	4	2
9	5	4	2	6	8	7	3	1
2	7	6	1	3	4	5	9	8
4	8	9	6	1	3	2	5	7
7	6	2	9	8	5	3	1	4
5	3	1	4	7	2	9	8	6
1	9	3	8	2	7	4	6	5
8	4	7	5	9	6	1	2	3
6	2	5	3	4	1	8	7	9

Solution # 387

7	2	5	3	8	4	6	1	9
4	3	6	1	9	2	8	7	5
8	1	9	5	7	6	2	3	4
1	8	7	9	4	5	3	2	6
2	5	3	8	6	7	4	9	1
6	9	4	2	3	1	5	8	7
9	4	2	7	5	3	1	6	8
3	6	8	4	1	9	7	5	2
5	7	1	6	2	8	9	4	3

Solution # 388

3	2	9	1	5	8	6	4	7
7	8	6	4	2	3	5	1	9
5	4	1	6	9	7	2	8	3
6	7	4	2	1	9	8	3	5
9	5	2	8	3	4	7	6	1
1	3	8	7	6	5	4	9	2
4	6	3	5	7	1	9	2	8
2	9	7	3	8	6	1	5	4
8	1	5	9	4	2	3	7	6

Solution # 389

6	3	8	2	5	1	7	4	9
7	2	1	4	9	8	5	6	3
4	9	5	3	7	6	8	2	1
5	8	3	7	4	9	6	1	2
1	7	4	6	2	3	9	5	8
9	6	2	1	8	5	3	7	4
3	5	7	9	1	4	2	8	6
8	4	6	5	3	2	1	9	7
2	1	9	8	6	7	4	3	5

Solution # 390

4	5	1	9	6	7	3	2	8
3	7	6	8	2	5	9	4	1
9	8	2	1	4	3	6	7	5
7	9	3	5	1	2	4	8	6
6	1	4	3	8	9	2	5	7
5	2	8	6	7	4	1	3	9
8	3	5	4	9	1	7	6	2
1	6	7	2	3	8	5	9	4
2	4	9	7	5	6	8	1	3

Solution # 391

1	9	7	6	4	5	2	8	3
6	2	8	1	9	3	5	7	4
4	5	3	7	8	2	6	1	9
8	3	2	5	6	7	4	9	1
7	1	5	4	3	9	8	6	2
9	4	6	8	2	1	7	3	5
3	7	1	2	5	6	9	4	8
5	8	9	3	7	4	1	2	6
2	6	4	9	1	8	3	5	7

Solution # 392

4	8	3	1	7	5	9	6	2
1	9	5	2	8	6	3	4	7
2	6	7	3	4	9	5	1	8
7	1	9	4	5	8	2	3	6
3	5	2	6	9	1	7	8	4
8	4	6	7	2	3	1	9	5
6	7	4	9	1	2	8	5	3
5	2	1	8	3	4	6	7	9
9	3	8	5	6	7	4	2	1

Solution # 393

1	8	4	3	7	6	2	5	9
2	6	9	5	1	8	3	7	4
5	3	7	9	2	4	8	1	6
7	9	8	4	5	3	1	6	2
6	1	5	2	8	9	7	4	3
4	2	3	1	6	7	5	9	8
9	5	1	6	3	2	4	8	7
3	7	6	8	4	5	9	2	1
8	4	2	7	9	1	6	3	5

Solution # 394

5	7	4	2	1	3	9	8	6
3	6	1	5	8	9	7	4	2
8	2	9	7	4	6	5	3	1
4	3	2	6	9	7	8	1	5
1	8	6	4	3	5	2	9	7
9	5	7	8	2	1	3	6	4
2	4	5	9	6	8	1	7	3
7	9	3	1	5	4	6	2	8
6	1	8	3	7	2	4	5	9

Solution # 395

6	5	3	8	9	4	2	7	1
4	1	8	3	7	2	9	6	5
7	2	9	6	5	1	3	4	8
2	4	1	9	3	6	8	5	7
9	3	5	1	8	7	6	2	4
8	6	7	4	2	5	1	9	3
3	9	6	7	4	8	5	1	2
5	8	4	2	1	9	7	3	6
1	7	2	5	6	3	4	8	9

Solution # 396

8	9	3	7	6	1	5	4	2
2	6	5	8	9	4	7	3	1
1	4	7	5	2	3	6	9	8
6	2	9	4	7	8	3	1	5
4	5	1	9	3	2	8	7	6
3	7	8	6	1	5	4	2	9
9	3	6	1	5	7	2	8	4
5	8	2	3	4	9	1	6	7
7	1	4	2	8	6	9	5	3

Solution # 397

4	9	6	3	5	7	2	8	1
8	3	2	4	9	1	5	7	6
5	7	1	8	2	6	3	4	9
7	2	9	1	4	8	6	5	3
6	8	5	9	7	3	4	1	2
1	4	3	2	6	5	7	9	8
9	6	4	5	1	2	8	3	7
3	5	7	6	8	9	1	2	4
2	1	8	7	3	4	9	6	5

Solution # 398

2	4	5	3	7	1	8	9	6
7	6	8	2	9	5	3	1	4
1	3	9	4	6	8	7	2	5
9	5	6	7	4	3	2	8	1
3	8	2	1	5	9	4	6	7
4	1	7	8	2	6	9	5	3
6	2	1	9	3	7	5	4	8
5	7	4	6	8	2	1	3	9
8	9	3	5	1	4	6	7	2

Solution # 399

6	5	9	4	2	3	7	1	8
3	4	7	9	1	8	2	5	6
2	1	8	6	7	5	3	9	4
8	9	5	3	4	7	1	6	2
4	2	3	1	5	6	8	7	9
1	7	6	8	9	2	4	3	5
5	8	2	7	6	1	9	4	3
9	3	1	5	8	4	6	2	7
7	6	4	2	3	9	5	8	1

Solution # 400

9	7	2	8	5	6	1	3	4
5	1	4	7	2	3	6	8	9
3	6	8	4	9	1	2	7	5
1	9	5	2	4	8	3	6	7
8	3	7	5	6	9	4	1	2
4	2	6	3	1	7	9	5	8
6	5	3	9	7	4	8	2	1
2	8	9	1	3	5	7	4	6
7	4	1	6	8	2	5	9	3

Solution # 401

3	4	7	2	8	6	9	1	5
8	2	6	9	5	1	7	4	3
5	1	9	4	7	3	6	8	2
4	8	2	1	3	7	5	6	9
7	3	1	5	6	9	8	2	4
6	9	5	8	2	4	1	3	7
2	5	3	7	1	8	4	9	6
9	7	8	6	4	2	3	5	1
1	6	4	3	9	5	2	7	8

Solution # 402

4	5	8	7	9	1	3	2	6
3	1	6	2	4	5	7	9	8
9	7	2	6	8	3	1	5	4
7	6	4	8	5	2	9	1	3
5	9	3	1	6	4	8	7	2
8	2	1	9	3	7	4	6	5
1	8	9	3	2	6	5	4	7
2	4	7	5	1	8	6	3	9
6	3	5	4	7	9	2	8	1

Solution # 403

8	2	4	5	1	6	9	3	7
1	6	3	7	2	9	8	4	5
5	9	7	3	8	4	1	6	2
6	4	5	9	7	2	3	1	8
9	3	8	6	5	1	7	2	4
2	7	1	4	3	8	6	5	9
7	5	9	1	4	3	2	8	6
3	8	6	2	9	5	4	7	1
4	1	2	8	6	7	5	9	3

Solution # 404

3	1	7	8	2	6	9	4	5
8	9	2	4	1	5	6	3	7
4	5	6	9	7	3	1	2	8
5	2	4	1	8	7	3	9	6
6	7	1	3	4	9	8	5	2
9	3	8	5	6	2	4	7	1
7	4	5	6	3	1	2	8	9
1	8	9	2	5	4	7	6	3
2	6	3	7	9	8	5	1	4

Solution # 405

4	8	9	5	7	3	6	1	2
1	2	5	4	9	6	3	7	8
6	3	7	1	2	8	9	4	5
9	7	3	8	6	4	2	5	1
2	6	1	7	5	9	8	3	4
8	5	4	3	1	2	7	9	6
7	9	2	6	4	1	5	8	3
5	4	8	2	3	7	1	6	9
3	1	6	9	8	5	4	2	7

Solution # 406

7	2	1	3	4	5	8	6	9
6	9	3	8	2	1	5	7	4
8	5	4	7	6	9	2	1	3
2	3	7	1	5	6	9	4	8
9	1	5	2	8	4	6	3	7
4	6	8	9	7	3	1	5	2
5	8	6	4	3	2	7	9	1
1	4	2	5	9	7	3	8	6
3	7	9	6	1	8	4	2	5

Solution # 407

7	1	6	9	4	2	3	8	5
2	9	5	6	3	8	7	4	1
3	4	8	7	1	5	6	9	2
6	8	4	2	9	7	1	5	3
5	7	2	3	8	1	9	6	4
1	3	9	4	5	6	2	7	8
8	6	1	5	7	3	4	2	9
9	2	3	8	6	4	5	1	7
4	5	7	1	2	9	8	3	6

Solution # 408

1	9	5	7	2	4	6	3	8
4	6	3	1	9	8	2	5	7
2	8	7	6	3	5	1	4	9
6	4	9	3	1	7	8	2	5
3	5	8	9	4	2	7	6	1
7	2	1	8	5	6	3	9	4
5	7	6	2	8	9	4	1	3
9	1	2	4	7	3	5	8	6
8	3	4	5	6	1	9	7	2

Solution # 409

8	4	2	6	9	7	1	5	3
1	9	6	5	3	2	7	8	4
3	7	5	1	8	4	9	6	2
4	8	3	7	2	5	6	1	9
6	2	9	8	1	3	5	4	7
5	1	7	4	6	9	2	3	8
2	6	4	3	7	1	8	9	5
7	5	1	9	4	8	3	2	6
9	3	8	2	5	6	4	7	1

Solution # 410

8	3	4	9	2	5	7	6	1
7	2	6	1	8	3	5	4	9
9	1	5	6	4	7	8	3	2
1	7	9	2	3	6	4	8	5
5	4	2	7	9	8	6	1	3
6	8	3	5	1	4	2	9	7
2	9	7	8	6	1	3	5	4
4	6	1	3	5	2	9	7	8
3	5	8	4	7	9	1	2	6

Solution # 411

2	9	1	6	4	8	7	3	5
5	4	6	3	7	2	8	1	9
7	8	3	9	1	5	4	2	6
6	1	9	5	3	4	2	8	7
3	5	2	8	9	7	6	4	1
8	7	4	1	2	6	5	9	3
1	6	7	2	8	3	9	5	4
4	3	8	7	5	9	1	6	2
9	2	5	4	6	1	3	7	8

Solution # 412

9	6	3	1	4	5	7	2	8
7	1	8	6	9	2	5	4	3
5	2	4	8	3	7	6	9	1
4	8	7	9	6	3	1	5	2
2	3	1	5	7	4	8	6	9
6	9	5	2	1	8	3	7	4
8	4	2	7	5	1	9	3	6
1	5	6	3	2	9	4	8	7
3	7	9	4	8	6	2	1	5

Solution # 413

4	5	8	3	9	6	1	7	2
9	7	2	4	5	1	8	3	6
1	3	6	7	2	8	5	9	4
7	9	3	5	6	2	4	8	1
5	6	4	1	8	7	9	2	3
2	8	1	9	3	4	6	5	7
8	1	7	2	4	9	3	6	5
3	4	9	6	7	5	2	1	8
6	2	5	8	1	3	7	4	9

Solution # 414

8	2	1	3	5	6	7	4	9
3	9	4	7	8	1	6	2	5
6	7	5	4	9	2	3	8	1
2	1	8	9	3	4	5	6	7
7	3	6	5	2	8	1	9	4
5	4	9	1	6	7	8	3	2
1	8	7	6	4	9	2	5	3
9	5	2	8	7	3	4	1	6
4	6	3	2	1	5	9	7	8

Solution # 415

5	1	2	4	6	3	8	7	9
4	7	6	8	1	9	5	2	3
9	3	8	7	2	5	1	4	6
7	6	9	1	4	8	2	3	5
3	2	1	5	9	7	4	6	8
8	5	4	6	3	2	7	9	1
2	4	5	3	8	6	9	1	7
6	9	7	2	5	1	3	8	4
1	8	3	9	7	4	6	5	2

Solution # 416

8	4	7	1	9	3	6	5	2
6	3	1	2	5	7	9	8	4
9	2	5	8	4	6	7	3	1
2	7	8	3	1	9	4	6	5
1	6	4	7	8	5	3	2	9
3	5	9	6	2	4	1	7	8
7	8	2	4	3	1	5	9	6
4	9	3	5	6	2	8	1	7
5	1	6	9	7	8	2	4	3

Solution # 417

7	8	3	4	5	2	1	6	9
5	9	1	6	7	3	8	4	2
4	6	2	8	1	9	7	5	3
9	3	7	2	8	6	4	1	5
2	1	4	7	9	5	6	3	8
8	5	6	1	3	4	9	2	7
3	7	5	9	4	1	2	8	6
6	4	9	3	2	8	5	7	1
1	2	8	5	6	7	3	9	4

Solution # 418

9	8	5	2	3	1	6	7	4
7	4	1	9	6	8	3	2	5
2	3	6	5	4	7	1	8	9
6	9	7	1	5	3	8	4	2
3	1	4	8	2	9	7	5	6
8	5	2	4	7	6	9	1	3
1	2	9	6	8	5	4	3	7
5	7	8	3	9	4	2	6	1
4	6	3	7	1	2	5	9	8

Solution # 419

4	2	8	5	3	1	9	6	7
6	3	1	9	7	2	4	5	8
5	7	9	6	4	8	2	3	1
2	5	7	1	8	6	3	4	9
3	8	6	2	9	4	1	7	5
1	9	4	7	5	3	8	2	6
8	6	3	4	1	5	7	9	2
7	4	5	8	2	9	6	1	3
9	1	2	3	6	7	5	8	4

Solution # 420

7	6	8	3	5	2	9	1	4
5	2	4	8	9	1	6	3	7
3	9	1	6	4	7	2	5	8
2	7	5	1	6	3	4	8	9
6	4	3	2	8	9	1	7	5
1	8	9	5	7	4	3	2	6
4	3	7	9	2	8	5	6	1
9	1	6	7	3	5	8	4	2
8	5	2	4	1	6	7	9	3

Solution # 421

9	4	6	5	2	8	1	7	3
8	7	3	9	6	1	5	2	4
1	5	2	4	3	7	8	9	6
2	3	9	1	7	4	6	8	5
7	6	4	8	5	2	9	3	1
5	1	8	3	9	6	2	4	7
3	9	7	6	8	5	4	1	2
6	8	1	2	4	3	7	5	9
4	2	5	7	1	9	3	6	8

Solution # 422

7	8	2	1	4	3	9	6	5
5	9	4	2	6	8	7	3	1
3	6	1	7	9	5	4	8	2
9	7	6	5	2	4	3	1	8
4	1	5	8	3	6	2	7	9
2	3	8	9	7	1	5	4	6
6	5	7	3	1	9	8	2	4
8	4	3	6	5	2	1	9	7
1	2	9	4	8	7	6	5	3

Solution # 423

9	8	6	1	2	7	4	3	5
3	7	1	5	4	6	9	2	8
4	5	2	3	8	9	1	7	6
5	6	3	2	1	4	7	8	9
2	9	8	7	3	5	6	4	1
7	1	4	9	6	8	3	5	2
8	4	7	6	9	2	5	1	3
6	3	5	8	7	1	2	9	4
1	2	9	4	5	3	8	6	7

Solution # 424

1	7	8	9	3	6	4	2	5
6	9	5	8	4	2	1	3	7
4	3	2	1	7	5	9	8	6
9	2	3	5	6	1	7	4	8
7	8	1	4	2	9	6	5	3
5	6	4	3	8	7	2	9	1
8	4	9	6	1	3	5	7	2
3	1	7	2	5	4	8	6	9
2	5	6	7	9	8	3	1	4

Solution # 425

7	3	6	8	2	1	5	9	4
8	4	2	5	3	9	7	1	6
1	9	5	6	7	4	8	3	2
9	5	8	2	6	7	1	4	3
6	1	4	3	9	5	2	8	7
2	7	3	1	4	8	9	6	5
4	8	7	9	5	6	3	2	1
3	6	9	7	1	2	4	5	8
5	2	1	4	8	3	6	7	9

Solution # 426

2	3	9	5	7	4	6	1	8
5	7	6	1	9	8	3	2	4
8	1	4	6	3	2	5	9	7
6	8	7	9	5	1	4	3	2
4	9	3	2	8	6	7	5	1
1	5	2	7	4	3	8	6	9
3	2	5	4	1	7	9	8	6
7	6	8	3	2	9	1	4	5
9	4	1	8	6	5	2	7	3

Solution # 427

3	5	1	8	9	4	2	7	6
6	2	7	1	3	5	8	9	4
9	4	8	7	2	6	3	1	5
8	3	2	5	1	9	4	6	7
4	1	9	6	7	3	5	2	8
5	7	6	2	4	8	1	3	9
7	9	3	4	5	2	6	8	1
1	6	4	3	8	7	9	5	2
2	8	5	9	6	1	7	4	3

Solution # 428

3	6	7	5	8	1	2	9	4
5	9	2	4	7	3	1	8	6
1	4	8	2	9	6	5	3	7
9	7	1	8	3	5	6	4	2
2	8	5	6	1	4	3	7	9
6	3	4	7	2	9	8	1	5
4	1	3	9	5	2	7	6	8
8	2	9	3	6	7	4	5	1
7	5	6	1	4	8	9	2	3

Solution # 429

9	7	2	4	8	3	6	5	1
1	4	8	6	5	7	2	9	3
3	5	6	2	9	1	8	7	4
6	8	7	3	1	9	4	2	5
5	2	1	8	4	6	9	3	7
4	3	9	5	7	2	1	6	8
2	1	5	9	3	8	7	4	6
7	6	3	1	2	4	5	8	9
8	9	4	7	6	5	3	1	2

Solution # 430

7	5	6	2	1	3	4	9	8
3	4	1	5	8	9	7	2	6
9	8	2	6	7	4	1	3	5
4	9	8	3	2	5	6	1	7
6	3	5	1	4	7	9	8	2
1	2	7	8	9	6	3	5	4
8	7	3	4	5	1	2	6	9
5	6	4	9	3	2	8	7	1
2	1	9	7	6	8	5	4	3

Solution # 431

5	4	2	8	9	7	6	3	1
6	9	7	3	1	4	5	2	8
3	1	8	6	2	5	7	4	9
8	5	4	7	3	2	1	9	6
1	2	9	5	4	6	8	7	3
7	6	3	1	8	9	4	5	2
4	8	1	9	7	3	2	6	5
2	3	5	4	6	8	9	1	7
9	7	6	2	5	1	3	8	4

Solution # 432

5	9	7	2	6	4	8	1	3
4	8	1	3	5	7	6	9	2
2	3	6	9	1	8	4	5	7
1	4	8	5	7	9	2	3	6
7	5	3	4	2	6	9	8	1
9	6	2	8	3	1	5	7	4
6	7	9	1	4	5	3	2	8
3	1	5	6	8	2	7	4	9
8	2	4	7	9	3	1	6	5

Solution # 433

6	7	8	9	3	2	1	4	5
4	5	1	6	7	8	3	2	9
3	2	9	4	5	1	6	8	7
7	9	3	2	4	5	8	1	6
1	4	2	8	9	6	7	5	3
5	8	6	3	1	7	2	9	4
9	6	7	1	2	4	5	3	8
8	1	4	5	6	3	9	7	2
2	3	5	7	8	9	4	6	1

Solution # 434

8	3	7	9	1	5	6	2	4
9	6	4	2	3	7	8	1	5
5	1	2	6	4	8	3	9	7
3	9	5	1	6	2	7	4	8
7	2	8	4	5	9	1	3	6
1	4	6	8	7	3	9	5	2
4	5	1	7	9	6	2	8	3
2	7	3	5	8	1	4	6	9
6	8	9	3	2	4	5	7	1

Solution # 435

5	3	6	1	2	4	8	9	7
9	2	1	7	8	3	4	6	5
8	4	7	9	5	6	1	3	2
7	8	9	3	1	5	6	2	4
1	6	2	4	7	8	9	5	3
4	5	3	2	6	9	7	1	8
3	7	5	6	4	1	2	8	9
2	1	8	5	9	7	3	4	6
6	9	4	8	3	2	5	7	1

Solution # 436

4	7	5	6	2	1	8	3	9
9	1	6	8	3	4	5	7	2
8	3	2	5	9	7	1	4	6
5	4	3	7	1	6	9	2	8
7	6	9	2	8	3	4	1	5
1	2	8	9	4	5	7	6	3
6	5	4	3	7	8	2	9	1
2	8	1	4	6	9	3	5	7
3	9	7	1	5	2	6	8	4

Solution # 437

8	6	4	5	1	2	9	3	7
1	5	9	6	3	7	2	8	4
3	7	2	9	8	4	5	6	1
5	9	1	4	6	3	7	2	8
2	8	6	7	5	1	3	4	9
7	4	3	2	9	8	1	5	6
9	1	8	3	4	5	6	7	2
4	2	5	1	7	6	8	9	3
6	3	7	8	2	9	4	1	5

Solution # 438

9	1	4	8	6	2	5	7	3
5	6	3	7	9	1	8	2	4
8	7	2	5	4	3	6	1	9
7	9	5	2	8	6	3	4	1
1	3	8	4	5	9	7	6	2
4	2	6	1	3	7	9	8	5
3	8	1	6	2	5	4	9	7
2	4	9	3	7	8	1	5	6
6	5	7	9	1	4	2	3	8

Solution # 439

1	4	6	7	2	9	8	5	3
9	8	5	3	1	6	7	2	4
2	3	7	4	5	8	9	1	6
8	1	4	9	7	3	2	6	5
3	6	2	5	8	4	1	9	7
7	5	9	1	6	2	3	4	8
4	2	1	6	3	7	5	8	9
6	7	8	2	9	5	4	3	1
5	9	3	8	4	1	6	7	2

Solution # 440

3	8	5	2	4	6	9	1	7
4	1	6	8	9	7	3	2	5
7	9	2	1	3	5	8	4	6
1	7	9	6	5	3	2	8	4
6	5	8	4	2	9	1	7	3
2	3	4	7	1	8	5	6	9
5	2	1	9	6	4	7	3	8
8	6	3	5	7	1	4	9	2
9	4	7	3	8	2	6	5	1

Solution # 441

7	3	6	4	1	8	5	9	2
2	1	4	9	5	7	8	3	6
5	8	9	6	3	2	1	7	4
9	4	1	3	7	6	2	8	5
6	2	3	8	4	5	7	1	9
8	5	7	2	9	1	4	6	3
1	9	8	5	2	3	6	4	7
3	7	2	1	6	4	9	5	8
4	6	5	7	8	9	3	2	1

Solution # 442

7	2	4	1	6	3	9	8	5
6	3	8	2	5	9	7	4	1
9	1	5	4	8	7	3	2	6
5	9	3	7	1	2	8	6	4
4	6	7	5	3	8	1	9	2
2	8	1	6	9	4	5	7	3
1	7	6	8	2	5	4	3	9
3	4	2	9	7	1	6	5	8
8	5	9	3	4	6	2	1	7

Solution # 443

2	6	4	1	8	7	5	9	3
1	9	3	6	5	2	4	8	7
5	7	8	9	3	4	1	2	6
3	2	6	5	9	8	7	4	1
8	4	5	7	6	1	9	3	2
7	1	9	2	4	3	6	5	8
4	3	7	8	1	9	2	6	5
9	5	1	3	2	6	8	7	4
6	8	2	4	7	5	3	1	9

Solution # 444

7	8	6	3	5	4	9	1	2
3	9	1	7	8	2	6	4	5
2	4	5	1	6	9	3	7	8
6	3	8	9	7	1	5	2	4
1	2	9	5	4	8	7	6	3
4	5	7	2	3	6	8	9	1
9	6	4	8	1	5	2	3	7
5	7	2	4	9	3	1	8	6
8	1	3	6	2	7	4	5	9

Solution # 445

5	1	2	8	7	4	3	9	6
8	4	6	3	9	5	7	2	1
9	3	7	6	1	2	5	8	4
4	9	1	7	3	6	2	5	8
6	2	3	5	8	1	4	7	9
7	8	5	2	4	9	1	6	3
1	6	4	9	5	7	8	3	2
3	7	9	1	2	8	6	4	5
2	5	8	4	6	3	9	1	7

Solution # 446

7	3	5	6	4	1	2	9	8
2	9	6	8	5	3	1	7	4
4	1	8	9	2	7	3	5	6
6	2	7	1	3	8	9	4	5
3	8	4	7	9	5	6	1	2
9	5	1	2	6	4	8	3	7
5	6	2	3	7	9	4	8	1
8	4	3	5	1	6	7	2	9
1	7	9	4	8	2	5	6	3

Solution # 447

7	1	4	6	9	5	2	8	3
6	9	3	8	2	7	1	4	5
8	5	2	3	4	1	7	6	9
2	4	1	5	7	8	9	3	6
9	7	6	4	3	2	5	1	8
5	3	8	1	6	9	4	2	7
1	6	7	2	5	3	8	9	4
3	8	5	9	1	4	6	7	2
4	2	9	7	8	6	3	5	1

Solution # 448

4	6	9	7	3	8	5	2	1
7	5	3	1	2	4	6	8	9
2	1	8	9	6	5	7	4	3
6	8	1	2	9	7	4	3	5
5	4	2	6	8	3	9	1	7
9	3	7	5	4	1	8	6	2
3	9	6	8	7	2	1	5	4
1	7	4	3	5	6	2	9	8
8	2	5	4	1	9	3	7	6

Solution # 449

5	9	1	7	3	8	2	6	4
7	4	8	6	2	5	9	3	1
2	6	3	1	4	9	5	7	8
1	8	6	9	5	4	7	2	3
4	5	2	3	7	1	6	8	9
9	3	7	2	8	6	1	4	5
8	2	4	5	1	7	3	9	6
6	7	5	4	9	3	8	1	2
3	1	9	8	6	2	4	5	7

Solution # 450

4	5	3	1	8	9	6	2	7
2	1	9	5	7	6	3	4	8
8	7	6	2	4	3	1	9	5
7	2	1	6	3	4	8	5	9
9	3	8	7	2	5	4	6	1
5	6	4	8	9	1	7	3	2
3	9	7	4	1	2	5	8	6
6	8	2	3	5	7	9	1	4
1	4	5	9	6	8	2	7	3

Solution # 451

2	9	8	4	7	3	6	5	1
7	3	1	6	9	5	2	4	8
6	4	5	2	1	8	9	3	7
4	7	2	5	6	9	8	1	3
3	8	6	1	2	7	4	9	5
1	5	9	3	8	4	7	6	2
8	6	3	9	5	2	1	7	4
9	2	4	7	3	1	5	8	6
5	1	7	8	4	6	3	2	9

Solution # 452

8	5	6	7	1	2	3	4	9
3	7	4	6	9	5	1	2	8
9	2	1	4	8	3	5	6	7
2	6	7	5	3	1	9	8	4
1	4	3	8	6	9	2	7	5
5	9	8	2	4	7	6	1	3
4	1	5	3	2	8	7	9	6
7	8	9	1	5	6	4	3	2
6	3	2	9	7	4	8	5	1

Solution # 453

1	5	3	7	9	4	2	6	8
4	7	6	2	8	1	3	9	5
8	9	2	5	6	3	4	7	1
7	6	8	4	3	2	5	1	9
2	1	4	9	5	6	7	8	3
9	3	5	1	7	8	6	4	2
5	8	9	6	2	7	1	3	4
6	2	1	3	4	9	8	5	7
3	4	7	8	1	5	9	2	6

Solution # 454

2	3	7	1	8	4	5	9	6
6	9	5	2	3	7	1	4	8
8	4	1	9	5	6	3	7	2
5	2	6	8	1	9	4	3	7
3	7	4	6	2	5	8	1	9
9	1	8	4	7	3	2	6	5
1	5	9	7	4	2	6	8	3
7	8	3	5	6	1	9	2	4
4	6	2	3	9	8	7	5	1

Solution # 455

4	3	9	2	6	8	7	5	1
6	5	2	3	1	7	4	9	8
1	8	7	9	5	4	3	2	6
9	4	5	6	3	2	8	1	7
8	2	3	5	7	1	9	6	4
7	6	1	8	4	9	5	3	2
3	7	4	1	9	6	2	8	5
5	1	8	4	2	3	6	7	9
2	9	6	7	8	5	1	4	3

Solution # 456

9	8	5	7	1	2	4	6	3
7	6	1	3	4	8	2	9	5
4	2	3	6	5	9	1	7	8
3	5	4	8	6	7	9	2	1
6	9	7	1	2	3	8	5	4
2	1	8	5	9	4	6	3	7
1	3	6	2	8	5	7	4	9
5	4	2	9	7	1	3	8	6
8	7	9	4	3	6	5	1	2

Solution # 457

9	7	5	6	3	2	1	4	8
1	4	8	9	5	7	2	3	6
3	6	2	1	8	4	7	9	5
5	3	6	2	1	8	4	7	9
4	2	1	7	6	9	8	5	3
8	9	7	3	4	5	6	2	1
6	5	9	4	2	1	3	8	7
7	1	4	8	9	3	5	6	2
2	8	3	5	7	6	9	1	4

Solution # 458

7	4	6	1	3	9	2	5	8
2	9	1	8	6	5	4	3	7
5	3	8	2	4	7	6	9	1
4	2	3	7	1	6	5	8	9
6	8	7	9	5	3	1	2	4
1	5	9	4	2	8	7	6	3
3	7	2	5	8	4	9	1	6
8	1	4	6	9	2	3	7	5
9	6	5	3	7	1	8	4	2

Solution # 459

4	3	6	7	2	9	5	8	1
5	2	7	8	1	4	6	3	9
1	8	9	6	5	3	7	2	4
2	5	1	4	6	8	9	7	3
7	4	3	1	9	2	8	5	6
6	9	8	5	3	7	4	1	2
3	6	4	2	8	5	1	9	7
8	7	2	9	4	1	3	6	5
9	1	5	3	7	6	2	4	8

Solution # 460

1	7	4	3	5	2	9	8	6
5	9	8	7	1	6	3	2	4
3	2	6	4	9	8	5	7	1
4	6	9	2	8	1	7	5	3
2	8	3	6	7	5	1	4	9
7	1	5	9	4	3	2	6	8
9	4	2	8	3	7	6	1	5
6	3	1	5	2	4	8	9	7
8	5	7	1	6	9	4	3	2

Solution # 461

7	3	8	5	2	9	4	6	1
1	6	2	8	4	7	3	9	5
9	5	4	6	3	1	7	2	8
3	7	6	4	5	2	8	1	9
5	4	9	7	1	8	2	3	6
8	2	1	3	9	6	5	4	7
2	9	3	1	7	5	6	8	4
4	8	7	9	6	3	1	5	2
6	1	5	2	8	4	9	7	3

Solution # 462

7	9	1	5	6	8	4	3	2
4	8	5	2	3	7	1	9	6
6	3	2	4	9	1	8	7	5
8	5	6	3	4	2	7	1	9
2	7	4	6	1	9	5	8	3
3	1	9	7	8	5	6	2	4
9	6	7	8	2	4	3	5	1
5	2	3	1	7	6	9	4	8
1	4	8	9	5	3	2	6	7

Solution # 463

1	4	9	2	3	8	5	6	7
8	7	3	4	6	5	2	1	9
6	5	2	9	1	7	3	8	4
3	8	5	1	4	9	7	2	6
2	9	4	8	7	6	1	5	3
7	6	1	5	2	3	4	9	8
5	2	6	3	9	4	8	7	1
9	3	8	7	5	1	6	4	2
4	1	7	6	8	2	9	3	5

Solution # 464

1	8	4	3	9	2	7	5	6
6	3	9	8	5	7	4	2	1
7	5	2	1	4	6	8	9	3
4	1	8	9	6	5	2	3	7
9	7	3	2	8	1	6	4	5
5	2	6	4	7	3	1	8	9
8	4	5	7	1	9	3	6	2
2	9	1	6	3	4	5	7	8
3	6	7	5	2	8	9	1	4

Solution # 465

1	9	8	6	5	4	7	2	3
4	7	6	3	2	9	5	8	1
2	3	5	7	1	8	4	6	9
7	6	9	5	4	3	2	1	8
5	2	1	8	9	6	3	4	7
3	8	4	2	7	1	6	9	5
6	4	3	9	8	7	1	5	2
9	5	7	1	6	2	8	3	4
8	1	2	4	3	5	9	7	6

Solution # 466

8	5	6	3	2	4	7	1	9
2	4	9	8	1	7	3	6	5
1	3	7	5	9	6	4	2	8
7	8	4	6	5	9	1	3	2
6	2	3	7	8	1	9	5	4
9	1	5	4	3	2	8	7	6
3	7	2	9	6	8	5	4	1
4	9	1	2	7	5	6	8	3
5	6	8	1	4	3	2	9	7

Solution # 467

9	6	4	3	2	8	5	7	1
5	8	1	7	6	9	2	4	3
3	2	7	5	4	1	9	8	6
1	5	2	4	3	6	7	9	8
8	4	6	9	5	7	1	3	2
7	3	9	1	8	2	6	5	4
4	7	3	2	1	5	8	6	9
6	1	5	8	9	4	3	2	7
2	9	8	6	7	3	4	1	5

Solution # 468

4	1	3	8	7	2	5	6	9
5	7	2	6	9	4	1	8	3
9	6	8	5	1	3	4	2	7
1	2	9	7	3	8	6	4	5
7	8	6	4	2	5	9	3	1
3	5	4	1	6	9	8	7	2
6	9	7	3	4	1	2	5	8
2	3	5	9	8	6	7	1	4
8	4	1	2	5	7	3	9	6

Solution # 469

9	3	4	8	6	2	7	5	1
7	2	1	4	5	9	3	8	6
8	6	5	7	1	3	9	2	4
3	1	9	2	4	5	6	7	8
2	4	6	1	7	8	5	3	9
5	7	8	3	9	6	4	1	2
1	5	3	6	2	4	8	9	7
6	9	2	5	8	7	1	4	3
4	8	7	9	3	1	2	6	5

Solution # 470

4	2	1	5	7	8	6	9	3
8	5	6	9	3	2	7	4	1
7	9	3	1	4	6	2	5	8
9	7	4	3	8	1	5	2	6
2	1	5	7	6	4	8	3	9
6	3	8	2	5	9	1	7	4
3	8	2	4	1	5	9	6	7
5	6	7	8	9	3	4	1	2
1	4	9	6	2	7	3	8	5

Solution # 471

2	1	5	4	8	3	6	7	9
4	7	9	6	1	2	8	5	3
3	6	8	5	9	7	2	4	1
1	5	2	7	3	4	9	8	6
6	8	4	1	2	9	7	3	5
7	9	3	8	6	5	4	1	2
9	3	7	2	5	8	1	6	4
8	2	1	3	4	6	5	9	7
5	4	6	9	7	1	3	2	8

Solution # 472

3	9	8	7	5	4	1	2	6
6	5	7	8	1	2	3	4	9
4	2	1	9	6	3	7	8	5
2	3	4	1	8	9	5	6	7
7	1	9	5	2	6	8	3	4
8	6	5	4	3	7	2	9	1
1	4	2	3	9	5	6	7	8
9	8	3	6	7	1	4	5	2
5	7	6	2	4	8	9	1	3

Solution # 473

8	7	1	6	3	5	4	2	9
2	3	6	9	4	8	1	7	5
9	4	5	7	1	2	8	6	3
1	5	4	2	7	9	6	3	8
7	8	9	4	6	3	2	5	1
3	6	2	8	5	1	7	9	4
6	1	7	3	9	4	5	8	2
5	2	3	1	8	6	9	4	7
4	9	8	5	2	7	3	1	6

Solution # 474

1	7	2	4	9	3	8	6	5
3	5	9	7	8	6	1	2	4
8	6	4	1	5	2	9	7	3
4	3	7	6	2	9	5	1	8
9	2	8	5	3	1	7	4	6
5	1	6	8	7	4	3	9	2
2	4	3	9	1	8	6	5	7
7	8	1	2	6	5	4	3	9
6	9	5	3	4	7	2	8	1

Solution # 475

5	8	1	7	2	4	6	9	3
2	7	3	1	6	9	5	4	8
4	6	9	3	5	8	2	7	1
9	1	5	4	7	3	8	2	6
3	2	8	6	9	1	4	5	7
7	4	6	5	8	2	1	3	9
8	9	7	2	4	6	3	1	5
1	5	2	8	3	7	9	6	4
6	3	4	9	1	5	7	8	2

Solution # 476

1	6	7	8	5	3	9	4	2
2	5	3	4	1	9	7	6	8
9	8	4	6	2	7	1	5	3
3	1	2	9	4	6	5	8	7
7	4	8	5	3	1	2	9	6
5	9	6	7	8	2	3	1	4
4	2	1	3	6	5	8	7	9
8	3	9	1	7	4	6	2	5
6	7	5	2	9	8	4	3	1

Solution # 477

1	6	9	3	7	8	4	2	5
2	5	3	1	6	4	7	8	9
8	7	4	5	9	2	6	1	3
7	1	6	4	8	5	3	9	2
4	9	5	6	2	3	8	7	1
3	8	2	7	1	9	5	4	6
9	3	7	8	5	1	2	6	4
6	4	1	2	3	7	9	5	8
5	2	8	9	4	6	1	3	7

Solution # 478

3	8	9	6	2	7	1	4	5
5	6	4	3	1	8	9	7	2
2	7	1	5	9	4	8	6	3
9	2	7	1	5	3	4	8	6
4	3	8	7	6	9	5	2	1
6	1	5	4	8	2	3	9	7
7	9	6	8	3	1	2	5	4
1	4	2	9	7	5	6	3	8
8	5	3	2	4	6	7	1	9

Solution # 479

5	1	7	4	9	8	2	6	3
9	4	3	2	6	5	7	8	1
8	2	6	3	7	1	9	4	5
6	3	2	5	4	7	8	1	9
1	9	4	8	2	6	5	3	7
7	5	8	9	1	3	6	2	4
2	7	5	1	8	4	3	9	6
4	6	9	7	3	2	1	5	8
3	8	1	6	5	9	4	7	2

Solution # 480

2	6	7	8	9	5	4	3	1
8	4	3	2	7	1	6	5	9
1	9	5	4	6	3	8	2	7
4	3	9	7	2	6	1	8	5
6	8	1	5	3	9	7	4	2
7	5	2	1	8	4	9	6	3
9	1	4	3	5	8	2	7	6
5	2	8	6	1	7	3	9	4
3	7	6	9	4	2	5	1	8

Solution # 481

7	5	2	1	4	6	9	3	8
3	9	4	7	5	8	2	6	1
6	8	1	2	3	9	4	7	5
9	6	8	4	2	7	1	5	3
4	2	5	8	1	3	7	9	6
1	7	3	6	9	5	8	4	2
5	1	6	9	8	4	3	2	7
2	4	7	3	6	1	5	8	9
8	3	9	5	7	2	6	1	4

Solution # 482

5	8	1	2	3	7	4	9	6
9	6	2	4	8	5	1	7	3
4	3	7	9	1	6	2	8	5
3	9	5	8	7	2	6	4	1
1	7	6	3	4	9	5	2	8
8	2	4	6	5	1	7	3	9
7	4	9	5	6	8	3	1	2
6	1	8	7	2	3	9	5	4
2	5	3	1	9	4	8	6	7

Solution # 483

5	3	2	4	6	7	9	8	1
9	6	8	1	2	5	4	3	7
1	4	7	9	8	3	5	6	2
2	8	1	3	7	4	6	9	5
4	7	9	2	5	6	8	1	3
6	5	3	8	1	9	7	2	4
7	9	5	6	3	2	1	4	8
3	1	4	7	9	8	2	5	6
8	2	6	5	4	1	3	7	9

Solution # 484

7	8	5	4	1	6	2	9	3
3	6	2	7	5	9	8	1	4
9	1	4	3	2	8	5	7	6
8	5	6	9	3	1	7	4	2
4	9	7	8	6	2	1	3	5
1	2	3	5	4	7	6	8	9
2	4	9	1	8	5	3	6	7
6	3	1	2	7	4	9	5	8
5	7	8	6	9	3	4	2	1

Solution # 485

9	7	1	5	2	6	4	3	8
8	2	4	3	1	9	5	7	6
5	6	3	4	7	8	1	9	2
7	9	2	6	5	4	3	8	1
3	8	5	1	9	2	6	4	7
4	1	6	8	3	7	2	5	9
2	5	7	9	4	1	8	6	3
1	4	8	7	6	3	9	2	5
6	3	9	2	8	5	7	1	4

Solution # 486

2	3	6	8	7	5	4	9	1
9	8	5	1	2	4	7	6	3
1	7	4	9	6	3	8	2	5
7	5	1	6	4	2	3	8	9
6	9	2	7	3	8	1	5	4
8	4	3	5	1	9	6	7	2
3	1	8	2	9	6	5	4	7
5	2	7	4	8	1	9	3	6
4	6	9	3	5	7	2	1	8

Solution # 487

1	3	7	9	5	2	8	4	6
6	5	2	4	1	8	9	7	3
9	4	8	6	7	3	2	1	5
8	6	9	3	4	5	1	2	7
4	2	5	7	6	1	3	9	8
3	7	1	2	8	9	5	6	4
2	1	6	8	3	7	4	5	9
5	8	4	1	9	6	7	3	2
7	9	3	5	2	4	6	8	1

Solution # 488

1	4	6	9	3	7	2	8	5
7	9	8	2	6	5	4	1	3
5	3	2	8	1	4	7	6	9
2	7	3	1	5	6	8	9	4
6	8	1	3	4	9	5	7	2
4	5	9	7	2	8	6	3	1
9	2	4	6	8	1	3	5	7
3	6	7	5	9	2	1	4	8
8	1	5	4	7	3	9	2	6

Solution # 489

5	7	2	8	9	4	1	3	6
3	1	4	2	7	6	8	5	9
6	9	8	1	3	5	7	4	2
7	4	1	9	5	3	2	6	8
2	8	5	4	6	1	3	9	7
9	3	6	7	8	2	4	1	5
4	6	7	3	2	9	5	8	1
8	5	3	6	1	7	9	2	4
1	2	9	5	4	8	6	7	3

Solution # 490

7	4	8	3	5	1	2	9	6
2	9	6	4	7	8	3	5	1
5	3	1	6	2	9	8	7	4
9	5	3	2	1	7	4	6	8
4	6	7	8	9	3	1	2	5
8	1	2	5	6	4	7	3	9
6	8	4	7	3	5	9	1	2
1	7	5	9	8	2	6	4	3
3	2	9	1	4	6	5	8	7

Solution # 491

5	3	2	8	7	4	6	9	1
4	1	7	3	6	9	8	2	5
8	6	9	1	2	5	7	4	3
1	2	3	6	5	7	4	8	9
7	4	8	9	3	2	5	1	6
6	9	5	4	1	8	3	7	2
2	7	6	5	8	1	9	3	4
3	8	4	2	9	6	1	5	7
9	5	1	7	4	3	2	6	8

Solution # 492

5	8	1	6	7	9	4	2	3
7	2	9	8	3	4	5	6	1
6	3	4	5	2	1	9	7	8
3	1	7	9	6	2	8	5	4
8	4	6	3	1	5	2	9	7
9	5	2	7	4	8	1	3	6
1	9	3	4	5	7	6	8	2
4	7	5	2	8	6	3	1	9
2	6	8	1	9	3	7	4	5

Solution # 493

7	5	6	2	4	1	3	8	9
9	8	4	3	5	7	2	6	1
1	2	3	8	9	6	4	7	5
3	1	2	5	6	8	7	9	4
5	9	7	4	3	2	8	1	6
6	4	8	7	1	9	5	3	2
2	3	9	6	8	5	1	4	7
4	6	5	1	7	3	9	2	8
8	7	1	9	2	4	6	5	3

Solution # 494

1	5	8	2	7	6	9	4	3
3	4	2	9	1	8	5	7	6
7	9	6	3	4	5	1	8	2
6	1	4	7	8	2	3	9	5
2	7	3	5	6	9	4	1	8
9	8	5	1	3	4	2	6	7
8	6	9	4	2	3	7	5	1
4	3	7	8	5	1	6	2	9
5	2	1	6	9	7	8	3	4

Solution # 495

4	5	2	1	8	7	3	6	9
3	1	7	9	2	6	5	8	4
9	6	8	3	4	5	2	7	1
8	2	5	7	1	9	6	4	3
1	3	6	4	5	8	9	2	7
7	9	4	6	3	2	1	5	8
6	4	9	2	7	3	8	1	5
2	8	1	5	9	4	7	3	6
5	7	3	8	6	1	4	9	2

Solution # 496

1	3	4	8	2	5	7	6	9
8	7	5	9	1	6	4	2	3
6	2	9	7	4	3	5	1	8
3	5	7	6	9	2	8	4	1
9	6	1	4	5	8	2	3	7
4	8	2	3	7	1	6	9	5
7	4	6	5	3	9	1	8	2
2	9	8	1	6	7	3	5	4
5	1	3	2	8	4	9	7	6

Solution # 497

9	1	3	8	6	4	2	7	5
7	4	6	5	9	2	1	3	8
5	8	2	7	3	1	4	9	6
3	2	1	4	5	7	6	8	9
6	5	8	9	2	3	7	4	1
4	7	9	1	8	6	5	2	3
8	9	7	6	4	5	3	1	2
1	3	5	2	7	8	9	6	4
2	6	4	3	1	9	8	5	7

Solution # 498

3	6	4	7	5	9	2	1	8
9	8	1	6	4	2	5	7	3
2	5	7	3	1	8	4	9	6
1	7	5	9	3	6	8	4	2
8	3	6	4	2	7	9	5	1
4	2	9	1	8	5	3	6	7
7	4	3	2	9	1	6	8	5
6	9	8	5	7	3	1	2	4
5	1	2	8	6	4	7	3	9

Solution # 499

9	5	2	3	1	4	8	6	7
6	4	1	9	7	8	3	2	5
8	7	3	2	5	6	1	9	4
4	1	8	5	9	2	6	7	3
2	3	9	4	6	7	5	1	8
7	6	5	8	3	1	9	4	2
3	8	7	1	2	9	4	5	6
1	2	4	6	8	5	7	3	9
5	9	6	7	4	3	2	8	1

Solution # 500

9	2	6	5	4	8	7	1	3
5	8	4	1	3	7	2	9	6
1	3	7	9	6	2	8	4	5
8	5	1	3	2	9	4	6	7
7	4	9	6	8	5	3	2	1
3	6	2	7	1	4	5	8	9
4	9	3	8	7	6	1	5	2
6	1	8	2	5	3	9	7	4
2	7	5	4	9	1	6	3	8

Solution # 501

4	2	1	6	9	8	5	3	7
8	9	3	5	4	7	6	2	1
7	5	6	3	1	2	9	4	8
6	3	2	9	5	1	7	8	4
9	7	4	8	6	3	2	1	5
1	8	5	2	7	4	3	9	6
3	6	9	1	8	5	4	7	2
2	4	8	7	3	6	1	5	9
5	1	7	4	2	9	8	6	3

Solution # 502

4	5	9	2	3	7	8	6	1
8	3	6	9	1	4	7	2	5
7	2	1	6	8	5	4	3	9
5	4	2	7	9	3	1	8	6
3	1	8	5	6	2	9	7	4
6	9	7	1	4	8	3	5	2
9	7	4	3	2	6	5	1	8
1	6	5	8	7	9	2	4	3
2	8	3	4	5	1	6	9	7

Solution # 503

1	2	7	6	5	8	3	9	4
6	3	4	1	9	2	7	5	8
5	9	8	3	4	7	6	2	1
4	6	5	8	2	1	9	3	7
9	7	2	5	3	4	1	8	6
3	8	1	7	6	9	2	4	5
8	5	6	9	7	3	4	1	2
7	4	3	2	1	5	8	6	9
2	1	9	4	8	6	5	7	3

Solution # 504

4	8	7	2	9	5	3	1	6
2	3	6	4	1	7	5	9	8
1	5	9	8	6	3	7	2	4
9	7	5	3	4	2	6	8	1
6	4	8	1	7	9	2	3	5
3	2	1	5	8	6	4	7	9
7	6	3	9	5	1	8	4	2
5	9	4	7	2	8	1	6	3
8	1	2	6	3	4	9	5	7

Solution # 505

9	5	2	3	6	1	8	4	7
1	3	4	7	8	5	6	2	9
7	8	6	9	2	4	5	1	3
6	4	3	8	7	9	2	5	1
8	9	1	4	5	2	3	7	6
2	7	5	1	3	6	4	9	8
3	1	8	2	4	7	9	6	5
4	6	9	5	1	8	7	3	2
5	2	7	6	9	3	1	8	4

Solution # 506

9	5	2	4	7	3	1	6	8
8	3	7	6	9	1	5	2	4
1	4	6	8	5	2	7	3	9
3	7	5	1	4	9	2	8	6
4	2	1	3	6	8	9	7	5
6	8	9	7	2	5	3	4	1
2	6	8	9	1	7	4	5	3
7	1	3	5	8	4	6	9	2
5	9	4	2	3	6	8	1	7

Solution # 507

2	6	7	5	3	9	8	1	4
8	9	4	7	1	2	6	5	3
5	3	1	8	4	6	2	9	7
6	4	2	9	7	1	3	8	5
9	7	3	4	8	5	1	6	2
1	8	5	6	2	3	7	4	9
4	1	6	3	9	7	5	2	8
3	2	9	1	5	8	4	7	6
7	5	8	2	6	4	9	3	1

Solution # 508

5	4	2	7	3	9	6	8	1
8	6	3	4	2	1	9	7	5
7	9	1	8	5	6	4	3	2
6	2	5	3	7	4	1	9	8
1	8	4	9	6	2	7	5	3
3	7	9	1	8	5	2	4	6
9	5	8	6	1	7	3	2	4
4	3	6	2	9	8	5	1	7
2	1	7	5	4	3	8	6	9

Solution # 509

1	5	4	6	8	9	7	3	2
3	9	2	4	7	1	8	6	5
6	7	8	2	5	3	1	9	4
8	1	6	3	4	7	5	2	9
9	4	7	8	2	5	6	1	3
2	3	5	1	9	6	4	7	8
5	6	3	9	1	4	2	8	7
7	8	1	5	3	2	9	4	6
4	2	9	7	6	8	3	5	1

Solution # 510

4	5	9	6	2	7	3	1	8
7	2	3	1	8	4	5	6	9
8	6	1	9	3	5	7	2	4
3	1	7	4	5	2	8	9	6
6	8	4	7	9	3	2	5	1
5	9	2	8	6	1	4	3	7
9	7	5	3	1	8	6	4	2
2	4	6	5	7	9	1	8	3
1	3	8	2	4	6	9	7	5

Solution # 511

4	6	1	5	7	9	2	8	3
9	7	5	3	2	8	1	6	4
2	3	8	4	6	1	9	7	5
5	9	4	8	1	2	6	3	7
1	8	6	7	9	3	4	5	2
3	2	7	6	5	4	8	9	1
6	5	9	1	4	7	3	2	8
8	1	2	9	3	5	7	4	6
7	4	3	2	8	6	5	1	9

Solution # 512

9	2	6	3	4	5	1	7	8
8	5	7	1	9	2	4	6	3
3	1	4	7	6	8	5	9	2
5	9	1	8	2	4	7	3	6
2	7	8	5	3	6	9	4	1
4	6	3	9	1	7	2	8	5
6	3	2	4	5	9	8	1	7
1	8	9	2	7	3	6	5	4
7	4	5	6	8	1	3	2	9

Solution # 513

5	3	4	6	2	8	7	1	9
7	8	9	1	3	4	6	5	2
1	2	6	5	9	7	8	4	3
3	7	5	8	6	1	2	9	4
4	1	2	9	7	5	3	6	8
9	6	8	3	4	2	1	7	5
2	5	3	4	1	6	9	8	7
6	4	7	2	8	9	5	3	1
8	9	1	7	5	3	4	2	6

Solution # 514

9	8	4	3	1	2	5	7	6
6	3	1	4	7	5	9	8	2
7	2	5	8	9	6	4	3	1
2	6	9	7	8	3	1	4	5
4	5	3	9	6	1	8	2	7
1	7	8	5	2	4	6	9	3
5	9	2	6	4	7	3	1	8
3	4	7	1	5	8	2	6	9
8	1	6	2	3	9	7	5	4

Solution # 515

1	6	5	4	3	2	8	9	7
2	8	9	6	1	7	4	3	5
3	4	7	8	9	5	2	6	1
6	5	2	3	8	1	7	4	9
7	9	1	5	4	6	3	2	8
8	3	4	7	2	9	1	5	6
4	7	3	9	6	8	5	1	2
5	1	6	2	7	4	9	8	3
9	2	8	1	5	3	6	7	4

Solution # 516

7	8	9	4	1	6	3	2	5
6	4	2	5	8	3	7	9	1
3	5	1	7	9	2	6	8	4
9	3	8	1	4	7	2	5	6
5	2	7	6	3	9	4	1	8
4	1	6	2	5	8	9	3	7
1	7	3	9	6	5	8	4	2
8	6	5	3	2	4	1	7	9
2	9	4	8	7	1	5	6	3

Solution # 517

7	8	1	4	5	6	9	2	3
3	4	2	7	9	8	1	5	6
9	5	6	2	1	3	4	8	7
8	1	3	5	4	2	6	7	9
4	2	5	9	6	7	3	1	8
6	7	9	3	8	1	5	4	2
2	6	4	8	3	5	7	9	1
1	9	7	6	2	4	8	3	5
5	3	8	1	7	9	2	6	4

Solution # 518

3	7	4	1	6	9	5	2	8
9	2	5	8	4	3	7	1	6
8	1	6	2	7	5	3	4	9
1	4	8	7	3	6	9	5	2
5	3	2	9	1	8	6	7	4
7	6	9	4	5	2	8	3	1
6	8	7	3	2	4	1	9	5
2	5	1	6	9	7	4	8	3
4	9	3	5	8	1	2	6	7

Solution # 519

2	5	7	8	3	9	4	6	1
6	1	8	2	7	4	9	3	5
9	4	3	6	5	1	8	2	7
4	8	2	5	9	3	1	7	6
1	3	9	7	6	2	5	8	4
5	7	6	1	4	8	3	9	2
3	9	5	4	2	6	7	1	8
8	2	4	9	1	7	6	5	3
7	6	1	3	8	5	2	4	9

Solution # 520

3	5	1	9	2	4	7	8	6
2	6	4	7	8	3	5	1	9
8	9	7	6	5	1	3	4	2
9	3	8	4	1	6	2	7	5
1	2	6	3	7	5	4	9	8
7	4	5	8	9	2	1	6	3
6	1	2	5	4	9	8	3	7
5	8	9	1	3	7	6	2	4
4	7	3	2	6	8	9	5	1

Solution # 521

4	9	5	3	8	2	1	6	7
1	8	2	6	7	4	3	5	9
6	7	3	9	5	1	8	4	2
7	1	6	2	4	8	9	3	5
9	2	8	5	3	7	4	1	6
5	3	4	1	6	9	2	7	8
2	6	9	4	1	5	7	8	3
8	5	1	7	9	3	6	2	4
3	4	7	8	2	6	5	9	1

Solution # 522

8	3	2	9	1	7	4	5	6
9	4	7	2	6	5	1	3	8
5	6	1	4	8	3	2	9	7
6	8	3	7	4	1	9	2	5
7	5	4	3	9	2	6	8	1
1	2	9	8	5	6	7	4	3
4	1	6	5	3	9	8	7	2
2	9	5	1	7	8	3	6	4
3	7	8	6	2	4	5	1	9

Solution # 523

9	5	3	4	1	6	7	2	8
6	1	8	2	7	3	9	4	5
4	2	7	5	9	8	6	3	1
5	4	1	8	6	7	3	9	2
7	6	9	3	2	1	8	5	4
3	8	2	9	5	4	1	7	6
1	3	6	7	4	2	5	8	9
2	7	5	1	8	9	4	6	3
8	9	4	6	3	5	2	1	7

Solution # 524

4	2	8	7	3	1	9	6	5
3	7	5	9	2	6	4	8	1
6	1	9	5	8	4	3	2	7
8	3	1	4	9	5	6	7	2
9	5	2	6	7	3	8	1	4
7	6	4	8	1	2	5	9	3
5	8	7	2	4	9	1	3	6
2	4	3	1	6	8	7	5	9
1	9	6	3	5	7	2	4	8

Solution # 525

7	9	6	3	1	2	4	8	5
8	1	4	6	7	5	3	9	2
3	2	5	8	4	9	1	6	7
2	5	7	4	6	8	9	1	3
1	6	9	7	5	3	2	4	8
4	8	3	9	2	1	5	7	6
6	7	1	2	3	4	8	5	9
9	4	2	5	8	6	7	3	1
5	3	8	1	9	7	6	2	4

Solution # 526

1	2	9	3	6	7	8	4	5
3	5	4	1	8	2	6	7	9
8	6	7	9	5	4	3	2	1
6	7	2	4	1	3	9	5	8
5	9	1	8	2	6	4	3	7
4	8	3	7	9	5	2	1	6
2	4	5	6	7	8	1	9	3
7	1	6	2	3	9	5	8	4
9	3	8	5	4	1	7	6	2

Solution # 527

2	1	5	3	9	6	4	7	8
4	6	9	2	7	8	1	5	3
7	8	3	4	1	5	6	2	9
1	3	2	5	6	7	8	9	4
8	5	6	1	4	9	7	3	2
9	7	4	8	2	3	5	6	1
5	9	1	6	8	2	3	4	7
6	4	7	9	3	1	2	8	5
3	2	8	7	5	4	9	1	6

Solution # 528

5	9	1	2	3	6	4	7	8
8	2	6	5	7	4	3	1	9
3	4	7	8	9	1	5	2	6
7	5	9	3	6	2	1	8	4
2	1	3	9	4	8	6	5	7
6	8	4	7	1	5	2	9	3
4	6	8	1	5	9	7	3	2
9	3	5	6	2	7	8	4	1
1	7	2	4	8	3	9	6	5

Solution # 529

6	2	5	9	1	8	3	4	7
7	9	4	5	3	2	8	1	6
3	1	8	4	6	7	5	9	2
9	6	2	8	5	4	1	7	3
1	5	7	3	2	6	4	8	9
4	8	3	1	7	9	2	6	5
8	3	6	2	9	1	7	5	4
5	4	9	7	8	3	6	2	1
2	7	1	6	4	5	9	3	8

Solution # 530

9	2	4	7	6	1	3	8	5
6	1	8	2	3	5	7	4	9
5	7	3	4	9	8	1	6	2
1	3	6	8	2	9	5	7	4
4	8	2	1	5	7	6	9	3
7	9	5	6	4	3	8	2	1
3	6	9	5	7	2	4	1	8
8	5	7	9	1	4	2	3	6
2	4	1	3	8	6	9	5	7

Solution # 531

2	1	7	9	5	8	6	4	3
4	9	5	6	3	1	2	7	8
3	8	6	2	7	4	1	9	5
5	3	4	1	6	9	7	8	2
9	6	8	5	2	7	4	3	1
1	7	2	8	4	3	9	5	6
8	5	9	4	1	6	3	2	7
6	2	3	7	9	5	8	1	4
7	4	1	3	8	2	5	6	9

Solution # 532

5	4	9	1	3	8	7	6	2
8	7	6	4	2	5	1	3	9
2	1	3	6	7	9	8	4	5
4	6	5	7	1	3	2	9	8
9	8	1	2	5	4	3	7	6
3	2	7	9	8	6	4	5	1
7	9	4	8	6	2	5	1	3
1	5	2	3	9	7	6	8	4
6	3	8	5	4	1	9	2	7

Solution # 533

3	5	1	8	7	6	9	2	4
6	8	4	9	1	2	7	5	3
2	9	7	3	5	4	1	6	8
1	7	2	5	9	8	3	4	6
4	3	8	7	6	1	5	9	2
5	6	9	4	2	3	8	1	7
8	1	5	2	4	7	6	3	9
7	4	6	1	3	9	2	8	5
9	2	3	6	8	5	4	7	1

Solution # 534

4	9	1	2	7	5	3	8	6
8	6	3	1	9	4	7	5	2
5	7	2	8	6	3	9	1	4
2	3	7	6	5	9	8	4	1
9	1	8	4	3	2	6	7	5
6	4	5	7	8	1	2	3	9
1	8	4	9	2	7	5	6	3
7	5	9	3	4	6	1	2	8
3	2	6	5	1	8	4	9	7

Solution # 535

8	6	4	3	1	9	5	2	7
7	3	1	4	2	5	9	8	6
9	5	2	8	7	6	4	3	1
1	4	3	6	5	8	7	9	2
5	7	9	2	3	1	6	4	8
2	8	6	7	9	4	3	1	5
4	9	5	1	6	2	8	7	3
6	2	7	9	8	3	1	5	4
3	1	8	5	4	7	2	6	9

Solution # 536

5	3	4	8	7	2	6	9	1
2	1	9	3	4	6	5	8	7
6	8	7	5	1	9	4	3	2
7	5	2	1	8	4	9	6	3
9	4	1	6	5	3	2	7	8
3	6	8	2	9	7	1	5	4
8	7	6	4	2	5	3	1	9
4	9	5	7	3	1	8	2	6
1	2	3	9	6	8	7	4	5

Solution # 537

8	5	1	9	4	2	6	7	3
6	7	9	5	1	3	4	8	2
4	3	2	7	6	8	9	1	5
3	4	6	1	2	7	5	9	8
2	8	7	3	9	5	1	6	4
1	9	5	6	8	4	3	2	7
9	2	8	4	5	1	7	3	6
5	1	3	2	7	6	8	4	9
7	6	4	8	3	9	2	5	1

Solution # 538

6	8	5	4	7	9	3	1	2
4	2	3	6	8	1	9	7	5
1	9	7	5	2	3	6	8	4
2	5	6	8	3	4	1	9	7
8	4	1	7	9	5	2	3	6
3	7	9	1	6	2	4	5	8
9	3	8	2	4	7	5	6	1
7	1	2	3	5	6	8	4	9
5	6	4	9	1	8	7	2	3

Solution # 539

9	3	6	5	2	4	8	1	7
1	5	2	6	8	7	4	3	9
4	8	7	9	1	3	5	6	2
2	9	5	8	3	1	7	4	6
8	6	3	7	4	9	1	2	5
7	4	1	2	6	5	9	8	3
6	2	9	4	7	8	3	5	1
5	1	8	3	9	2	6	7	4
3	7	4	1	5	6	2	9	8

Solution # 540

1	4	3	7	9	6	5	2	8
8	7	9	2	5	3	6	4	1
6	5	2	4	8	1	7	9	3
4	2	1	5	7	9	8	3	6
3	6	7	8	1	4	9	5	2
9	8	5	3	6	2	4	1	7
7	1	8	9	3	5	2	6	4
5	3	4	6	2	8	1	7	9
2	9	6	1	4	7	3	8	5

Solution # 541

5	2	8	9	6	3	4	1	7
7	1	6	2	8	4	5	3	9
9	3	4	7	5	1	2	6	8
1	9	3	6	4	7	8	2	5
8	6	2	3	9	5	1	7	4
4	7	5	8	1	2	3	9	6
3	8	7	5	2	6	9	4	1
2	4	9	1	7	8	6	5	3
6	5	1	4	3	9	7	8	2

Solution # 542

7	2	8	5	1	3	4	6	9
1	6	9	4	8	2	5	7	3
3	4	5	6	7	9	1	2	8
5	3	2	1	6	8	9	4	7
8	7	1	9	4	5	6	3	2
6	9	4	2	3	7	8	5	1
9	8	3	7	5	6	2	1	4
2	1	6	3	9	4	7	8	5
4	5	7	8	2	1	3	9	6

Solution # 543

8	6	7	3	9	5	4	2	1
3	2	1	8	4	6	7	5	9
9	4	5	7	2	1	8	3	6
6	7	4	1	5	2	3	9	8
1	8	2	6	3	9	5	7	4
5	3	9	4	8	7	1	6	2
7	1	3	2	6	8	9	4	5
2	5	8	9	7	4	6	1	3
4	9	6	5	1	3	2	8	7

Solution # 544

4	3	2	6	7	1	8	5	9
8	7	5	9	2	4	6	3	1
1	9	6	3	8	5	7	2	4
9	1	3	5	6	2	4	8	7
2	6	4	8	3	7	9	1	5
7	5	8	4	1	9	3	6	2
3	4	7	1	5	8	2	9	6
5	8	9	2	4	6	1	7	3
6	2	1	7	9	3	5	4	8

Solution # 545

3	8	4	7	2	5	6	1	9
1	7	5	9	6	3	4	8	2
2	9	6	4	1	8	3	7	5
9	1	8	5	3	6	2	4	7
7	4	2	8	9	1	5	3	6
6	5	3	2	4	7	8	9	1
4	6	7	3	5	9	1	2	8
5	3	9	1	8	2	7	6	4
8	2	1	6	7	4	9	5	3

Solution # 546

8	1	3	4	6	7	5	2	9
4	9	5	2	3	8	7	6	1
7	2	6	1	9	5	8	4	3
9	5	2	7	8	3	6	1	4
1	8	7	6	4	9	3	5	2
6	3	4	5	2	1	9	7	8
3	7	1	8	5	2	4	9	6
2	4	8	9	7	6	1	3	5
5	6	9	3	1	4	2	8	7

Solution # 547

6	7	8	3	2	1	9	5	4
5	1	2	9	4	8	6	7	3
3	4	9	5	6	7	8	1	2
9	8	6	7	1	4	2	3	5
4	5	1	6	3	2	7	9	8
7	2	3	8	5	9	1	4	6
8	9	4	2	7	5	3	6	1
2	3	5	1	9	6	4	8	7
1	6	7	4	8	3	5	2	9

Solution # 548

2	9	4	1	5	8	6	7	3
5	1	3	9	7	6	2	4	8
8	7	6	4	2	3	9	5	1
3	6	7	8	4	2	5	1	9
1	8	5	7	6	9	4	3	2
4	2	9	5	3	1	8	6	7
9	4	8	6	1	7	3	2	5
7	5	2	3	8	4	1	9	6
6	3	1	2	9	5	7	8	4

Solution # 549

1	6	8	2	4	9	3	7	5
4	3	5	7	8	6	2	9	1
7	9	2	1	3	5	4	6	8
9	7	1	5	2	8	6	4	3
2	8	6	3	1	4	7	5	9
5	4	3	6	9	7	1	8	2
6	1	4	9	5	3	8	2	7
3	5	7	8	6	2	9	1	4
8	2	9	4	7	1	5	3	6

Solution # 550

1	6	8	2	7	9	5	4	3
9	2	5	1	3	4	8	7	6
3	7	4	6	8	5	2	9	1
6	4	7	5	9	3	1	8	2
8	5	1	4	2	7	6	3	9
2	3	9	8	1	6	4	5	7
5	1	3	9	4	2	7	6	8
7	8	6	3	5	1	9	2	4
4	9	2	7	6	8	3	1	5

Solution # 551

1	3	9	6	2	4	7	5	8
7	6	2	8	9	5	1	4	3
4	5	8	7	3	1	9	2	6
9	8	1	4	5	6	3	7	2
6	2	7	3	1	8	4	9	5
3	4	5	9	7	2	8	6	1
2	9	6	1	4	3	5	8	7
5	1	4	2	8	7	6	3	9
8	7	3	5	6	9	2	1	4

Solution # 552

4	5	2	7	6	1	3	9	8
3	6	7	2	9	8	5	1	4
9	1	8	3	5	4	6	7	2
1	8	3	4	7	9	2	5	6
5	9	6	8	1	2	7	4	3
7	2	4	6	3	5	9	8	1
2	4	5	9	8	6	1	3	7
6	3	1	5	4	7	8	2	9
8	7	9	1	2	3	4	6	5

Solution # 553

6	1	2	4	8	5	3	7	9
8	9	7	2	6	3	4	1	5
4	5	3	9	1	7	2	8	6
1	2	8	5	9	4	6	3	7
7	3	5	8	2	6	9	4	1
9	4	6	3	7	1	8	5	2
2	7	9	1	4	8	5	6	3
3	8	1	6	5	2	7	9	4
5	6	4	7	3	9	1	2	8

Solution # 554

4	3	1	9	5	8	2	7	6
6	7	8	3	1	2	4	5	9
5	2	9	7	4	6	8	3	1
3	4	5	1	6	9	7	2	8
2	9	7	4	8	5	6	1	3
8	1	6	2	7	3	9	4	5
9	5	3	8	2	4	1	6	7
7	8	4	6	3	1	5	9	2
1	6	2	5	9	7	3	8	4

Solution # 555

8	5	4	6	2	7	1	3	9
7	3	6	9	1	8	2	4	5
2	1	9	3	5	4	7	8	6
3	9	8	2	4	5	6	7	1
5	4	7	1	6	9	8	2	3
1	6	2	7	8	3	9	5	4
6	7	1	5	3	2	4	9	8
9	8	3	4	7	1	5	6	2
4	2	5	8	9	6	3	1	7

Solution # 556

3	1	9	7	2	6	8	4	5
8	4	2	3	1	5	9	6	7
5	6	7	8	9	4	2	1	3
7	5	8	6	4	2	3	9	1
1	2	4	9	5	3	7	8	6
9	3	6	1	7	8	4	5	2
4	8	1	2	6	7	5	3	9
2	9	3	5	8	1	6	7	4
6	7	5	4	3	9	1	2	8

Solution # 557

9	6	3	1	8	7	5	4	2
2	4	1	5	9	3	6	8	7
8	5	7	2	4	6	9	3	1
7	9	4	8	3	1	2	5	6
1	8	2	6	7	5	4	9	3
5	3	6	9	2	4	1	7	8
3	7	5	4	1	2	8	6	9
4	2	8	3	6	9	7	1	5
6	1	9	7	5	8	3	2	4

Solution # 558

5	7	1	9	3	4	6	8	2
4	9	6	2	5	8	7	3	1
8	3	2	6	1	7	9	4	5
7	1	8	4	9	3	5	2	6
2	6	4	7	8	5	3	1	9
9	5	3	1	2	6	8	7	4
6	8	9	3	4	2	1	5	7
3	4	7	5	6	1	2	9	8
1	2	5	8	7	9	4	6	3

Solution # 559

1	6	5	2	7	4	8	9	3
3	7	9	8	6	1	2	4	5
4	2	8	5	3	9	7	1	6
6	9	1	4	5	8	3	7	2
2	4	3	6	9	7	1	5	8
8	5	7	3	1	2	4	6	9
7	1	6	9	2	3	5	8	4
9	8	2	1	4	5	6	3	7
5	3	4	7	8	6	9	2	1

Solution # 560

8	5	9	4	2	6	3	1	7
4	3	2	8	7	1	5	9	6
7	6	1	3	5	9	4	8	2
5	9	6	1	8	3	2	7	4
3	1	4	7	6	2	8	5	9
2	8	7	5	9	4	6	3	1
6	4	3	9	1	8	7	2	5
9	7	8	2	4	5	1	6	3
1	2	5	6	3	7	9	4	8

Solution # 561

7	1	9	3	5	4	6	8	2
5	3	2	6	8	7	4	1	9
4	6	8	1	9	2	3	5	7
2	5	4	8	6	3	9	7	1
1	8	7	9	4	5	2	3	6
3	9	6	2	7	1	5	4	8
9	4	5	7	1	6	8	2	3
8	2	1	5	3	9	7	6	4
6	7	3	4	2	8	1	9	5

Solution # 562

4	5	9	8	2	6	7	1	3
2	7	3	4	5	1	8	9	6
1	6	8	9	3	7	5	4	2
5	3	1	7	8	9	6	2	4
9	8	2	6	4	5	3	7	1
6	4	7	3	1	2	9	5	8
3	9	4	2	7	8	1	6	5
8	1	6	5	9	4	2	3	7
7	2	5	1	6	3	4	8	9

Solution # 563

2	3	5	1	7	9	6	8	4
4	8	6	3	5	2	7	9	1
1	7	9	6	4	8	5	3	2
5	1	2	4	6	3	8	7	9
7	9	8	2	1	5	4	6	3
3	6	4	8	9	7	1	2	5
8	2	1	7	3	4	9	5	6
6	5	3	9	8	1	2	4	7
9	4	7	5	2	6	3	1	8

Solution # 564

2	1	6	9	4	3	5	8	7
7	4	9	2	5	8	1	3	6
5	3	8	1	7	6	4	9	2
8	9	7	6	1	2	3	4	5
3	2	5	7	9	4	8	6	1
4	6	1	8	3	5	2	7	9
1	8	2	3	6	9	7	5	4
6	5	3	4	2	7	9	1	8
9	7	4	5	8	1	6	2	3

Solution # 565

4	9	1	3	6	2	7	5	8
7	5	3	9	1	8	6	4	2
6	8	2	5	7	4	3	1	9
8	6	7	1	2	5	9	3	4
2	4	5	8	3	9	1	7	6
1	3	9	7	4	6	8	2	5
3	2	4	6	9	1	5	8	7
9	1	8	2	5	7	4	6	3
5	7	6	4	8	3	2	9	1

Solution # 566

6	7	4	3	8	1	5	9	2
3	9	5	7	2	4	6	1	8
1	2	8	9	6	5	4	7	3
8	1	6	5	3	2	9	4	7
9	5	3	6	4	7	8	2	1
7	4	2	1	9	8	3	5	6
4	3	9	2	1	6	7	8	5
5	8	1	4	7	3	2	6	9
2	6	7	8	5	9	1	3	4

Solution # 567

8	6	1	2	9	7	3	5	4
2	9	7	4	3	5	1	8	6
3	5	4	8	6	1	2	7	9
9	4	3	5	8	6	7	1	2
1	8	5	7	4	2	9	6	3
6	7	2	3	1	9	5	4	8
4	2	9	1	5	8	6	3	7
5	3	6	9	7	4	8	2	1
7	1	8	6	2	3	4	9	5

Solution # 568

7	8	5	3	6	9	4	2	1
4	9	1	2	7	8	5	3	6
3	2	6	4	1	5	8	9	7
6	4	2	1	5	3	9	7	8
9	1	3	7	8	4	6	5	2
5	7	8	6	9	2	3	1	4
1	6	9	5	4	7	2	8	3
8	3	7	9	2	6	1	4	5
2	5	4	8	3	1	7	6	9

Solution # 569

3	7	8	6	2	9	4	5	1
5	4	6	8	3	1	2	9	7
2	1	9	4	5	7	6	3	8
1	2	3	5	8	6	9	7	4
7	9	5	1	4	2	8	6	3
6	8	4	7	9	3	5	1	2
8	6	7	9	1	4	3	2	5
9	5	2	3	7	8	1	4	6
4	3	1	2	6	5	7	8	9

Solution # 570

8	3	6	2	7	1	9	5	4
9	4	2	5	8	6	3	1	7
5	7	1	3	9	4	2	8	6
3	5	9	8	4	2	6	7	1
6	8	7	9	1	5	4	2	3
2	1	4	6	3	7	5	9	8
7	6	3	1	2	9	8	4	5
4	9	5	7	6	8	1	3	2
1	2	8	4	5	3	7	6	9

Solution # 571

9	4	7	6	3	1	5	8	2
5	8	2	4	9	7	6	3	1
3	1	6	8	2	5	4	9	7
7	6	4	3	1	8	2	5	9
2	3	9	7	5	4	8	1	6
1	5	8	9	6	2	7	4	3
6	7	1	5	4	9	3	2	8
8	9	5	2	7	3	1	6	4
4	2	3	1	8	6	9	7	5

Solution # 572

1	2	3	5	6	9	8	7	4
7	5	8	4	2	3	1	9	6
4	9	6	1	7	8	2	5	3
3	7	2	9	8	6	5	4	1
9	4	1	3	5	2	6	8	7
8	6	5	7	4	1	3	2	9
5	8	7	6	3	4	9	1	2
2	3	9	8	1	7	4	6	5
6	1	4	2	9	5	7	3	8

Solution # 573

6	1	5	7	8	4	9	3	2
8	7	9	3	2	5	1	6	4
3	2	4	9	6	1	7	5	8
2	3	6	5	9	7	4	8	1
7	9	1	8	4	3	6	2	5
5	4	8	6	1	2	3	7	9
9	8	3	4	5	6	2	1	7
4	6	2	1	7	8	5	9	3
1	5	7	2	3	9	8	4	6

Solution # 574

8	1	3	6	7	4	9	2	5
2	9	6	5	1	3	4	8	7
7	5	4	9	2	8	1	3	6
4	2	8	1	3	7	5	6	9
1	3	5	2	6	9	8	7	4
9	6	7	4	8	5	2	1	3
6	8	9	3	5	1	7	4	2
5	7	2	8	4	6	3	9	1
3	4	1	7	9	2	6	5	8

Solution # 575

5	9	7	4	3	1	6	2	8
6	2	8	9	5	7	1	3	4
1	4	3	2	8	6	7	5	9
3	5	2	1	4	8	9	6	7
4	8	6	7	9	3	2	1	5
9	7	1	5	6	2	8	4	3
2	1	5	8	7	4	3	9	6
8	3	9	6	1	5	4	7	2
7	6	4	3	2	9	5	8	1

Solution # 576

9	8	3	5	6	1	7	4	2
4	7	6	2	8	3	5	9	1
1	5	2	9	4	7	8	3	6
7	1	4	3	5	9	6	2	8
5	6	9	7	2	8	3	1	4
3	2	8	4	1	6	9	7	5
2	4	7	6	3	5	1	8	9
8	3	5	1	9	4	2	6	7
6	9	1	8	7	2	4	5	3

Solution # 577

2	4	7	9	5	8	1	3	6
9	6	8	1	3	4	7	2	5
5	1	3	6	7	2	4	9	8
4	9	6	5	8	1	2	7	3
7	3	5	2	4	9	6	8	1
8	2	1	3	6	7	5	4	9
1	8	9	4	2	6	3	5	7
6	5	4	7	9	3	8	1	2
3	7	2	8	1	5	9	6	4

Solution # 578

1	5	8	6	7	2	9	3	4
4	7	3	9	1	8	2	6	5
6	2	9	5	4	3	7	8	1
8	3	7	4	9	5	6	1	2
2	9	4	1	8	6	5	7	3
5	6	1	2	3	7	8	4	9
7	8	2	3	5	1	4	9	6
9	1	6	8	2	4	3	5	7
3	4	5	7	6	9	1	2	8

Solution # 579

1	8	3	9	5	2	7	4	6
9	2	6	7	8	4	5	3	1
7	5	4	1	6	3	9	8	2
8	1	9	2	3	6	4	5	7
4	7	5	8	1	9	2	6	3
3	6	2	4	7	5	1	9	8
2	3	7	5	9	8	6	1	4
6	9	1	3	4	7	8	2	5
5	4	8	6	2	1	3	7	9

Solution # 580

7	1	6	5	2	9	8	4	3
4	9	3	8	7	1	5	6	2
2	5	8	6	3	4	9	7	1
3	6	4	7	8	5	1	2	9
5	2	9	1	4	3	7	8	6
8	7	1	2	9	6	4	3	5
1	4	2	9	6	8	3	5	7
9	8	7	3	5	2	6	1	4
6	3	5	4	1	7	2	9	8

Solution # 581

9	3	4	2	1	8	7	5	6
2	6	5	4	7	3	8	9	1
8	1	7	6	5	9	2	3	4
7	4	8	9	3	6	1	2	5
6	5	2	1	8	4	9	7	3
1	9	3	7	2	5	4	6	8
4	2	1	3	6	7	5	8	9
3	8	9	5	4	2	6	1	7
5	7	6	8	9	1	3	4	2

Solution # 582

8	9	4	3	6	1	7	5	2
3	2	1	7	4	5	8	9	6
6	5	7	2	9	8	1	3	4
1	6	5	9	2	3	4	7	8
9	3	2	4	8	7	5	6	1
4	7	8	1	5	6	3	2	9
5	1	9	6	7	4	2	8	3
7	4	6	8	3	2	9	1	5
2	8	3	5	1	9	6	4	7

Solution # 583

9	4	3	8	7	2	5	1	6
6	1	2	4	5	9	8	3	7
8	7	5	1	6	3	2	9	4
3	5	4	2	8	6	1	7	9
2	6	1	9	3	7	4	5	8
7	9	8	5	1	4	3	6	2
1	3	9	7	2	8	6	4	5
5	2	7	6	4	1	9	8	3
4	8	6	3	9	5	7	2	1

Solution # 584

2	7	3	6	4	1	5	8	9
6	1	5	8	7	9	3	4	2
8	9	4	5	2	3	1	6	7
5	8	2	1	6	4	7	9	3
3	6	9	7	8	2	4	5	1
7	4	1	3	9	5	6	2	8
1	2	7	9	5	6	8	3	4
9	3	6	4	1	8	2	7	5
4	5	8	2	3	7	9	1	6

Solution # 585

6	4	2	9	1	7	5	8	3
1	8	7	3	4	5	2	9	6
5	3	9	6	2	8	7	4	1
3	6	4	1	5	9	8	7	2
7	2	8	4	6	3	9	1	5
9	5	1	8	7	2	6	3	4
2	9	5	7	3	4	1	6	8
8	1	3	5	9	6	4	2	7
4	7	6	2	8	1	3	5	9

Solution # 586

6	7	1	4	5	2	8	9	3
2	4	9	8	3	6	1	7	5
5	3	8	9	1	7	2	6	4
7	2	6	1	9	4	3	5	8
8	1	4	3	7	5	9	2	6
9	5	3	6	2	8	7	4	1
4	9	2	5	8	1	6	3	7
3	8	5	7	6	9	4	1	2
1	6	7	2	4	3	5	8	9

Solution # 587

1	8	4	9	6	7	3	5	2
5	2	3	8	4	1	7	6	9
9	7	6	3	5	2	1	4	8
7	9	2	6	8	5	4	3	1
8	4	5	1	3	9	2	7	6
6	3	1	7	2	4	8	9	5
3	5	8	4	1	6	9	2	7
4	6	9	2	7	8	5	1	3
2	1	7	5	9	3	6	8	4

Solution # 588

3	6	5	1	9	4	8	7	2
4	8	9	3	2	7	5	6	1
1	2	7	6	5	8	4	3	9
8	9	4	2	7	1	6	5	3
2	7	3	8	6	5	1	9	4
5	1	6	4	3	9	2	8	7
6	4	8	7	1	3	9	2	5
9	3	2	5	4	6	7	1	8
7	5	1	9	8	2	3	4	6

Solution # 589

9	3	5	7	1	2	6	8	4
8	6	1	5	3	4	7	2	9
7	2	4	6	8	9	3	5	1
5	9	6	2	7	1	8	4	3
3	4	7	9	5	8	1	6	2
1	8	2	4	6	3	5	9	7
2	1	3	8	4	6	9	7	5
6	5	9	3	2	7	4	1	8
4	7	8	1	9	5	2	3	6

Solution # 590

4	9	5	7	3	2	8	6	1
6	3	8	4	1	5	9	2	7
7	2	1	8	6	9	5	4	3
1	4	9	6	5	3	7	8	2
8	5	7	2	9	4	1	3	6
3	6	2	1	7	8	4	9	5
2	1	6	9	4	7	3	5	8
9	8	3	5	2	1	6	7	4
5	7	4	3	8	6	2	1	9

Solution # 591

7	5	6	1	8	4	2	3	9
8	9	2	5	7	3	4	6	1
1	3	4	2	9	6	7	5	8
3	8	7	9	4	2	5	1	6
2	1	9	6	5	8	3	7	4
6	4	5	7	3	1	9	8	2
9	6	1	3	2	5	8	4	7
5	7	8	4	6	9	1	2	3
4	2	3	8	1	7	6	9	5

Solution # 592

2	1	7	6	8	9	3	5	4
5	6	3	2	1	4	8	9	7
8	9	4	5	3	7	6	2	1
9	3	1	4	2	5	7	8	6
7	5	8	1	9	6	2	4	3
4	2	6	3	7	8	9	1	5
1	7	2	8	4	3	5	6	9
6	8	9	7	5	1	4	3	2
3	4	5	9	6	2	1	7	8

Solution # 593

6	2	8	1	7	5	9	4	3
4	7	3	6	9	8	2	1	5
9	1	5	4	3	2	6	7	8
1	8	2	5	6	7	4	3	9
3	9	7	2	4	1	8	5	6
5	6	4	9	8	3	1	2	7
8	5	6	7	1	4	3	9	2
2	4	9	3	5	6	7	8	1
7	3	1	8	2	9	5	6	4

Solution # 594

9	7	2	4	5	3	8	6	1
5	1	4	6	9	8	3	7	2
3	8	6	7	1	2	4	5	9
8	4	1	9	6	7	5	2	3
6	5	9	2	3	4	1	8	7
7	2	3	5	8	1	9	4	6
1	9	5	8	7	6	2	3	4
4	3	7	1	2	5	6	9	8
2	6	8	3	4	9	7	1	5

Solution # 595

4	2	7	5	8	1	9	3	6
3	9	5	2	4	6	8	7	1
6	8	1	7	9	3	4	5	2
2	7	3	6	1	9	5	4	8
1	6	8	4	3	5	7	2	9
9	5	4	8	2	7	6	1	3
8	4	9	3	5	2	1	6	7
7	1	2	9	6	4	3	8	5
5	3	6	1	7	8	2	9	4

Solution # 596

1	3	7	6	2	9	4	8	5
4	6	2	7	5	8	3	9	1
8	5	9	1	3	4	6	7	2
3	2	4	8	1	7	5	6	9
9	1	8	4	6	5	2	3	7
6	7	5	2	9	3	1	4	8
7	9	3	5	4	2	8	1	6
2	8	6	3	7	1	9	5	4
5	4	1	9	8	6	7	2	3

Solution # 597

7	8	1	6	3	2	4	5	9
4	9	6	5	8	1	7	3	2
3	2	5	4	7	9	6	8	1
1	3	2	9	5	6	8	4	7
9	4	8	3	1	7	2	6	5
5	6	7	8	2	4	9	1	3
8	5	9	7	6	3	1	2	4
2	7	3	1	4	8	5	9	6
6	1	4	2	9	5	3	7	8

Solution # 598

9	5	4	3	1	6	7	8	2
7	1	6	2	9	8	3	5	4
8	3	2	5	4	7	6	1	9
6	7	8	1	3	2	9	4	5
5	2	1	4	6	9	8	3	7
3	4	9	7	8	5	2	6	1
4	9	5	6	7	3	1	2	8
1	6	7	8	2	4	5	9	3
2	8	3	9	5	1	4	7	6

Solution # 599

6	7	3	8	9	1	2	4	5
4	8	5	6	2	7	9	3	1
2	9	1	3	4	5	8	7	6
3	5	7	4	1	2	6	8	9
8	2	9	5	7	6	4	1	3
1	6	4	9	3	8	7	5	2
7	1	6	2	5	4	3	9	8
9	4	8	1	6	3	5	2	7
5	3	2	7	8	9	1	6	4

Solution # 600

8	1	6	7	5	9	2	3	4
3	4	9	1	8	2	6	7	5
2	7	5	3	6	4	8	1	9
1	2	4	6	9	7	3	5	8
5	6	7	4	3	8	1	9	2
9	3	8	2	1	5	4	6	7
6	9	2	5	4	1	7	8	3
7	8	3	9	2	6	5	4	1
4	5	1	8	7	3	9	2	6

Solution # 601

6	8	2	7	1	3	5	4	9
4	1	9	2	5	6	3	8	7
7	3	5	4	8	9	2	1	6
1	7	4	9	3	2	8	6	5
9	2	6	5	4	8	7	3	1
8	5	3	6	7	1	4	9	2
3	6	7	1	2	4	9	5	8
2	4	1	8	9	5	6	7	3
5	9	8	3	6	7	1	2	4

Solution # 602

5	9	3	1	6	2	4	7	8
6	8	1	4	7	3	5	2	9
2	4	7	5	9	8	6	3	1
4	2	9	7	5	1	8	6	3
7	3	8	9	2	6	1	4	5
1	6	5	8	3	4	7	9	2
9	5	6	2	1	7	3	8	4
3	1	4	6	8	9	2	5	7
8	7	2	3	4	5	9	1	6

Solution # 603

6	3	4	7	1	8	9	2	5
1	9	8	2	5	3	4	6	7
5	7	2	4	6	9	1	8	3
4	6	9	5	8	7	2	3	1
7	1	3	9	2	6	8	5	4
8	2	5	1	3	4	7	9	6
9	5	7	6	4	2	3	1	8
3	4	1	8	9	5	6	7	2
2	8	6	3	7	1	5	4	9

Solution # 604

5	6	9	3	1	4	7	8	2
2	4	1	9	8	7	6	5	3
3	8	7	6	2	5	4	1	9
7	1	8	5	9	2	3	4	6
9	5	4	7	6	3	8	2	1
6	2	3	8	4	1	9	7	5
8	7	5	1	3	9	2	6	4
1	9	2	4	7	6	5	3	8
4	3	6	2	5	8	1	9	7

Solution # 605

1	4	6	9	2	7	8	5	3
5	7	9	3	1	8	4	2	6
8	2	3	5	6	4	1	7	9
4	6	1	2	8	9	7	3	5
3	9	8	7	4	5	6	1	2
7	5	2	6	3	1	9	8	4
6	1	5	4	7	2	3	9	8
2	3	7	8	9	6	5	4	1
9	8	4	1	5	3	2	6	7

Solution # 606

9	2	8	5	3	1	6	4	7
6	1	3	4	7	2	5	9	8
4	7	5	9	6	8	1	3	2
1	9	7	3	4	5	8	2	6
5	4	6	8	2	9	7	1	3
8	3	2	7	1	6	9	5	4
2	8	9	6	5	3	4	7	1
3	5	4	1	8	7	2	6	9
7	6	1	2	9	4	3	8	5

Solution # 607

9	2	6	3	8	7	5	1	4
7	5	3	1	6	4	8	9	2
8	4	1	2	5	9	6	7	3
3	8	4	6	1	5	9	2	7
5	9	2	7	4	8	3	6	1
1	6	7	9	2	3	4	8	5
6	3	5	8	7	1	2	4	9
4	7	8	5	9	2	1	3	6
2	1	9	4	3	6	7	5	8

Solution # 608

6	3	2	8	5	7	9	1	4
8	7	4	3	1	9	2	5	6
9	1	5	6	4	2	8	7	3
2	8	6	4	7	5	3	9	1
5	4	1	9	3	6	7	2	8
3	9	7	2	8	1	6	4	5
1	2	3	7	6	4	5	8	9
4	6	9	5	2	8	1	3	7
7	5	8	1	9	3	4	6	2

Solution # 609

9	6	8	2	4	5	7	1	3
4	1	3	6	9	7	8	5	2
7	5	2	1	8	3	6	9	4
8	3	5	9	6	1	4	2	7
2	7	1	8	5	4	9	3	6
6	9	4	3	7	2	5	8	1
3	4	9	5	2	6	1	7	8
5	2	6	7	1	8	3	4	9
1	8	7	4	3	9	2	6	5

Solution # 610

9	8	5	3	2	1	6	4	7
7	4	2	8	6	5	3	1	9
1	3	6	7	9	4	2	8	5
3	6	8	1	5	7	4	9	2
4	1	7	9	3	2	5	6	8
5	2	9	4	8	6	1	7	3
6	7	3	2	1	9	8	5	4
2	9	1	5	4	8	7	3	6
8	5	4	6	7	3	9	2	1

Solution # 611

2	6	4	1	5	9	3	7	8
1	7	3	2	4	8	9	6	5
5	9	8	7	3	6	1	2	4
7	3	1	4	6	5	8	9	2
9	4	2	8	1	7	6	5	3
8	5	6	3	9	2	7	4	1
3	1	5	9	7	4	2	8	6
6	2	9	5	8	1	4	3	7
4	8	7	6	2	3	5	1	9

Solution # 612

4	8	1	2	9	5	3	7	6
5	6	3	7	8	4	1	9	2
7	9	2	3	1	6	8	4	5
8	7	5	6	2	1	4	3	9
9	1	4	8	5	3	2	6	7
2	3	6	4	7	9	5	1	8
6	5	8	1	3	7	9	2	4
3	4	9	5	6	2	7	8	1
1	2	7	9	4	8	6	5	3

Solution # 613

3	4	1	9	8	2	5	6	7
6	9	5	1	7	3	4	8	2
2	8	7	5	6	4	1	9	3
7	1	6	8	2	9	3	4	5
8	5	4	6	3	1	7	2	9
9	3	2	7	4	5	8	1	6
5	6	8	2	1	7	9	3	4
1	7	3	4	9	6	2	5	8
4	2	9	3	5	8	6	7	1

Solution # 614

2	5	3	1	8	9	6	4	7
1	7	8	6	3	4	9	5	2
6	9	4	7	5	2	1	8	3
7	8	2	5	9	6	4	3	1
4	1	9	3	2	7	8	6	5
3	6	5	4	1	8	2	7	9
5	3	6	9	4	1	7	2	8
9	2	7	8	6	5	3	1	4
8	4	1	2	7	3	5	9	6

Solution # 615

5	2	4	1	3	9	7	6	8
8	3	9	7	6	4	1	5	2
7	6	1	8	2	5	4	9	3
1	7	6	9	8	2	5	3	4
3	4	8	5	7	6	2	1	9
9	5	2	4	1	3	6	8	7
4	9	7	6	5	8	3	2	1
2	8	5	3	4	1	9	7	6
6	1	3	2	9	7	8	4	5

Solution # 616

7	9	2	6	3	4	1	8	5
4	8	6	2	5	1	3	7	9
1	3	5	9	7	8	4	6	2
2	6	3	8	1	7	9	5	4
9	5	7	4	2	6	8	3	1
8	4	1	3	9	5	6	2	7
6	7	8	1	4	2	5	9	3
3	2	4	5	6	9	7	1	8
5	1	9	7	8	3	2	4	6

Solution # 617

7	1	5	6	9	8	2	3	4
6	2	8	7	3	4	9	1	5
3	4	9	1	5	2	8	7	6
1	9	6	2	7	5	4	8	3
5	7	4	9	8	3	6	2	1
8	3	2	4	1	6	7	5	9
4	5	1	8	6	7	3	9	2
9	6	7	3	2	1	5	4	8
2	8	3	5	4	9	1	6	7

Solution # 618

5	1	7	6	9	8	2	4	3
2	8	6	7	3	4	9	5	1
3	9	4	5	2	1	7	6	8
9	7	8	2	1	6	5	3	4
4	6	5	9	8	3	1	7	2
1	3	2	4	7	5	6	8	9
7	4	1	8	6	9	3	2	5
6	5	9	3	4	2	8	1	7
8	2	3	1	5	7	4	9	6

Solution # 619

2	6	8	4	3	7	5	1	9
4	3	9	6	5	1	7	2	8
5	1	7	9	8	2	6	4	3
8	5	2	1	7	3	9	6	4
1	9	6	2	4	5	8	3	7
7	4	3	8	9	6	1	5	2
6	2	4	7	1	8	3	9	5
3	8	1	5	2	9	4	7	6
9	7	5	3	6	4	2	8	1

Solution # 620

8	2	3	5	4	1	6	9	7
6	5	9	2	7	8	3	4	1
7	4	1	6	3	9	2	8	5
5	1	2	8	9	7	4	3	6
4	9	6	3	1	2	5	7	8
3	7	8	4	5	6	1	2	9
2	8	5	7	6	4	9	1	3
1	6	4	9	8	3	7	5	2
9	3	7	1	2	5	8	6	4

Solution # 621

8	9	1	2	6	3	5	4	7
4	7	2	1	9	5	6	8	3
5	3	6	8	4	7	9	1	2
7	5	4	3	2	8	1	9	6
3	1	8	6	5	9	7	2	4
2	6	9	4	7	1	3	5	8
1	8	5	7	3	2	4	6	9
9	4	3	5	8	6	2	7	1
6	2	7	9	1	4	8	3	5

Solution # 622

2	5	3	9	8	4	1	7	6
9	7	4	1	5	6	8	2	3
1	8	6	3	7	2	9	4	5
6	1	9	5	2	7	4	3	8
7	4	2	8	3	9	5	6	1
8	3	5	4	6	1	2	9	7
4	2	7	6	1	5	3	8	9
3	6	1	2	9	8	7	5	4
5	9	8	7	4	3	6	1	2

Solution # 623

6	9	7	2	1	5	3	4	8
3	2	5	4	8	9	6	1	7
8	1	4	3	7	6	9	5	2
4	7	8	6	2	1	5	3	9
9	6	2	5	3	4	7	8	1
1	5	3	8	9	7	4	2	6
2	4	6	7	5	8	1	9	3
7	8	1	9	4	3	2	6	5
5	3	9	1	6	2	8	7	4

Solution # 624

7	9	4	1	5	6	2	8	3
1	8	5	3	2	4	6	9	7
2	6	3	7	8	9	1	4	5
3	7	9	2	1	5	8	6	4
6	4	1	8	9	7	5	3	2
8	5	2	4	6	3	7	1	9
9	1	8	5	4	2	3	7	6
4	2	7	6	3	8	9	5	1
5	3	6	9	7	1	4	2	8

Solution # 625

4	1	3	7	6	5	2	8	9
2	8	6	9	1	3	4	5	7
9	5	7	2	4	8	6	3	1
3	2	1	6	9	4	5	7	8
6	7	5	3	8	2	9	1	4
8	4	9	1	5	7	3	2	6
7	6	8	5	2	9	1	4	3
5	9	4	8	3	1	7	6	2
1	3	2	4	7	6	8	9	5

Solution # 626

3	1	9	5	8	4	2	6	7
7	2	5	6	9	3	4	8	1
4	8	6	7	2	1	3	5	9
2	5	1	4	3	6	7	9	8
6	3	8	1	7	9	5	2	4
9	7	4	2	5	8	6	1	3
5	9	2	3	1	7	8	4	6
8	6	3	9	4	5	1	7	2
1	4	7	8	6	2	9	3	5

Solution # 627

6	5	4	1	8	7	9	2	3
3	9	1	5	4	2	6	7	8
7	8	2	3	9	6	5	1	4
9	4	6	7	3	1	8	5	2
2	3	7	6	5	8	4	9	1
5	1	8	4	2	9	3	6	7
1	2	9	8	6	4	7	3	5
8	6	3	2	7	5	1	4	9
4	7	5	9	1	3	2	8	6

Solution # 628

7	2	6	3	8	1	4	9	5
8	5	1	4	2	9	3	7	6
9	4	3	7	5	6	1	2	8
4	3	8	6	9	7	5	1	2
2	6	9	8	1	5	7	3	4
1	7	5	2	4	3	8	6	9
5	1	2	9	3	4	6	8	7
3	8	7	5	6	2	9	4	1
6	9	4	1	7	8	2	5	3

Solution # 629

8	3	1	4	6	9	7	2	5
2	4	6	5	3	7	1	9	8
9	7	5	1	2	8	3	4	6
1	6	3	7	4	2	8	5	9
4	9	7	6	8	5	2	3	1
5	8	2	3	9	1	4	6	7
7	2	9	8	5	3	6	1	4
3	1	4	9	7	6	5	8	2
6	5	8	2	1	4	9	7	3

Solution # 630

3	7	8	2	4	1	5	9	6
2	9	5	7	3	6	1	8	4
1	4	6	9	5	8	3	2	7
7	2	4	6	1	5	8	3	9
5	1	3	8	9	4	6	7	2
8	6	9	3	7	2	4	5	1
9	5	7	4	6	3	2	1	8
4	8	1	5	2	9	7	6	3
6	3	2	1	8	7	9	4	5

Solution # 631

1	5	8	2	3	4	7	6	9
2	7	9	1	8	6	4	5	3
4	6	3	5	9	7	8	1	2
5	8	1	7	2	3	9	4	6
9	2	6	4	1	8	5	3	7
7	3	4	6	5	9	2	8	1
8	9	5	3	7	1	6	2	4
3	4	2	9	6	5	1	7	8
6	1	7	8	4	2	3	9	5

Solution # 632

8	1	2	7	6	4	9	5	3
7	9	5	2	1	3	4	6	8
6	4	3	8	9	5	7	1	2
4	6	7	9	8	2	1	3	5
3	2	1	5	4	6	8	7	9
5	8	9	1	3	7	6	2	4
9	3	8	6	2	1	5	4	7
2	7	6	4	5	9	3	8	1
1	5	4	3	7	8	2	9	6

Solution # 633

3	7	9	8	6	2	1	5	4
2	8	1	5	7	4	9	6	3
6	5	4	3	9	1	8	7	2
9	6	5	4	2	3	7	1	8
4	1	3	7	8	6	2	9	5
8	2	7	1	5	9	3	4	6
1	3	6	2	4	7	5	8	9
5	4	2	9	1	8	6	3	7
7	9	8	6	3	5	4	2	1

Solution # 634

6	5	7	2	9	3	1	8	4
4	9	1	6	5	8	3	2	7
2	3	8	7	1	4	5	6	9
7	4	5	1	8	2	9	3	6
3	8	6	9	4	7	2	5	1
9	1	2	3	6	5	7	4	8
1	2	9	8	3	6	4	7	5
5	6	3	4	7	9	8	1	2
8	7	4	5	2	1	6	9	3

Solution # 635

3	4	9	6	5	2	7	1	8
8	5	7	1	3	9	6	4	2
1	2	6	7	4	8	9	5	3
6	7	3	9	1	5	8	2	4
5	9	2	4	8	6	3	7	1
4	1	8	2	7	3	5	6	9
9	6	5	8	2	4	1	3	7
2	3	1	5	9	7	4	8	6
7	8	4	3	6	1	2	9	5

Solution # 636

6	3	9	5	4	8	7	1	2
4	2	5	1	7	6	8	3	9
1	7	8	3	9	2	5	4	6
2	8	6	4	3	5	9	7	1
9	4	3	7	8	1	6	2	5
5	1	7	2	6	9	4	8	3
7	5	2	6	1	4	3	9	8
8	6	4	9	2	3	1	5	7
3	9	1	8	5	7	2	6	4

Solution # 637

5	4	1	9	8	6	3	7	2
2	9	3	5	4	7	8	1	6
7	6	8	3	2	1	4	5	9
3	7	6	8	1	9	5	2	4
4	8	2	7	3	5	6	9	1
9	1	5	2	6	4	7	8	3
1	2	4	6	5	8	9	3	7
8	3	7	4	9	2	1	6	5
6	5	9	1	7	3	2	4	8

Solution # 638

7	5	1	3	4	8	2	6	9
8	6	2	9	7	1	5	3	4
4	3	9	6	5	2	1	7	8
9	1	6	8	3	4	7	5	2
3	8	5	2	6	7	9	4	1
2	7	4	1	9	5	3	8	6
1	4	3	7	8	9	6	2	5
6	9	8	5	2	3	4	1	7
5	2	7	4	1	6	8	9	3

Solution # 639

3	9	6	2	4	1	8	5	7
4	2	7	9	5	8	3	1	6
8	5	1	7	6	3	9	4	2
1	6	3	8	9	7	5	2	4
2	7	8	5	1	4	6	3	9
5	4	9	6	3	2	7	8	1
7	1	2	3	8	9	4	6	5
9	3	5	4	2	6	1	7	8
6	8	4	1	7	5	2	9	3

Solution # 640

4	7	3	1	6	8	9	5	2
2	8	1	5	3	9	7	4	6
6	9	5	7	2	4	1	3	8
9	1	6	3	8	5	2	7	4
7	5	4	6	1	2	3	8	9
8	3	2	4	9	7	6	1	5
1	4	7	9	5	6	8	2	3
5	2	9	8	7	3	4	6	1
3	6	8	2	4	1	5	9	7

Solution # 641

2	7	5	1	6	4	8	9	3
9	3	6	2	8	7	4	1	5
8	1	4	9	5	3	7	2	6
6	2	1	4	9	5	3	8	7
7	8	3	6	2	1	5	4	9
4	5	9	7	3	8	1	6	2
3	4	2	5	1	9	6	7	8
5	9	7	8	4	6	2	3	1
1	6	8	3	7	2	9	5	4

Solution # 642

4	7	2	3	8	5	1	6	9
8	1	9	4	6	7	2	3	5
6	5	3	9	1	2	7	4	8
1	9	6	5	7	3	4	8	2
2	3	5	6	4	8	9	1	7
7	4	8	2	9	1	3	5	6
3	6	4	8	2	9	5	7	1
5	2	1	7	3	6	8	9	4
9	8	7	1	5	4	6	2	3

Solution # 643

6	1	9	2	3	5	7	8	4
8	2	3	7	6	4	9	1	5
7	5	4	8	9	1	6	3	2
9	7	2	6	1	8	4	5	3
4	3	1	5	7	9	8	2	6
5	8	6	3	4	2	1	7	9
2	9	8	1	5	6	3	4	7
3	6	5	4	8	7	2	9	1
1	4	7	9	2	3	5	6	8

Solution # 644

8	5	3	1	6	9	7	4	2
9	6	2	4	8	7	3	5	1
7	1	4	3	5	2	9	6	8
5	3	1	9	2	6	8	7	4
6	4	7	5	3	8	2	1	9
2	9	8	7	4	1	5	3	6
3	7	6	2	9	4	1	8	5
1	8	9	6	7	5	4	2	3
4	2	5	8	1	3	6	9	7

Solution # 645

3	2	1	5	9	4	7	8	6
4	7	6	1	2	8	5	3	9
8	9	5	3	7	6	4	2	1
9	5	4	2	6	3	1	7	8
7	3	8	4	1	9	6	5	2
1	6	2	7	8	5	3	9	4
6	1	9	8	3	7	2	4	5
5	8	7	6	4	2	9	1	3
2	4	3	9	5	1	8	6	7

Solution # 646

9	6	3	5	1	4	7	8	2
2	5	7	9	8	6	4	3	1
4	8	1	3	2	7	9	5	6
1	7	2	6	3	8	5	9	4
6	3	4	7	9	5	2	1	8
8	9	5	2	4	1	3	6	7
3	1	8	4	7	9	6	2	5
7	2	6	1	5	3	8	4	9
5	4	9	8	6	2	1	7	3

Solution # 647

5	7	1	4	8	6	9	2	3
9	4	3	2	5	1	8	7	6
6	8	2	7	9	3	4	5	1
7	6	8	1	3	9	2	4	5
1	3	4	5	2	8	7	6	9
2	9	5	6	7	4	3	1	8
3	2	7	8	1	5	6	9	4
8	5	6	9	4	2	1	3	7
4	1	9	3	6	7	5	8	2

Solution # 648

1	4	3	6	7	5	2	8	9
9	8	5	4	2	1	3	7	6
6	7	2	8	3	9	4	5	1
8	3	4	2	5	6	9	1	7
7	5	9	1	8	3	6	2	4
2	6	1	7	9	4	5	3	8
5	9	7	3	4	8	1	6	2
4	2	6	5	1	7	8	9	3
3	1	8	9	6	2	7	4	5

Solution # 649

3	8	4	5	7	9	2	6	1
6	2	9	8	3	1	4	5	7
5	7	1	4	6	2	8	9	3
7	5	6	9	4	8	1	3	2
8	4	2	3	1	5	6	7	9
9	1	3	6	2	7	5	8	4
4	3	5	1	9	6	7	2	8
2	9	8	7	5	4	3	1	6
1	6	7	2	8	3	9	4	5

Solution # 650

1	5	3	2	7	4	9	6	8
6	8	4	9	3	1	5	7	2
2	9	7	6	5	8	4	1	3
8	7	6	5	2	3	1	9	4
4	3	5	1	6	9	2	8	7
9	1	2	8	4	7	6	3	5
5	2	9	7	8	6	3	4	1
3	6	8	4	1	5	7	2	9
7	4	1	3	9	2	8	5	6

Solution # 651

1	7	8	6	2	5	9	3	4
4	3	5	1	7	9	2	6	8
9	6	2	4	8	3	5	7	1
2	8	7	5	4	6	3	1	9
3	5	1	2	9	7	4	8	6
6	9	4	8	3	1	7	2	5
5	4	3	7	6	8	1	9	2
8	2	9	3	1	4	6	5	7
7	1	6	9	5	2	8	4	3

Solution # 652

8	9	1	5	6	4	2	7	3
3	5	6	8	2	7	9	1	4
7	2	4	1	3	9	8	6	5
2	6	9	3	5	8	1	4	7
5	7	8	4	1	2	6	3	9
4	1	3	7	9	6	5	8	2
1	4	7	9	8	5	3	2	6
9	8	2	6	4	3	7	5	1
6	3	5	2	7	1	4	9	8

Solution # 653

3	1	9	2	6	5	4	7	8
5	7	4	9	1	8	6	3	2
6	8	2	4	3	7	1	5	9
8	3	1	5	7	9	2	6	4
9	2	5	6	4	3	8	1	7
4	6	7	1	8	2	3	9	5
1	9	3	8	5	4	7	2	6
7	5	8	3	2	6	9	4	1
2	4	6	7	9	1	5	8	3

Solution # 654

6	1	8	3	7	4	5	9	2
3	2	5	9	6	8	1	4	7
7	4	9	1	5	2	3	8	6
8	7	2	5	4	9	6	3	1
5	6	4	7	1	3	9	2	8
1	9	3	8	2	6	7	5	4
2	5	7	4	9	1	8	6	3
9	3	6	2	8	7	4	1	5
4	8	1	6	3	5	2	7	9

Solution # 655

7	1	8	6	2	5	9	4	3
9	4	5	7	8	3	6	2	1
2	6	3	4	9	1	5	7	8
1	7	2	9	3	8	4	6	5
4	3	6	1	5	7	2	8	9
8	5	9	2	4	6	3	1	7
6	9	7	3	1	4	8	5	2
5	2	1	8	6	9	7	3	4
3	8	4	5	7	2	1	9	6

Solution # 656

3	4	8	9	6	2	1	5	7
5	6	2	1	3	7	9	8	4
1	7	9	4	5	8	2	6	3
9	1	6	8	7	4	5	3	2
8	3	7	6	2	5	4	1	9
4	2	5	3	1	9	6	7	8
2	5	3	7	9	1	8	4	6
6	9	4	5	8	3	7	2	1
7	8	1	2	4	6	3	9	5

Solution # 657

8	7	9	2	3	1	4	5	6
5	3	4	9	6	7	1	2	8
2	1	6	8	4	5	7	9	3
1	8	3	6	7	2	5	4	9
9	2	5	3	1	4	6	8	7
6	4	7	5	8	9	3	1	2
3	9	2	4	5	6	8	7	1
7	5	8	1	2	3	9	6	4
4	6	1	7	9	8	2	3	5

Solution # 658

6	2	5	8	4	3	9	7	1
3	9	4	7	2	1	8	6	5
7	1	8	6	5	9	2	3	4
2	7	9	1	6	8	4	5	3
5	3	6	9	7	4	1	8	2
8	4	1	2	3	5	7	9	6
9	5	3	4	8	2	6	1	7
4	8	7	3	1	6	5	2	9
1	6	2	5	9	7	3	4	8

Solution # 659

3	1	8	5	4	7	2	6	9
6	4	5	9	2	3	8	1	7
9	2	7	1	8	6	3	5	4
8	7	2	6	1	9	5	4	3
1	3	9	2	5	4	6	7	8
4	5	6	3	7	8	1	9	2
5	8	1	4	9	2	7	3	6
7	6	4	8	3	5	9	2	1
2	9	3	7	6	1	4	8	5

Solution # 660

9	3	6	1	7	4	5	2	8
5	7	4	9	8	2	3	6	1
8	2	1	5	3	6	9	4	7
3	4	9	8	6	7	1	5	2
6	1	7	2	4	5	8	3	9
2	5	8	3	1	9	4	7	6
1	8	5	6	2	3	7	9	4
7	6	3	4	9	1	2	8	5
4	9	2	7	5	8	6	1	3

Solution # 661

3	7	1	5	4	9	6	2	8
9	6	5	8	2	3	4	1	7
8	2	4	7	6	1	9	5	3
6	4	8	1	5	7	3	9	2
1	9	3	6	8	2	7	4	5
2	5	7	3	9	4	1	8	6
7	1	2	9	3	5	8	6	4
4	3	6	2	1	8	5	7	9
5	8	9	4	7	6	2	3	1

Solution # 662

7	1	9	5	4	2	8	3	6
2	5	8	6	3	7	4	9	1
4	6	3	8	9	1	7	2	5
5	4	7	2	6	8	9	1	3
1	3	6	7	5	9	2	4	8
9	8	2	4	1	3	5	6	7
6	2	4	3	7	5	1	8	9
3	7	1	9	8	4	6	5	2
8	9	5	1	2	6	3	7	4

Solution # 663

2	5	7	1	6	9	3	4	8
4	3	1	7	8	2	5	6	9
9	8	6	3	4	5	2	7	1
1	9	5	6	3	4	7	8	2
6	2	3	5	7	8	9	1	4
8	7	4	9	2	1	6	5	3
5	1	2	4	9	6	8	3	7
7	6	8	2	1	3	4	9	5
3	4	9	8	5	7	1	2	6

Solution # 664

4	3	2	6	1	5	8	9	7
7	8	5	3	9	2	4	1	6
1	6	9	8	7	4	5	2	3
2	9	7	5	4	3	6	8	1
5	4	6	9	8	1	7	3	2
8	1	3	2	6	7	9	4	5
3	5	4	7	2	8	1	6	9
6	2	8	1	5	9	3	7	4
9	7	1	4	3	6	2	5	8

Solution # 665

2	9	4	7	8	6	3	5	1
8	3	1	5	4	9	6	7	2
5	6	7	3	1	2	9	8	4
1	4	3	2	9	7	8	6	5
7	2	8	1	6	5	4	3	9
6	5	9	4	3	8	2	1	7
3	1	2	6	5	4	7	9	8
9	7	6	8	2	1	5	4	3
4	8	5	9	7	3	1	2	6

Solution # 666

6	7	4	9	1	5	3	8	2
8	9	2	4	7	3	5	1	6
3	5	1	8	6	2	9	7	4
1	4	7	5	8	6	2	3	9
5	3	8	2	9	4	1	6	7
2	6	9	7	3	1	4	5	8
9	2	3	6	5	8	7	4	1
7	1	6	3	4	9	8	2	5
4	8	5	1	2	7	6	9	3

Solution # 667

1	6	5	7	4	8	9	3	2
4	7	9	2	3	6	8	5	1
2	3	8	1	5	9	7	6	4
6	4	2	8	9	5	3	1	7
9	1	3	4	7	2	6	8	5
5	8	7	3	6	1	4	2	9
3	9	1	5	8	4	2	7	6
8	2	6	9	1	7	5	4	3
7	5	4	6	2	3	1	9	8

Solution # 668

4	8	7	1	3	2	9	6	5
1	2	6	7	9	5	8	4	3
3	9	5	8	4	6	1	2	7
5	1	3	6	2	8	4	7	9
7	4	8	9	5	1	2	3	6
2	6	9	3	7	4	5	8	1
9	7	1	2	8	3	6	5	4
8	3	4	5	6	9	7	1	2
6	5	2	4	1	7	3	9	8

Solution # 669

5	3	9	7	4	6	2	8	1
7	4	2	1	5	8	6	3	9
1	8	6	2	9	3	7	5	4
2	1	8	9	7	5	3	4	6
9	5	4	6	3	2	1	7	8
6	7	3	8	1	4	5	9	2
8	9	5	3	6	1	4	2	7
3	2	1	4	8	7	9	6	5
4	6	7	5	2	9	8	1	3

Solution # 670

1	7	2	9	8	6	5	4	3
5	3	9	1	4	2	8	6	7
6	4	8	3	5	7	1	2	9
3	6	5	8	1	9	2	7	4
9	8	1	2	7	4	3	5	6
7	2	4	6	3	5	9	8	1
8	1	6	4	2	3	7	9	5
4	5	3	7	9	8	6	1	2
2	9	7	5	6	1	4	3	8

Solution # 671

2	7	8	5	1	3	6	4	9
3	1	4	7	6	9	8	5	2
5	9	6	4	8	2	3	1	7
8	3	9	1	2	7	5	6	4
6	2	1	8	5	4	9	7	3
7	4	5	9	3	6	2	8	1
9	5	2	6	4	1	7	3	8
4	6	7	3	9	8	1	2	5
1	8	3	2	7	5	4	9	6

Solution # 672

6	9	8	5	2	3	1	4	7
4	2	3	1	7	6	8	5	9
5	7	1	4	9	8	6	2	3
9	8	7	2	1	5	4	3	6
1	6	4	7	3	9	2	8	5
2	3	5	6	8	4	9	7	1
7	4	6	8	5	1	3	9	2
3	1	2	9	4	7	5	6	8
8	5	9	3	6	2	7	1	4

Solution # 673

4	3	5	9	6	2	7	8	1
8	2	1	7	4	5	3	6	9
6	9	7	8	1	3	4	5	2
3	6	9	5	2	4	8	1	7
5	8	4	1	7	6	2	9	3
7	1	2	3	9	8	5	4	6
9	5	8	2	3	1	6	7	4
2	7	6	4	5	9	1	3	8
1	4	3	6	8	7	9	2	5

Solution # 674

2	7	3	4	9	6	8	1	5
9	6	1	8	5	7	4	2	3
8	4	5	3	2	1	6	7	9
4	8	9	6	7	5	1	3	2
1	5	7	2	3	8	9	4	6
6	3	2	1	4	9	5	8	7
5	2	6	7	1	4	3	9	8
3	1	8	9	6	2	7	5	4
7	9	4	5	8	3	2	6	1

Solution # 675

8	3	5	9	2	1	4	7	6
9	2	4	7	8	6	5	1	3
1	6	7	5	3	4	8	2	9
3	8	6	2	7	5	9	4	1
5	4	1	6	9	8	7	3	2
7	9	2	1	4	3	6	5	8
2	1	9	8	5	7	3	6	4
4	5	8	3	6	2	1	9	7
6	7	3	4	1	9	2	8	5

Solution # 676

6	9	3	1	5	7	2	8	4
5	8	1	4	9	2	7	6	3
2	7	4	6	3	8	1	9	5
7	4	9	8	6	5	3	1	2
8	1	5	2	7	3	9	4	6
3	6	2	9	1	4	5	7	8
4	5	6	7	2	9	8	3	1
9	3	8	5	4	1	6	2	7
1	2	7	3	8	6	4	5	9

Solution # 677

9	2	3	1	5	4	6	7	8
8	5	7	9	6	3	1	2	4
1	6	4	7	8	2	9	3	5
7	4	8	3	1	5	2	9	6
5	1	2	8	9	6	3	4	7
6	3	9	4	2	7	8	5	1
2	9	1	5	7	8	4	6	3
3	7	6	2	4	1	5	8	9
4	8	5	6	3	9	7	1	2

Solution # 678

9	1	7	2	3	5	6	4	8
8	5	2	6	7	4	3	1	9
4	3	6	1	9	8	7	5	2
6	9	3	4	5	1	8	2	7
1	4	8	7	2	3	9	6	5
7	2	5	8	6	9	4	3	1
3	8	1	5	4	7	2	9	6
5	6	4	9	8	2	1	7	3
2	7	9	3	1	6	5	8	4

Solution # 679

2	3	8	5	4	9	1	6	7
9	1	6	7	2	3	5	4	8
5	4	7	1	6	8	9	2	3
1	5	4	3	8	6	7	9	2
7	8	9	4	5	2	6	3	1
6	2	3	9	7	1	8	5	4
8	7	2	6	3	5	4	1	9
3	6	1	8	9	4	2	7	5
4	9	5	2	1	7	3	8	6

Solution # 680

6	8	5	7	9	3	1	4	2
7	4	9	8	1	2	5	6	3
1	3	2	6	4	5	9	8	7
5	6	8	1	3	7	4	2	9
9	2	1	4	8	6	3	7	5
4	7	3	2	5	9	6	1	8
2	5	7	9	6	4	8	3	1
3	1	4	5	2	8	7	9	6
8	9	6	3	7	1	2	5	4

Solution # 681

7	9	8	6	1	2	5	3	4
1	4	3	5	7	8	6	9	2
6	5	2	4	3	9	1	7	8
2	6	7	3	4	1	9	8	5
5	1	9	8	2	7	3	4	6
8	3	4	9	5	6	7	2	1
4	8	5	1	9	3	2	6	7
3	2	1	7	6	4	8	5	9
9	7	6	2	8	5	4	1	3

Solution # 682

7	4	5	1	6	3	2	9	8
3	6	8	9	2	7	1	5	4
1	9	2	8	4	5	7	6	3
8	2	1	3	9	4	5	7	6
4	5	9	7	8	6	3	2	1
6	3	7	5	1	2	8	4	9
9	8	6	2	5	1	4	3	7
2	7	4	6	3	8	9	1	5
5	1	3	4	7	9	6	8	2

Solution # 683

3	9	2	7	4	5	6	1	8
7	6	5	3	8	1	9	2	4
4	1	8	2	9	6	5	3	7
2	7	9	8	6	3	1	4	5
5	3	1	9	7	4	8	6	2
6	8	4	5	1	2	7	9	3
8	2	7	1	3	9	4	5	6
9	4	3	6	5	7	2	8	1
1	5	6	4	2	8	3	7	9

Solution # 684

6	8	9	5	3	1	2	7	4
1	3	2	7	4	6	8	5	9
4	5	7	9	2	8	1	3	6
8	6	3	4	9	5	7	2	1
9	1	4	2	8	7	5	6	3
2	7	5	1	6	3	4	9	8
5	4	6	3	1	2	9	8	7
7	9	8	6	5	4	3	1	2
3	2	1	8	7	9	6	4	5

Solution # 685

8	5	3	1	2	7	6	9	4
1	9	7	5	4	6	8	2	3
6	2	4	9	8	3	1	5	7
4	8	6	2	3	9	7	1	5
3	7	2	6	5	1	4	8	9
9	1	5	8	7	4	2	3	6
2	4	1	3	6	5	9	7	8
7	3	8	4	9	2	5	6	1
5	6	9	7	1	8	3	4	2

Solution # 686

8	5	7	3	6	9	1	2	4
4	1	6	7	2	8	9	3	5
9	2	3	1	4	5	7	6	8
5	7	4	8	9	3	2	1	6
3	8	2	6	1	4	5	7	9
6	9	1	2	5	7	8	4	3
1	6	9	5	3	2	4	8	7
7	3	5	4	8	1	6	9	2
2	4	8	9	7	6	3	5	1

Solution # 687

4	8	9	2	7	6	3	5	1
2	5	6	3	1	8	4	7	9
1	7	3	4	5	9	8	2	6
5	6	7	1	2	4	9	3	8
3	4	8	5	9	7	6	1	2
9	2	1	8	6	3	5	4	7
6	3	2	9	4	1	7	8	5
7	1	4	6	8	5	2	9	3
8	9	5	7	3	2	1	6	4

Solution # 688

9	3	6	2	1	7	8	4	5
1	8	5	9	3	4	2	6	7
4	7	2	6	5	8	3	9	1
3	5	4	8	2	6	7	1	9
6	2	1	7	9	3	5	8	4
7	9	8	5	4	1	6	3	2
5	6	9	1	8	2	4	7	3
2	4	7	3	6	9	1	5	8
8	1	3	4	7	5	9	2	6

Solution # 689

2	3	4	8	9	5	7	1	6
5	7	8	6	3	1	2	9	4
1	6	9	4	7	2	8	3	5
9	8	1	3	4	7	5	6	2
4	2	3	1	5	6	9	7	8
6	5	7	9	2	8	3	4	1
8	4	5	7	6	3	1	2	9
3	1	6	2	8	9	4	5	7
7	9	2	5	1	4	6	8	3

Solution # 690

2	4	5	7	1	6	3	8	9
8	1	6	9	3	4	7	5	2
9	3	7	2	5	8	6	4	1
4	8	3	1	9	2	5	6	7
6	7	2	3	4	5	9	1	8
1	5	9	6	8	7	4	2	3
5	9	4	8	2	3	1	7	6
3	6	8	4	7	1	2	9	5
7	2	1	5	6	9	8	3	4

Solution # 691

9	2	1	5	7	3	8	4	6
7	4	6	9	1	8	2	3	5
5	3	8	2	6	4	1	7	9
4	5	2	8	9	1	3	6	7
8	7	9	3	5	6	4	2	1
6	1	3	4	2	7	5	9	8
2	8	5	7	3	9	6	1	4
3	6	7	1	4	5	9	8	2
1	9	4	6	8	2	7	5	3

Solution # 692

2	9	3	1	6	5	8	7	4
8	6	7	9	3	4	5	1	2
4	1	5	2	7	8	6	9	3
9	7	2	6	8	3	4	5	1
5	8	1	4	9	2	3	6	7
6	3	4	7	5	1	2	8	9
1	5	9	3	4	6	7	2	8
7	4	8	5	2	9	1	3	6
3	2	6	8	1	7	9	4	5

Solution # 693

3	7	5	1	4	8	2	6	9
8	9	1	6	2	5	3	4	7
6	4	2	7	9	3	1	5	8
9	2	8	4	3	7	5	1	6
5	6	4	8	1	2	9	7	3
1	3	7	9	5	6	4	8	2
2	1	6	5	7	9	8	3	4
4	8	9	3	6	1	7	2	5
7	5	3	2	8	4	6	9	1

Solution # 694

1	4	2	9	8	7	3	6	5
7	3	5	1	2	6	9	8	4
6	8	9	3	5	4	7	2	1
8	2	4	7	9	1	5	3	6
9	1	6	2	3	5	4	7	8
5	7	3	6	4	8	1	9	2
2	6	1	4	7	9	8	5	3
4	5	7	8	6	3	2	1	9
3	9	8	5	1	2	6	4	7

Solution # 695

1	5	9	4	6	8	7	2	3
4	6	2	7	5	3	9	8	1
3	8	7	9	1	2	5	6	4
5	1	6	8	7	4	2	3	9
8	9	3	6	2	1	4	7	5
2	7	4	5	3	9	8	1	6
6	2	5	3	9	7	1	4	8
7	3	8	1	4	5	6	9	2
9	4	1	2	8	6	3	5	7

Solution # 696

8	1	4	5	9	3	7	6	2
7	5	9	6	1	2	4	8	3
6	3	2	7	8	4	1	9	5
3	4	7	8	5	9	2	1	6
5	2	8	3	6	1	9	7	4
9	6	1	4	2	7	5	3	8
4	9	5	1	3	6	8	2	7
1	8	6	2	7	5	3	4	9
2	7	3	9	4	8	6	5	1

Solution # 697

9	4	8	5	2	6	7	1	3
5	6	7	3	1	8	2	9	4
3	1	2	7	9	4	5	6	8
6	7	5	9	8	3	4	2	1
1	9	4	2	7	5	3	8	6
2	8	3	4	6	1	9	5	7
8	2	9	6	4	7	1	3	5
7	5	1	8	3	9	6	4	2
4	3	6	1	5	2	8	7	9

Solution # 698

5	1	3	2	8	6	7	9	4
2	4	8	7	9	3	6	1	5
6	7	9	4	1	5	3	2	8
1	5	6	3	2	8	4	7	9
9	3	7	1	5	4	8	6	2
4	8	2	9	6	7	1	5	3
8	9	1	6	3	2	5	4	7
7	6	5	8	4	9	2	3	1
3	2	4	5	7	1	9	8	6

Solution # 699

5	3	7	1	2	4	6	9	8
4	2	9	3	6	8	5	7	1
6	1	8	9	7	5	3	4	2
9	6	1	8	5	2	4	3	7
8	4	5	6	3	7	2	1	9
2	7	3	4	1	9	8	6	5
1	9	4	2	8	3	7	5	6
7	8	6	5	4	1	9	2	3
3	5	2	7	9	6	1	8	4

Solution # 700

9	1	7	8	6	3	2	4	5
5	6	2	7	1	4	3	9	8
8	4	3	5	9	2	6	7	1
2	3	8	6	4	5	7	1	9
1	7	6	3	8	9	5	2	4
4	9	5	1	2	7	8	3	6
3	5	1	9	7	6	4	8	2
7	8	4	2	5	1	9	6	3
6	2	9	4	3	8	1	5	7

Solution # 701

6	1	5	7	3	8	9	4	2
7	9	3	4	2	5	6	1	8
4	2	8	9	1	6	5	3	7
9	4	2	5	7	1	3	8	6
3	7	6	8	4	2	1	5	9
5	8	1	3	6	9	7	2	4
1	5	4	6	8	7	2	9	3
2	3	7	1	9	4	8	6	5
8	6	9	2	5	3	4	7	1

Solution # 702

3	9	6	2	7	4	8	1	5
7	8	2	1	5	9	4	3	6
5	1	4	6	8	3	2	7	9
8	7	3	5	6	2	1	9	4
4	2	9	3	1	8	6	5	7
1	6	5	9	4	7	3	8	2
2	4	8	7	3	5	9	6	1
6	3	7	4	9	1	5	2	8
9	5	1	8	2	6	7	4	3

Solution # 703

9	5	8	6	7	2	4	1	3
1	2	7	3	4	8	9	6	5
4	3	6	5	1	9	7	2	8
2	6	4	7	8	5	1	3	9
8	9	3	1	2	6	5	7	4
7	1	5	9	3	4	6	8	2
6	8	2	4	5	7	3	9	1
3	4	9	2	6	1	8	5	7
5	7	1	8	9	3	2	4	6

Solution # 704

4	1	8	2	5	7	9	6	3
9	3	5	6	1	8	4	2	7
2	6	7	4	3	9	8	5	1
6	8	9	7	2	5	1	3	4
3	7	4	8	6	1	5	9	2
1	5	2	9	4	3	6	7	8
7	2	6	1	9	4	3	8	5
5	9	1	3	8	2	7	4	6
8	4	3	5	7	6	2	1	9

Solution # 705

3	9	7	1	2	6	4	8	5
8	5	2	4	3	7	9	6	1
6	1	4	9	8	5	3	2	7
7	3	5	2	6	4	8	1	9
1	2	9	7	5	8	6	3	4
4	6	8	3	1	9	5	7	2
5	7	1	6	4	3	2	9	8
2	4	3	8	9	1	7	5	6
9	8	6	5	7	2	1	4	3

Solution # 706

9	5	2	7	1	3	4	6	8
6	8	3	5	4	2	9	7	1
7	1	4	8	9	6	3	5	2
4	3	8	9	7	1	6	2	5
2	9	7	6	5	8	1	4	3
1	6	5	2	3	4	8	9	7
8	4	9	1	2	7	5	3	6
5	2	1	3	6	9	7	8	4
3	7	6	4	8	5	2	1	9

Solution # 707

8	5	1	3	7	6	9	4	2
6	7	2	1	9	4	3	8	5
3	4	9	2	5	8	7	6	1
1	6	5	9	8	7	4	2	3
7	3	8	4	6	2	1	5	9
9	2	4	5	3	1	8	7	6
2	9	3	8	4	5	6	1	7
5	8	7	6	1	3	2	9	4
4	1	6	7	2	9	5	3	8

Solution # 708

3	9	6	5	4	1	2	7	8
1	4	7	8	2	9	3	5	6
8	2	5	7	3	6	4	1	9
7	6	4	9	8	5	1	3	2
2	5	1	3	6	7	9	8	4
9	8	3	4	1	2	7	6	5
4	7	2	1	5	8	6	9	3
5	3	9	6	7	4	8	2	1
6	1	8	2	9	3	5	4	7

Solution # 709

4	1	9	8	6	2	5	3	7
3	5	2	7	9	1	8	4	6
6	7	8	3	5	4	9	2	1
8	4	1	9	2	7	3	6	5
5	9	6	4	8	3	1	7	2
7	2	3	6	1	5	4	9	8
9	8	4	5	7	6	2	1	3
1	3	7	2	4	8	6	5	9
2	6	5	1	3	9	7	8	4

Solution # 710

5	1	2	3	7	9	8	4	6
7	8	6	2	4	5	1	9	3
9	4	3	1	8	6	7	5	2
3	5	1	6	9	7	4	2	8
4	7	9	8	2	3	6	1	5
6	2	8	4	5	1	3	7	9
1	9	4	5	3	8	2	6	7
8	6	7	9	1	2	5	3	4
2	3	5	7	6	4	9	8	1

Solution # 711

8	9	3	1	7	5	4	6	2
1	7	6	8	4	2	3	5	9
2	4	5	9	3	6	7	8	1
5	8	9	6	2	3	1	7	4
3	6	1	7	9	4	8	2	5
7	2	4	5	1	8	9	3	6
4	5	2	3	8	9	6	1	7
6	1	8	4	5	7	2	9	3
9	3	7	2	6	1	5	4	8

Solution # 712

6	3	7	5	2	4	9	8	1
8	2	9	1	6	7	3	4	5
5	4	1	9	3	8	7	6	2
7	5	4	2	9	6	1	3	8
2	6	3	8	1	5	4	7	9
1	9	8	7	4	3	5	2	6
4	8	6	3	5	9	2	1	7
9	7	2	4	8	1	6	5	3
3	1	5	6	7	2	8	9	4

Solution # 713

6	7	4	3	1	9	8	5	2
8	5	2	6	4	7	3	1	9
9	3	1	8	2	5	7	6	4
4	6	9	1	7	2	5	8	3
3	2	8	5	9	6	4	7	1
7	1	5	4	3	8	9	2	6
2	4	7	9	5	1	6	3	8
1	9	6	7	8	3	2	4	5
5	8	3	2	6	4	1	9	7

Solution # 714

3	8	1	6	9	7	5	2	4
9	4	7	5	2	8	3	6	1
5	6	2	1	4	3	7	9	8
8	5	6	2	7	4	1	3	9
1	7	9	8	3	5	6	4	2
2	3	4	9	6	1	8	7	5
6	2	5	3	1	9	4	8	7
7	9	8	4	5	6	2	1	3
4	1	3	7	8	2	9	5	6

Solution # 715

7	9	1	2	5	6	3	4	8
4	3	8	7	1	9	6	2	5
2	5	6	4	8	3	9	1	7
9	7	2	5	3	4	1	8	6
6	1	4	8	9	7	5	3	2
5	8	3	6	2	1	7	9	4
1	4	5	9	6	8	2	7	3
3	6	7	1	4	2	8	5	9
8	2	9	3	7	5	4	6	1

Solution # 716

5	7	8	9	4	2	3	1	6
9	1	4	6	7	3	5	8	2
3	6	2	1	5	8	9	7	4
8	2	3	7	1	9	4	6	5
7	4	1	2	6	5	8	9	3
6	5	9	3	8	4	7	2	1
4	3	6	8	2	7	1	5	9
2	8	5	4	9	1	6	3	7
1	9	7	5	3	6	2	4	8

Solution # 717

8	5	9	6	1	2	7	3	4
3	7	1	5	4	9	2	6	8
2	6	4	7	3	8	1	5	9
5	2	6	4	8	1	3	9	7
4	3	8	9	5	7	6	2	1
1	9	7	2	6	3	8	4	5
9	1	2	3	7	5	4	8	6
7	4	3	8	9	6	5	1	2
6	8	5	1	2	4	9	7	3

Solution # 718

6	5	4	7	3	8	1	2	9
2	8	7	5	1	9	4	3	6
1	3	9	2	4	6	8	5	7
4	6	1	8	2	3	9	7	5
9	2	8	6	7	5	3	1	4
5	7	3	4	9	1	6	8	2
3	1	2	9	6	7	5	4	8
7	9	5	3	8	4	2	6	1
8	4	6	1	5	2	7	9	3

Solution # 719

7	9	3	2	8	6	4	5	1
1	8	6	7	4	5	9	3	2
5	4	2	3	1	9	6	8	7
6	3	4	8	5	7	1	2	9
9	5	1	4	2	3	8	7	6
2	7	8	9	6	1	3	4	5
3	2	5	6	9	8	7	1	4
4	6	7	1	3	2	5	9	8
8	1	9	5	7	4	2	6	3

Solution # 720

5	3	2	6	7	4	1	8	9
7	1	4	9	8	3	5	6	2
8	9	6	1	5	2	3	4	7
6	8	3	5	1	9	7	2	4
9	2	5	8	4	7	6	3	1
4	7	1	2	3	6	9	5	8
2	6	7	4	9	5	8	1	3
3	4	8	7	6	1	2	9	5
1	5	9	3	2	8	4	7	6

Solution # 721

5	6	8	2	4	3	7	1	9
4	9	2	1	8	7	3	5	6
1	7	3	9	6	5	8	4	2
6	4	7	5	3	8	9	2	1
2	1	5	4	7	9	6	8	3
8	3	9	6	2	1	4	7	5
9	2	6	7	1	4	5	3	8
7	8	1	3	5	6	2	9	4
3	5	4	8	9	2	1	6	7

Solution # 722

6	1	7	8	2	9	4	3	5
4	5	8	1	7	3	2	9	6
3	2	9	4	6	5	7	8	1
2	7	1	9	4	6	3	5	8
9	4	3	5	8	1	6	2	7
5	8	6	7	3	2	9	1	4
8	3	5	6	9	4	1	7	2
7	6	2	3	1	8	5	4	9
1	9	4	2	5	7	8	6	3

Solution # 723

9	8	1	3	7	2	5	6	4
3	7	4	1	5	6	9	2	8
5	2	6	8	9	4	3	7	1
2	1	9	6	4	5	7	8	3
8	5	7	9	1	3	2	4	6
6	4	3	7	2	8	1	5	9
4	9	2	5	8	1	6	3	7
1	3	5	4	6	7	8	9	2
7	6	8	2	3	9	4	1	5

Solution # 724

7	4	1	5	9	3	6	8	2
9	6	8	4	1	2	5	3	7
3	5	2	6	8	7	1	9	4
1	9	3	7	2	5	8	4	6
6	2	7	9	4	8	3	5	1
5	8	4	3	6	1	7	2	9
8	1	9	2	5	6	4	7	3
2	7	5	1	3	4	9	6	8
4	3	6	8	7	9	2	1	5

Solution # 725

1	6	7	8	5	4	2	9	3
5	8	2	9	6	3	4	1	7
3	4	9	1	7	2	6	5	8
2	1	8	3	9	6	5	7	4
4	9	5	2	8	7	3	6	1
7	3	6	4	1	5	9	8	2
6	5	3	7	4	1	8	2	9
9	7	4	6	2	8	1	3	5
8	2	1	5	3	9	7	4	6

Solution # 726

4	6	9	8	1	3	5	2	7
5	1	2	7	6	4	8	3	9
3	7	8	5	9	2	1	6	4
9	4	5	1	3	7	2	8	6
6	8	7	4	2	5	9	1	3
2	3	1	6	8	9	7	4	5
8	2	3	9	5	6	4	7	1
1	9	4	3	7	8	6	5	2
7	5	6	2	4	1	3	9	8

Solution # 727

5	7	6	9	4	3	1	2	8
8	2	3	7	6	1	5	4	9
4	1	9	2	5	8	3	7	6
7	5	4	6	8	9	2	1	3
1	9	2	5	3	4	8	6	7
3	6	8	1	2	7	9	5	4
9	4	7	3	1	5	6	8	2
2	3	5	8	7	6	4	9	1
6	8	1	4	9	2	7	3	5

Solution # 728

9	5	3	4	6	1	7	8	2
4	7	2	5	9	8	3	6	1
6	1	8	7	2	3	9	5	4
7	8	4	2	3	5	6	1	9
5	3	6	9	1	7	2	4	8
1	2	9	6	8	4	5	7	3
8	6	1	3	5	2	4	9	7
2	4	5	1	7	9	8	3	6
3	9	7	8	4	6	1	2	5

Solution # 729

4	3	7	8	1	5	2	9	6
1	2	9	4	3	6	8	7	5
6	8	5	7	9	2	3	4	1
5	9	3	2	6	1	4	8	7
2	4	8	3	5	7	1	6	9
7	1	6	9	4	8	5	2	3
9	7	4	5	8	3	6	1	2
3	6	2	1	7	4	9	5	8
8	5	1	6	2	9	7	3	4

Solution # 730

9	3	7	6	1	5	4	8	2
5	8	6	7	4	2	3	1	9
4	2	1	3	8	9	7	6	5
6	4	2	8	7	1	5	9	3
3	9	8	2	5	6	1	7	4
1	7	5	4	9	3	8	2	6
2	6	4	1	3	8	9	5	7
7	1	9	5	2	4	6	3	8
8	5	3	9	6	7	2	4	1

Solution # 731

6	1	2	5	9	8	4	3	7
5	3	7	2	4	1	9	8	6
8	9	4	7	3	6	2	5	1
9	5	6	1	7	4	8	2	3
3	4	8	9	6	2	7	1	5
7	2	1	8	5	3	6	4	9
4	7	5	3	8	9	1	6	2
1	6	9	4	2	5	3	7	8
2	8	3	6	1	7	5	9	4

Solution # 732

1	5	4	2	6	3	7	8	9
6	3	8	5	9	7	2	1	4
2	7	9	8	1	4	6	5	3
8	1	2	6	4	5	9	3	7
3	9	6	1	7	2	5	4	8
5	4	7	3	8	9	1	2	6
9	8	3	7	5	1	4	6	2
7	6	5	4	2	8	3	9	1
4	2	1	9	3	6	8	7	5

Solution # 733

8	1	5	4	3	9	7	6	2
6	7	4	2	5	1	3	9	8
9	2	3	6	7	8	1	5	4
4	3	6	9	8	2	5	1	7
7	5	2	1	4	3	9	8	6
1	9	8	5	6	7	2	4	3
5	6	9	7	2	4	8	3	1
2	8	1	3	9	6	4	7	5
3	4	7	8	1	5	6	2	9

Solution # 734

5	4	3	1	7	8	2	6	9
9	2	1	6	5	4	8	7	3
8	6	7	3	2	9	5	1	4
4	9	6	7	1	2	3	5	8
7	5	2	8	6	3	9	4	1
3	1	8	4	9	5	7	2	6
6	7	5	9	3	1	4	8	2
2	3	4	5	8	6	1	9	7
1	8	9	2	4	7	6	3	5

Solution # 735

4	9	8	1	6	5	7	3	2
7	2	1	8	3	9	5	6	4
5	3	6	2	7	4	8	9	1
6	1	7	5	9	8	4	2	3
9	4	3	6	2	7	1	5	8
8	5	2	4	1	3	6	7	9
1	6	5	3	4	2	9	8	7
2	7	4	9	8	6	3	1	5
3	8	9	7	5	1	2	4	6

Solution # 736

8	3	9	4	7	1	2	6	5
1	2	6	5	8	9	4	7	3
7	4	5	3	2	6	9	1	8
9	7	1	2	6	8	3	5	4
2	6	3	7	5	4	1	8	9
4	5	8	9	1	3	7	2	6
3	8	7	1	9	5	6	4	2
6	9	2	8	4	7	5	3	1
5	1	4	6	3	2	8	9	7

Solution # 737

3	4	8	2	7	9	6	5	1
9	7	6	5	8	1	4	3	2
2	1	5	3	4	6	8	9	7
7	2	4	6	1	5	3	8	9
5	3	1	7	9	8	2	4	6
8	6	9	4	2	3	7	1	5
1	9	3	8	6	7	5	2	4
4	8	7	9	5	2	1	6	3
6	5	2	1	3	4	9	7	8

Solution # 738

4	7	1	6	8	9	3	2	5
8	9	2	1	3	5	6	4	7
5	6	3	2	4	7	8	1	9
9	1	4	3	7	8	5	6	2
6	8	7	5	9	2	1	3	4
2	3	5	4	6	1	7	9	8
3	2	6	7	5	4	9	8	1
7	4	8	9	1	6	2	5	3
1	5	9	8	2	3	4	7	6

Solution # 739

1	6	7	4	2	5	9	3	8
5	9	2	3	8	7	4	1	6
8	4	3	1	6	9	5	7	2
3	8	1	5	9	6	2	4	7
9	7	4	8	3	2	1	6	5
2	5	6	7	4	1	8	9	3
4	2	9	6	5	3	7	8	1
7	3	8	2	1	4	6	5	9
6	1	5	9	7	8	3	2	4

Solution # 740

9	3	1	5	4	8	2	6	7
8	5	2	7	6	9	1	4	3
7	4	6	1	3	2	9	5	8
3	2	4	6	8	7	5	9	1
6	7	5	9	1	4	3	8	2
1	9	8	3	2	5	4	7	6
4	1	3	8	9	6	7	2	5
5	6	9	2	7	1	8	3	4
2	8	7	4	5	3	6	1	9

Solution # 741

7	6	4	1	5	8	2	9	3
1	2	8	6	3	9	7	4	5
3	9	5	7	2	4	1	6	8
6	3	1	4	7	2	8	5	9
9	8	7	5	1	3	6	2	4
5	4	2	9	8	6	3	7	1
2	1	6	3	9	5	4	8	7
4	5	3	8	6	7	9	1	2
8	7	9	2	4	1	5	3	6

Solution # 742

8	7	9	6	2	3	4	1	5
1	2	3	4	8	5	6	7	9
5	6	4	9	1	7	8	2	3
4	9	5	8	7	1	2	3	6
2	8	1	3	6	4	9	5	7
7	3	6	5	9	2	1	8	4
9	1	7	2	3	6	5	4	8
6	5	2	7	4	8	3	9	1
3	4	8	1	5	9	7	6	2

Solution # 743

2	6	4	3	7	1	8	5	9
1	9	7	5	4	8	6	3	2
8	3	5	9	2	6	1	4	7
7	1	6	8	9	3	4	2	5
5	2	8	7	6	4	9	1	3
3	4	9	2	1	5	7	8	6
4	7	1	6	5	2	3	9	8
9	8	2	4	3	7	5	6	1
6	5	3	1	8	9	2	7	4

Solution # 744

1	6	3	2	5	9	8	4	7
5	9	8	1	4	7	2	6	3
4	2	7	8	6	3	5	1	9
8	1	5	3	2	6	9	7	4
2	4	9	5	7	1	3	8	6
7	3	6	4	9	8	1	5	2
3	7	2	6	8	5	4	9	1
6	8	4	9	1	2	7	3	5
9	5	1	7	3	4	6	2	8

Solution # 745

9	3	2	4	1	8	7	6	5
6	1	8	5	7	3	4	2	9
7	4	5	6	2	9	1	3	8
2	8	1	9	3	7	5	4	6
4	6	9	8	5	2	3	1	7
5	7	3	1	4	6	8	9	2
1	9	7	2	8	4	6	5	3
3	2	4	7	6	5	9	8	1
8	5	6	3	9	1	2	7	4

Solution # 746

6	3	4	9	7	2	1	8	5
5	9	8	4	3	1	6	2	7
1	2	7	8	5	6	3	4	9
3	5	2	1	4	8	9	7	6
7	8	6	3	2	9	5	1	4
9	4	1	5	6	7	8	3	2
4	6	3	7	1	5	2	9	8
2	7	9	6	8	3	4	5	1
8	1	5	2	9	4	7	6	3

Solution # 747

7	6	2	4	5	3	9	1	8
1	3	8	7	6	9	5	2	4
4	5	9	1	8	2	3	7	6
8	2	3	6	7	1	4	5	9
6	9	4	5	2	8	7	3	1
5	7	1	3	9	4	6	8	2
9	1	7	2	3	6	8	4	5
3	4	6	8	1	5	2	9	7
2	8	5	9	4	7	1	6	3

Solution # 748

3	7	2	9	5	8	6	4	1
6	1	5	2	3	4	9	7	8
8	4	9	7	6	1	2	3	5
7	2	6	5	1	3	8	9	4
9	8	1	4	2	7	5	6	3
4	5	3	6	8	9	1	2	7
2	9	8	3	4	5	7	1	6
1	6	4	8	7	2	3	5	9
5	3	7	1	9	6	4	8	2

Solution # 749

5	8	7	6	9	4	3	2	1
3	1	4	7	5	2	6	9	8
2	6	9	3	1	8	5	4	7
6	2	1	9	3	7	8	5	4
9	4	8	1	2	5	7	6	3
7	5	3	8	4	6	9	1	2
1	3	5	2	7	9	4	8	6
8	9	2	4	6	3	1	7	5
4	7	6	5	8	1	2	3	9

Solution # 750

5	3	2	6	4	1	9	8	7
1	9	8	3	2	7	4	6	5
4	6	7	5	8	9	3	1	2
7	1	9	8	5	6	2	4	3
2	8	5	4	9	3	6	7	1
6	4	3	1	7	2	5	9	8
3	2	6	9	1	8	7	5	4
8	7	4	2	6	5	1	3	9
9	5	1	7	3	4	8	2	6

Solution # 751

2	4	9	7	5	3	1	6	8
7	1	3	6	2	8	9	5	4
6	5	8	4	9	1	3	2	7
9	7	1	3	8	6	2	4	5
5	3	6	2	4	9	8	7	1
4	8	2	5	1	7	6	9	3
8	2	4	1	6	5	7	3	9
3	9	5	8	7	2	4	1	6
1	6	7	9	3	4	5	8	2

Solution # 752

9	7	4	2	3	8	1	6	5
8	2	1	5	6	9	3	4	7
3	6	5	7	1	4	8	9	2
5	9	8	1	4	6	7	2	3
7	3	6	8	9	2	4	5	1
1	4	2	3	7	5	9	8	6
4	8	3	6	2	1	5	7	9
6	5	7	9	8	3	2	1	4
2	1	9	4	5	7	6	3	8

Solution # 753

2	9	5	1	4	3	7	8	6
8	3	4	9	6	7	5	1	2
1	7	6	2	8	5	3	4	9
4	5	3	7	2	1	6	9	8
6	2	9	8	5	4	1	7	3
7	1	8	3	9	6	2	5	4
3	4	2	5	1	9	8	6	7
5	6	7	4	3	8	9	2	1
9	8	1	6	7	2	4	3	5

Solution # 754

6	5	1	4	3	9	7	8	2
8	2	9	7	1	6	5	3	4
7	4	3	5	2	8	1	9	6
9	3	6	2	8	1	4	7	5
5	8	2	3	7	4	6	1	9
1	7	4	6	9	5	8	2	3
3	1	5	9	6	7	2	4	8
2	6	8	1	4	3	9	5	7
4	9	7	8	5	2	3	6	1

Solution # 755

5	1	6	2	7	4	3	9	8
2	9	4	3	8	6	5	7	1
8	3	7	5	9	1	2	4	6
6	4	3	9	1	8	7	5	2
7	8	5	4	6	2	9	1	3
1	2	9	7	5	3	8	6	4
4	7	1	8	2	9	6	3	5
3	5	8	6	4	7	1	2	9
9	6	2	1	3	5	4	8	7

Solution # 756

6	7	1	9	2	8	5	3	4
2	9	4	3	6	5	8	1	7
8	3	5	7	1	4	6	2	9
4	1	6	8	5	7	2	9	3
7	8	9	2	3	6	1	4	5
3	5	2	1	4	9	7	6	8
5	4	8	6	9	2	3	7	1
1	6	7	4	8	3	9	5	2
9	2	3	5	7	1	4	8	6

Solution # 757

8	6	9	4	2	7	3	1	5
4	3	1	6	8	5	2	7	9
2	5	7	1	3	9	4	8	6
3	7	6	5	4	1	9	2	8
9	4	2	8	7	6	5	3	1
5	1	8	2	9	3	6	4	7
7	8	5	3	6	2	1	9	4
6	2	4	9	1	8	7	5	3
1	9	3	7	5	4	8	6	2

Solution # 758

4	7	1	5	8	9	6	2	3
2	6	9	3	7	1	4	5	8
8	5	3	4	6	2	1	7	9
3	8	4	1	5	6	7	9	2
5	2	6	9	4	7	3	8	1
9	1	7	2	3	8	5	6	4
7	4	5	8	2	3	9	1	6
1	3	2	6	9	5	8	4	7
6	9	8	7	1	4	2	3	5

Solution # 759

5	9	1	4	8	6	7	3	2
8	4	2	3	7	9	1	5	6
3	7	6	1	2	5	9	8	4
2	6	8	9	5	4	3	1	7
4	1	7	8	6	3	5	2	9
9	3	5	2	1	7	4	6	8
7	5	4	6	3	8	2	9	1
1	8	9	5	4	2	6	7	3
6	2	3	7	9	1	8	4	5

Solution # 760

6	8	2	9	7	3	5	4	1
7	5	4	1	2	8	3	6	9
9	1	3	6	5	4	8	2	7
1	7	6	2	8	5	4	9	3
4	2	5	3	9	7	1	8	6
8	3	9	4	1	6	7	5	2
2	9	8	7	4	1	6	3	5
5	6	7	8	3	9	2	1	4
3	4	1	5	6	2	9	7	8

Solution # 761

1	5	9	6	7	2	4	8	3
4	7	3	8	1	9	5	6	2
2	6	8	5	4	3	7	1	9
3	4	2	9	6	5	8	7	1
8	1	6	7	2	4	9	3	5
7	9	5	3	8	1	2	4	6
5	8	7	1	9	6	3	2	4
9	2	1	4	3	7	6	5	8
6	3	4	2	5	8	1	9	7

Solution # 762

7	9	8	5	3	6	1	2	4
5	4	2	7	1	9	8	3	6
3	1	6	2	8	4	9	7	5
9	8	5	4	7	2	3	6	1
6	3	7	8	9	1	5	4	2
4	2	1	3	6	5	7	9	8
1	5	4	9	2	3	6	8	7
8	6	9	1	4	7	2	5	3
2	7	3	6	5	8	4	1	9

Solution # 763

6	3	4	8	7	1	9	5	2
1	7	2	3	5	9	8	4	6
5	8	9	2	6	4	3	1	7
8	5	3	7	9	6	1	2	4
2	4	6	1	8	5	7	3	9
9	1	7	4	3	2	5	6	8
3	9	1	6	4	8	2	7	5
7	6	5	9	2	3	4	8	1
4	2	8	5	1	7	6	9	3

Solution # 764

4	1	6	5	7	3	2	8	9
2	8	5	9	1	6	7	4	3
7	3	9	8	2	4	1	6	5
6	9	7	4	8	5	3	2	1
5	2	3	1	6	9	4	7	8
8	4	1	7	3	2	9	5	6
1	5	4	6	9	7	8	3	2
9	7	2	3	5	8	6	1	4
3	6	8	2	4	1	5	9	7

Solution # 765

7	5	1	8	9	2	3	6	4
3	8	9	4	6	1	2	7	5
2	4	6	5	7	3	9	8	1
1	3	8	6	5	9	7	4	2
9	7	2	3	4	8	5	1	6
4	6	5	2	1	7	8	3	9
6	2	4	7	8	5	1	9	3
5	1	7	9	3	6	4	2	8
8	9	3	1	2	4	6	5	7

Solution # 766

4	8	5	9	2	7	6	3	1
9	7	1	6	8	3	5	4	2
3	6	2	1	4	5	8	7	9
1	9	3	2	5	6	4	8	7
7	2	8	3	9	4	1	6	5
6	5	4	7	1	8	9	2	3
8	1	6	5	7	2	3	9	4
2	3	9	4	6	1	7	5	8
5	4	7	8	3	9	2	1	6

Solution # 767

1	3	6	9	5	2	4	8	7
7	8	9	4	1	3	2	6	5
5	4	2	7	8	6	1	9	3
8	9	4	2	7	1	5	3	6
3	7	1	6	4	5	9	2	8
2	6	5	3	9	8	7	4	1
6	1	8	5	2	4	3	7	9
4	5	7	8	3	9	6	1	2
9	2	3	1	6	7	8	5	4

Solution # 768

1	9	5	8	6	3	7	4	2
6	3	2	4	7	9	1	5	8
4	7	8	2	1	5	6	9	3
8	4	3	9	5	6	2	1	7
9	5	1	7	3	2	8	6	4
2	6	7	1	4	8	9	3	5
3	2	9	5	8	1	4	7	6
5	8	4	6	9	7	3	2	1
7	1	6	3	2	4	5	8	9

Solution # 769

1	7	5	9	2	6	4	3	8
4	6	9	7	8	3	1	2	5
3	8	2	5	1	4	6	7	9
9	1	3	8	5	7	2	6	4
5	2	7	6	4	9	8	1	3
8	4	6	1	3	2	5	9	7
7	5	8	3	6	1	9	4	2
2	3	1	4	9	5	7	8	6
6	9	4	2	7	8	3	5	1

Solution # 770

6	7	8	5	9	1	4	3	2
3	9	4	7	8	2	6	5	1
5	2	1	6	3	4	9	8	7
2	4	6	3	7	8	1	9	5
7	1	3	2	5	9	8	6	4
8	5	9	4	1	6	7	2	3
4	8	7	9	2	5	3	1	6
1	6	2	8	4	3	5	7	9
9	3	5	1	6	7	2	4	8

Solution # 771

4	9	5	7	1	3	6	2	8
1	7	6	5	2	8	3	9	4
2	8	3	9	6	4	1	5	7
9	6	8	3	5	1	4	7	2
3	1	2	4	7	9	8	6	5
5	4	7	2	8	6	9	1	3
8	5	4	1	9	2	7	3	6
7	3	9	6	4	5	2	8	1
6	2	1	8	3	7	5	4	9

Solution # 772

5	9	1	6	7	3	2	4	8
2	8	7	9	1	4	5	3	6
4	3	6	5	8	2	7	1	9
7	6	2	4	3	9	8	5	1
9	1	5	8	2	6	4	7	3
8	4	3	7	5	1	6	9	2
1	7	9	2	4	8	3	6	5
6	5	8	3	9	7	1	2	4
3	2	4	1	6	5	9	8	7

Solution # 773

9	1	5	7	8	6	4	3	2
8	2	3	9	4	5	1	6	7
6	4	7	3	2	1	5	8	9
3	6	1	4	9	2	8	7	5
5	7	9	6	3	8	2	1	4
2	8	4	1	5	7	6	9	3
7	9	8	2	1	4	3	5	6
1	3	2	5	6	9	7	4	8
4	5	6	8	7	3	9	2	1

Solution # 774

6	2	9	3	5	8	4	7	1
5	7	8	4	2	1	6	3	9
4	1	3	7	9	6	8	2	5
9	8	5	1	3	2	7	4	6
1	3	6	8	4	7	5	9	2
2	4	7	5	6	9	3	1	8
7	6	4	2	1	5	9	8	3
8	5	1	9	7	3	2	6	4
3	9	2	6	8	4	1	5	7

Solution # 775

7	6	4	5	9	8	1	3	2
1	8	5	3	2	7	9	6	4
3	2	9	1	6	4	8	7	5
5	1	7	9	8	3	2	4	6
2	9	6	4	7	5	3	1	8
4	3	8	2	1	6	5	9	7
8	5	3	7	4	1	6	2	9
9	7	1	6	5	2	4	8	3
6	4	2	8	3	9	7	5	1

Solution # 776

7	4	8	6	2	1	3	9	5
5	9	1	4	3	7	6	2	8
6	3	2	9	8	5	1	4	7
9	7	5	8	1	4	2	3	6
8	2	6	5	9	3	7	1	4
4	1	3	2	7	6	5	8	9
3	8	7	1	5	9	4	6	2
2	5	4	3	6	8	9	7	1
1	6	9	7	4	2	8	5	3

Solution # 777

3	7	2	9	1	8	6	4	5
6	5	8	3	7	4	1	2	9
9	1	4	2	5	6	3	7	8
8	9	1	7	6	2	4	5	3
4	2	5	8	3	1	7	9	6
7	6	3	5	4	9	8	1	2
1	3	6	4	2	5	9	8	7
5	4	9	6	8	7	2	3	1
2	8	7	1	9	3	5	6	4

Solution # 778

9	3	1	8	5	7	2	6	4
8	6	4	9	2	3	5	1	7
7	5	2	4	1	6	8	9	3
5	4	8	6	3	2	9	7	1
1	9	6	7	8	4	3	5	2
2	7	3	1	9	5	6	4	8
4	8	7	3	6	9	1	2	5
3	2	9	5	4	1	7	8	6
6	1	5	2	7	8	4	3	9

Solution # 779

6	7	8	3	2	9	5	1	4
9	4	3	7	1	5	2	8	6
1	5	2	8	6	4	7	9	3
3	2	7	4	9	6	8	5	1
5	8	1	2	7	3	6	4	9
4	9	6	1	5	8	3	7	2
2	6	5	9	8	1	4	3	7
7	1	4	5	3	2	9	6	8
8	3	9	6	4	7	1	2	5

Solution # 780

1	4	8	7	6	5	2	9	3
3	6	7	2	1	9	5	8	4
2	5	9	3	8	4	6	7	1
5	9	4	1	3	7	8	6	2
6	3	1	5	2	8	9	4	7
8	7	2	9	4	6	3	1	5
9	2	6	4	5	1	7	3	8
4	8	5	6	7	3	1	2	9
7	1	3	8	9	2	4	5	6

Solution # 781

6	3	7	4	2	9	5	1	8
2	8	4	1	7	5	6	3	9
1	5	9	8	3	6	7	2	4
4	2	6	3	1	8	9	5	7
8	1	3	5	9	7	2	4	6
9	7	5	6	4	2	3	8	1
5	9	8	2	6	1	4	7	3
7	4	2	9	8	3	1	6	5
3	6	1	7	5	4	8	9	2

Solution # 782

7	5	8	1	4	2	3	9	6
1	9	6	3	7	5	8	4	2
4	3	2	8	9	6	7	1	5
8	1	7	6	3	9	5	2	4
3	4	5	7	2	1	6	8	9
2	6	9	4	5	8	1	7	3
6	2	3	9	8	7	4	5	1
5	7	1	2	6	4	9	3	8
9	8	4	5	1	3	2	6	7

Solution # 783

9	4	8	1	3	2	6	7	5
1	5	3	4	7	6	2	9	8
7	6	2	8	9	5	3	1	4
2	9	4	6	5	7	8	3	1
8	3	7	9	4	1	5	2	6
6	1	5	3	2	8	7	4	9
5	2	9	7	8	4	1	6	3
4	7	1	5	6	3	9	8	2
3	8	6	2	1	9	4	5	7

Solution # 784

1	2	5	3	7	6	8	4	9
8	4	7	5	9	1	2	3	6
9	6	3	2	8	4	7	5	1
2	8	9	4	5	7	6	1	3
3	7	4	1	6	9	5	8	2
6	5	1	8	2	3	4	9	7
4	9	2	6	3	8	1	7	5
7	1	6	9	4	5	3	2	8
5	3	8	7	1	2	9	6	4

Solution # 785

7	3	5	4	8	2	1	6	9
4	2	6	9	1	7	5	3	8
9	1	8	5	3	6	2	4	7
2	8	1	3	4	9	6	7	5
6	7	9	2	5	8	3	1	4
3	5	4	6	7	1	8	9	2
1	4	2	7	6	5	9	8	3
5	6	3	8	9	4	7	2	1
8	9	7	1	2	3	4	5	6

Solution # 786

8	5	9	6	2	3	7	4	1
1	6	3	8	4	7	9	5	2
7	4	2	9	1	5	6	8	3
5	8	7	3	6	1	2	9	4
3	2	6	4	5	9	8	1	7
9	1	4	7	8	2	5	3	6
2	7	5	1	9	4	3	6	8
4	3	8	5	7	6	1	2	9
6	9	1	2	3	8	4	7	5

Solution # 787

7	8	6	3	2	5	9	4	1
3	9	1	7	8	4	6	2	5
4	2	5	1	6	9	3	7	8
1	3	4	6	9	7	8	5	2
8	6	2	5	4	1	7	3	9
5	7	9	8	3	2	1	6	4
2	5	8	9	7	6	4	1	3
6	4	3	2	1	8	5	9	7
9	1	7	4	5	3	2	8	6

Solution # 788

2	9	7	3	5	4	8	6	1
3	6	5	2	1	8	4	7	9
4	1	8	7	6	9	5	2	3
6	8	1	9	4	5	2	3	7
7	2	4	6	3	1	9	8	5
9	5	3	8	7	2	1	4	6
8	4	6	1	9	7	3	5	2
5	7	9	4	2	3	6	1	8
1	3	2	5	8	6	7	9	4

Solution # 789

8	2	3	6	7	4	9	5	1
1	9	6	5	8	3	4	7	2
5	7	4	9	1	2	6	8	3
3	5	9	1	6	8	7	2	4
7	6	1	4	2	5	8	3	9
2	4	8	3	9	7	1	6	5
9	1	2	7	3	6	5	4	8
4	8	7	2	5	1	3	9	6
6	3	5	8	4	9	2	1	7

Solution # 790

9	6	7	5	2	8	3	4	1
3	8	1	4	9	7	5	2	6
5	4	2	1	6	3	7	9	8
8	5	4	2	7	6	1	3	9
1	7	9	8	3	5	4	6	2
2	3	6	9	4	1	8	7	5
4	2	5	3	1	9	6	8	7
6	1	3	7	8	2	9	5	4
7	9	8	6	5	4	2	1	3

Solution # 791

8	5	6	4	3	1	2	7	9
4	2	9	7	5	8	6	1	3
7	3	1	2	9	6	8	4	5
1	4	2	6	8	3	5	9	7
5	6	8	9	1	7	3	2	4
9	7	3	5	2	4	1	8	6
2	8	4	3	6	9	7	5	1
6	1	7	8	4	5	9	3	2
3	9	5	1	7	2	4	6	8

Solution # 792

5	8	4	9	2	7	1	3	6
3	2	7	1	6	5	8	4	9
6	1	9	8	4	3	7	5	2
4	7	6	3	8	1	2	9	5
2	5	1	4	7	9	3	6	8
9	3	8	6	5	2	4	1	7
7	9	2	5	1	4	6	8	3
8	4	5	2	3	6	9	7	1
1	6	3	7	9	8	5	2	4

Solution # 793

3	9	7	2	4	1	8	5	6
4	5	1	6	9	8	3	2	7
6	8	2	7	3	5	1	9	4
9	7	4	1	2	3	6	8	5
2	1	6	8	5	9	4	7	3
8	3	5	4	6	7	9	1	2
1	4	8	5	7	6	2	3	9
5	2	3	9	8	4	7	6	1
7	6	9	3	1	2	5	4	8

Solution # 794

9	4	6	5	7	8	1	2	3
1	8	3	6	9	2	5	4	7
2	7	5	1	4	3	6	9	8
6	3	2	4	8	5	7	1	9
7	5	4	3	1	9	8	6	2
8	9	1	2	6	7	4	3	5
4	2	8	7	3	6	9	5	1
5	1	9	8	2	4	3	7	6
3	6	7	9	5	1	2	8	4

Solution # 795

6	9	2	8	3	1	7	5	4
4	8	3	5	7	6	9	2	1
7	5	1	4	2	9	8	6	3
1	6	9	2	4	5	3	7	8
3	2	4	7	9	8	5	1	6
5	7	8	6	1	3	4	9	2
8	3	5	9	6	2	1	4	7
9	4	6	1	8	7	2	3	5
2	1	7	3	5	4	6	8	9

Solution # 796

8	5	4	3	6	2	7	1	9
3	1	9	4	7	8	2	5	6
6	7	2	5	9	1	3	8	4
5	2	3	1	8	6	9	4	7
9	8	6	7	2	4	1	3	5
1	4	7	9	5	3	8	6	2
4	9	5	8	1	7	6	2	3
2	3	8	6	4	9	5	7	1
7	6	1	2	3	5	4	9	8

Solution # 797

3	9	6	5	4	8	7	1	2
2	7	8	3	6	1	9	4	5
4	1	5	9	2	7	3	8	6
9	3	2	7	1	6	8	5	4
8	6	1	4	5	3	2	9	7
7	5	4	8	9	2	6	3	1
1	2	9	6	3	5	4	7	8
5	8	3	2	7	4	1	6	9
6	4	7	1	8	9	5	2	3

Solution # 798

5	2	3	1	8	9	4	6	7
7	8	1	4	5	6	9	2	3
4	6	9	3	7	2	1	8	5
3	9	4	6	2	5	8	7	1
2	5	8	7	9	1	6	3	4
6	1	7	8	4	3	5	9	2
8	3	2	5	6	4	7	1	9
9	4	6	2	1	7	3	5	8
1	7	5	9	3	8	2	4	6

Solution # 799

6	8	2	1	5	3	7	9	4
1	5	9	6	4	7	8	3	2
4	7	3	9	8	2	5	1	6
2	4	7	5	9	1	6	8	3
9	6	1	7	3	8	2	4	5
5	3	8	2	6	4	1	7	9
7	1	4	3	2	5	9	6	8
3	9	5	8	7	6	4	2	1
8	2	6	4	1	9	3	5	7

Solution # 800

6	1	9	8	2	7	4	5	3
3	8	5	4	6	9	2	1	7
2	7	4	5	3	1	6	8	9
5	3	8	7	4	6	1	9	2
1	4	2	9	5	3	7	6	8
9	6	7	1	8	2	5	3	4
4	9	6	2	1	8	3	7	5
7	5	1	3	9	4	8	2	6
8	2	3	6	7	5	9	4	1

Solution # 801

8	3	6	2	4	7	9	5	1
2	1	9	6	5	8	4	3	7
7	5	4	9	3	1	6	8	2
4	7	3	5	9	6	2	1	8
6	8	1	7	2	3	5	9	4
5	9	2	1	8	4	7	6	3
3	2	7	8	6	5	1	4	9
1	4	5	3	7	9	8	2	6
9	6	8	4	1	2	3	7	5

Solution # 802

1	8	2	7	9	3	6	5	4
9	7	4	6	5	1	8	2	3
5	6	3	8	2	4	1	7	9
7	4	1	5	8	9	3	6	2
3	2	9	4	6	7	5	1	8
8	5	6	3	1	2	9	4	7
6	3	5	2	7	8	4	9	1
2	1	8	9	4	5	7	3	6
4	9	7	1	3	6	2	8	5

Solution # 803

6	2	7	3	5	4	8	9	1
3	4	1	9	8	2	6	7	5
5	9	8	7	1	6	3	2	4
9	1	5	6	3	7	4	8	2
2	6	4	1	9	8	5	3	7
7	8	3	2	4	5	9	1	6
8	7	6	4	2	3	1	5	9
1	5	2	8	6	9	7	4	3
4	3	9	5	7	1	2	6	8

Solution # 804

6	8	3	9	4	5	2	1	7
2	1	9	3	8	7	6	5	4
5	4	7	6	1	2	8	3	9
7	6	8	4	9	3	5	2	1
4	3	5	8	2	1	7	9	6
9	2	1	5	7	6	4	8	3
3	9	4	7	5	8	1	6	2
8	7	2	1	6	9	3	4	5
1	5	6	2	3	4	9	7	8

Solution # 805

8	9	3	1	5	7	6	2	4
5	7	2	6	3	4	9	1	8
1	6	4	8	9	2	3	7	5
2	4	7	5	1	6	8	3	9
6	5	8	9	7	3	2	4	1
9	3	1	2	4	8	5	6	7
3	8	9	7	2	1	4	5	6
7	2	5	4	6	9	1	8	3
4	1	6	3	8	5	7	9	2

Solution # 806

2	9	5	3	8	7	1	4	6
8	6	7	1	2	4	9	5	3
3	4	1	5	9	6	8	7	2
1	7	9	8	6	3	4	2	5
5	2	6	4	7	1	3	8	9
4	8	3	2	5	9	6	1	7
9	5	8	6	1	2	7	3	4
7	3	2	9	4	8	5	6	1
6	1	4	7	3	5	2	9	8

Solution # 807

9	8	2	7	3	6	1	5	4
4	3	7	1	9	5	2	6	8
5	1	6	2	8	4	9	7	3
3	6	8	9	5	1	7	4	2
7	5	9	8	4	2	6	3	1
1	2	4	3	6	7	5	8	9
6	4	1	5	2	3	8	9	7
8	7	5	4	1	9	3	2	6
2	9	3	6	7	8	4	1	5

Solution # 808

7	8	2	6	3	5	4	9	1
6	1	9	7	2	4	8	5	3
3	4	5	8	9	1	7	2	6
1	5	3	2	8	7	9	6	4
8	7	6	5	4	9	3	1	2
2	9	4	1	6	3	5	7	8
5	3	7	4	1	2	6	8	9
9	6	1	3	7	8	2	4	5
4	2	8	9	5	6	1	3	7

Solution # 809

4	7	1	8	2	3	5	9	6
3	6	9	4	5	7	8	1	2
8	2	5	9	1	6	7	4	3
1	8	2	7	6	4	3	5	9
7	4	3	1	9	5	6	2	8
5	9	6	3	8	2	1	7	4
2	3	7	6	4	1	9	8	5
6	5	8	2	7	9	4	3	1
9	1	4	5	3	8	2	6	7

Solution # 810

3	7	2	8	4	9	6	1	5
6	5	8	3	1	2	4	9	7
9	1	4	5	6	7	3	2	8
5	9	6	4	3	8	2	7	1
8	4	3	2	7	1	5	6	9
7	2	1	9	5	6	8	4	3
4	3	9	1	2	5	7	8	6
1	6	5	7	8	4	9	3	2
2	8	7	6	9	3	1	5	4

Solution # 811

8	5	6	3	7	9	2	1	4
4	3	2	8	6	1	5	7	9
7	1	9	4	2	5	3	6	8
3	6	1	7	8	2	9	4	5
2	9	7	5	3	4	1	8	6
5	4	8	1	9	6	7	2	3
9	2	3	6	4	7	8	5	1
1	7	4	9	5	8	6	3	2
6	8	5	2	1	3	4	9	7

Solution # 812

5	7	9	6	8	3	1	4	2
2	8	4	1	5	9	3	6	7
6	3	1	7	2	4	8	9	5
9	4	8	2	1	7	6	5	3
3	1	2	5	9	6	7	8	4
7	6	5	3	4	8	9	2	1
8	9	3	4	7	5	2	1	6
1	5	6	9	3	2	4	7	8
4	2	7	8	6	1	5	3	9

Solution # 813

7	2	8	3	4	9	6	1	5
4	3	6	1	2	5	9	8	7
1	5	9	7	6	8	2	4	3
9	6	1	2	3	4	5	7	8
2	8	3	5	7	1	4	9	6
5	4	7	8	9	6	1	3	2
8	7	4	9	5	2	3	6	1
6	1	2	4	8	3	7	5	9
3	9	5	6	1	7	8	2	4

Solution # 814

7	2	3	9	4	6	5	8	1
6	9	8	2	5	1	7	3	4
4	1	5	3	7	8	9	2	6
3	5	6	4	1	9	8	7	2
1	8	9	6	2	7	4	5	3
2	7	4	5	8	3	6	1	9
8	4	7	1	9	2	3	6	5
9	6	2	7	3	5	1	4	8
5	3	1	8	6	4	2	9	7

Solution # 815

9	3	5	7	1	8	4	2	6
8	4	1	6	2	9	3	7	5
6	2	7	5	3	4	1	8	9
2	7	6	3	9	1	5	4	8
3	5	8	4	7	6	2	9	1
1	9	4	2	8	5	7	6	3
5	8	9	1	4	7	6	3	2
7	1	2	8	6	3	9	5	4
4	6	3	9	5	2	8	1	7

Solution # 816

1	8	7	5	3	4	9	6	2
5	4	9	2	7	6	3	1	8
6	2	3	9	1	8	5	4	7
4	5	2	1	8	7	6	9	3
9	3	1	6	4	2	8	7	5
8	7	6	3	9	5	4	2	1
7	1	5	4	6	3	2	8	9
2	9	4	8	5	1	7	3	6
3	6	8	7	2	9	1	5	4

Solution # 817

9	6	3	2	8	1	4	5	7
8	5	2	7	6	4	3	9	1
7	1	4	9	5	3	2	6	8
6	9	7	5	1	2	8	4	3
1	4	8	3	7	6	9	2	5
2	3	5	8	4	9	7	1	6
5	2	9	1	3	8	6	7	4
4	8	1	6	9	7	5	3	2
3	7	6	4	2	5	1	8	9

Solution # 818

4	5	6	3	9	7	8	2	1
3	9	2	4	1	8	5	7	6
1	7	8	6	5	2	4	3	9
8	2	1	7	6	5	9	4	3
7	3	9	1	8	4	6	5	2
5	6	4	9	2	3	7	1	8
6	4	7	8	3	1	2	9	5
9	1	5	2	7	6	3	8	4
2	8	3	5	4	9	1	6	7

Solution # 819

5	9	1	4	7	6	2	8	3
4	3	2	1	5	8	6	7	9
8	7	6	2	9	3	1	5	4
7	6	9	8	2	5	3	4	1
2	4	8	7	3	1	5	9	6
1	5	3	6	4	9	7	2	8
9	8	5	3	6	2	4	1	7
6	1	4	5	8	7	9	3	2
3	2	7	9	1	4	8	6	5

Solution # 820

3	5	7	4	8	2	1	9	6
9	6	2	1	3	7	8	4	5
8	1	4	6	5	9	7	2	3
4	3	6	5	7	1	2	8	9
7	9	8	2	6	3	4	5	1
5	2	1	8	9	4	6	3	7
2	8	5	9	1	6	3	7	4
1	4	3	7	2	5	9	6	8
6	7	9	3	4	8	5	1	2

Solution # 821

8	2	5	7	9	4	1	6	3
7	4	1	6	3	5	9	8	2
3	6	9	8	1	2	7	5	4
1	9	4	3	7	8	6	2	5
6	5	8	1	2	9	3	4	7
2	3	7	5	4	6	8	1	9
4	7	6	2	8	3	5	9	1
9	8	3	4	5	1	2	7	6
5	1	2	9	6	7	4	3	8

Solution # 822

9	5	2	6	4	7	3	8	1
4	1	6	5	8	3	2	7	9
7	8	3	9	2	1	4	6	5
2	6	5	4	3	9	8	1	7
1	9	8	2	7	5	6	3	4
3	7	4	8	1	6	9	5	2
8	2	7	3	5	4	1	9	6
6	3	1	7	9	2	5	4	8
5	4	9	1	6	8	7	2	3

Solution # 823

8	5	3	4	1	2	6	7	9
2	6	7	5	3	9	8	1	4
1	9	4	8	7	6	2	5	3
9	3	1	6	5	8	4	2	7
6	2	5	9	4	7	3	8	1
7	4	8	1	2	3	9	6	5
4	1	6	3	8	5	7	9	2
3	8	2	7	9	1	5	4	6
5	7	9	2	6	4	1	3	8

Solution # 824

4	8	3	2	7	6	9	5	1
9	2	7	1	5	3	8	4	6
6	1	5	8	4	9	7	3	2
8	5	1	7	9	2	3	6	4
2	7	6	3	1	4	5	9	8
3	9	4	6	8	5	2	1	7
7	6	9	5	2	1	4	8	3
1	4	2	9	3	8	6	7	5
5	3	8	4	6	7	1	2	9

Solution # 825

9	7	3	4	2	8	5	6	1
6	4	1	9	5	3	8	7	2
5	2	8	6	1	7	3	4	9
4	3	9	2	6	5	1	8	7
2	1	5	7	8	4	6	9	3
7	8	6	3	9	1	2	5	4
1	6	7	5	4	2	9	3	8
8	9	4	1	3	6	7	2	5
3	5	2	8	7	9	4	1	6

Solution # 826

2	9	8	3	7	4	6	5	1
1	6	4	5	8	2	9	3	7
3	7	5	1	9	6	8	4	2
9	2	6	8	3	1	4	7	5
4	8	7	9	2	5	1	6	3
5	1	3	6	4	7	2	9	8
6	4	1	7	5	8	3	2	9
7	3	2	4	1	9	5	8	6
8	5	9	2	6	3	7	1	4

Solution # 827

4	7	9	3	2	6	1	8	5
6	3	8	9	1	5	4	7	2
5	2	1	7	4	8	9	3	6
2	4	3	1	5	9	8	6	7
1	8	7	4	6	3	5	2	9
9	5	6	2	8	7	3	1	4
7	1	4	5	3	2	6	9	8
8	9	5	6	7	1	2	4	3
3	6	2	8	9	4	7	5	1

Solution # 828

2	4	6	8	5	3	7	1	9
3	7	9	2	1	6	5	8	4
5	1	8	4	9	7	2	3	6
6	9	5	1	8	4	3	7	2
4	8	7	3	2	9	1	6	5
1	3	2	7	6	5	4	9	8
9	6	4	5	3	1	8	2	7
7	2	3	6	4	8	9	5	1
8	5	1	9	7	2	6	4	3

Solution # 829

4	8	7	3	2	1	9	5	6
5	9	3	4	6	8	2	7	1
1	6	2	5	9	7	8	4	3
3	1	9	2	7	4	6	8	5
8	7	5	6	1	3	4	9	2
6	2	4	8	5	9	3	1	7
9	5	8	1	3	6	7	2	4
7	3	1	9	4	2	5	6	8
2	4	6	7	8	5	1	3	9

Solution # 830

7	4	6	1	5	9	2	8	3
3	5	1	4	2	8	7	9	6
9	2	8	7	3	6	1	4	5
5	1	4	6	8	7	9	3	2
8	9	3	2	1	5	4	6	7
2	6	7	3	9	4	5	1	8
4	8	9	5	6	2	3	7	1
6	3	2	9	7	1	8	5	4
1	7	5	8	4	3	6	2	9

Solution # 831

5	9	4	8	3	6	2	7	1
7	6	8	5	1	2	4	9	3
2	1	3	4	7	9	8	6	5
1	7	2	9	5	3	6	4	8
4	5	9	7	6	8	3	1	2
8	3	6	2	4	1	9	5	7
6	4	5	3	2	7	1	8	9
3	8	7	1	9	4	5	2	6
9	2	1	6	8	5	7	3	4

Solution # 832

2	5	8	4	9	6	7	1	3
3	7	9	8	1	5	2	4	6
4	6	1	2	7	3	9	5	8
7	8	2	5	6	9	4	3	1
1	4	5	3	2	8	6	7	9
9	3	6	1	4	7	5	8	2
5	1	7	9	3	2	8	6	4
8	2	3	6	5	4	1	9	7
6	9	4	7	8	1	3	2	5

Solution # 833

3	1	4	2	9	5	7	6	8
6	8	5	1	7	4	3	9	2
9	2	7	6	8	3	4	1	5
4	6	1	8	2	9	5	3	7
2	5	3	7	4	1	6	8	9
8	7	9	3	5	6	2	4	1
1	3	8	5	6	7	9	2	4
7	4	2	9	3	8	1	5	6
5	9	6	4	1	2	8	7	3

Solution # 834

9	2	8	7	6	4	1	3	5
4	6	5	1	3	2	8	7	9
7	3	1	8	5	9	6	2	4
8	7	4	2	1	3	5	9	6
5	1	2	9	8	6	3	4	7
3	9	6	4	7	5	2	8	1
1	4	3	6	2	7	9	5	8
6	5	9	3	4	8	7	1	2
2	8	7	5	9	1	4	6	3

Solution # 835

1	6	4	8	5	7	9	2	3
3	5	9	2	1	6	7	4	8
7	8	2	9	4	3	6	1	5
9	7	5	1	3	4	2	8	6
4	2	1	5	6	8	3	7	9
6	3	8	7	9	2	4	5	1
5	4	7	3	8	9	1	6	2
8	9	6	4	2	1	5	3	7
2	1	3	6	7	5	8	9	4

Solution # 836

9	1	3	8	2	4	5	7	6
7	6	8	1	5	9	3	4	2
2	5	4	3	6	7	1	9	8
6	4	7	9	8	3	2	5	1
1	9	2	6	4	5	7	8	3
3	8	5	2	7	1	9	6	4
4	3	1	7	9	8	6	2	5
8	2	9	5	3	6	4	1	7
5	7	6	4	1	2	8	3	9

Solution # 837

5	9	6	4	1	3	2	7	8
8	1	3	7	5	2	4	6	9
4	7	2	8	6	9	3	5	1
2	8	4	9	7	1	5	3	6
7	3	1	6	8	5	9	4	2
9	6	5	3	2	4	1	8	7
3	5	8	2	9	6	7	1	4
1	2	7	5	4	8	6	9	3
6	4	9	1	3	7	8	2	5

Solution # 838

5	1	6	8	2	9	3	7	4
9	4	7	1	5	3	8	6	2
2	8	3	7	6	4	1	9	5
7	5	4	9	3	2	6	8	1
8	6	9	4	1	5	2	3	7
3	2	1	6	8	7	5	4	9
1	7	5	3	4	6	9	2	8
6	9	8	2	7	1	4	5	3
4	3	2	5	9	8	7	1	6

Solution # 839

6	5	8	9	4	2	3	1	7
2	9	3	5	7	1	6	8	4
4	7	1	6	8	3	5	9	2
5	8	7	1	2	9	4	3	6
3	4	2	7	6	8	9	5	1
1	6	9	3	5	4	2	7	8
7	1	5	4	3	6	8	2	9
9	2	4	8	1	5	7	6	3
8	3	6	2	9	7	1	4	5

Solution # 840

7	5	2	3	9	6	4	8	1
8	4	3	2	7	1	9	6	5
9	1	6	4	8	5	3	7	2
1	9	7	8	2	3	6	5	4
5	6	8	1	4	9	2	3	7
3	2	4	5	6	7	8	1	9
2	3	5	6	1	4	7	9	8
6	8	9	7	5	2	1	4	3
4	7	1	9	3	8	5	2	6

Solution # 841

6	7	8	4	3	2	5	9	1
2	4	5	1	8	9	6	7	3
3	1	9	6	5	7	8	4	2
8	3	4	2	6	1	7	5	9
7	5	6	8	9	3	2	1	4
9	2	1	5	7	4	3	8	6
5	9	2	7	4	6	1	3	8
1	8	3	9	2	5	4	6	7
4	6	7	3	1	8	9	2	5

Solution # 842

2	4	5	8	7	9	1	3	6
8	6	9	1	5	3	4	7	2
3	7	1	2	4	6	8	5	9
5	9	2	7	8	1	3	6	4
7	8	4	3	6	2	9	1	5
6	1	3	4	9	5	2	8	7
9	3	8	6	2	7	5	4	1
4	5	7	9	1	8	6	2	3
1	2	6	5	3	4	7	9	8

Solution # 843

4	7	9	5	6	1	3	8	2
1	8	5	3	9	2	6	7	4
3	2	6	8	4	7	1	5	9
5	4	8	2	1	3	7	9	6
6	1	2	4	7	9	8	3	5
7	9	3	6	5	8	2	4	1
2	5	1	7	3	4	9	6	8
9	6	7	1	8	5	4	2	3
8	3	4	9	2	6	5	1	7

Solution # 844

3	4	2	1	6	5	9	8	7
7	8	5	4	2	9	1	6	3
1	6	9	8	7	3	5	4	2
6	3	8	5	9	2	7	1	4
4	9	7	6	1	8	3	2	5
5	2	1	7	3	4	6	9	8
8	5	3	9	4	1	2	7	6
9	7	4	2	5	6	8	3	1
2	1	6	3	8	7	4	5	9

Solution # 845

3	7	9	5	1	8	6	2	4
6	4	8	3	9	2	1	5	7
5	2	1	4	7	6	9	3	8
9	6	4	8	2	1	3	7	5
7	8	5	6	3	9	4	1	2
1	3	2	7	5	4	8	9	6
4	9	3	2	6	7	5	8	1
8	5	7	1	4	3	2	6	9
2	1	6	9	8	5	7	4	3

Solution # 846

6	5	7	8	4	1	2	9	3
4	9	8	2	3	6	7	5	1
2	1	3	7	9	5	4	6	8
5	7	2	3	8	4	6	1	9
8	6	1	9	5	7	3	2	4
3	4	9	6	1	2	5	8	7
9	8	5	4	6	3	1	7	2
7	3	6	1	2	9	8	4	5
1	2	4	5	7	8	9	3	6

Solution # 847

3	9	8	6	7	5	4	1	2
5	2	7	4	1	8	9	6	3
4	6	1	3	9	2	5	8	7
2	8	5	7	6	3	1	4	9
7	1	6	9	5	4	3	2	8
9	3	4	2	8	1	6	7	5
6	5	2	8	4	9	7	3	1
8	7	9	1	3	6	2	5	4
1	4	3	5	2	7	8	9	6

Solution # 848

7	8	4	6	3	1	9	5	2
3	2	5	7	4	9	6	1	8
6	9	1	8	5	2	4	7	3
8	3	6	9	1	4	5	2	7
5	7	2	3	8	6	1	9	4
4	1	9	5	2	7	3	8	6
9	5	8	2	6	3	7	4	1
2	4	3	1	7	5	8	6	9
1	6	7	4	9	8	2	3	5

Solution # 849

9	2	7	8	6	3	5	4	1
3	6	1	4	5	2	9	8	7
8	4	5	9	1	7	6	3	2
6	8	9	2	4	1	3	7	5
1	3	2	5	7	8	4	6	9
5	7	4	3	9	6	1	2	8
4	5	3	7	2	9	8	1	6
7	9	6	1	8	4	2	5	3
2	1	8	6	3	5	7	9	4

Solution # 850

5	6	7	3	9	1	4	8	2
8	2	3	7	4	6	9	5	1
1	9	4	2	8	5	7	6	3
7	8	2	9	5	3	6	1	4
3	1	6	8	7	4	2	9	5
9	4	5	6	1	2	3	7	8
2	7	9	1	3	8	5	4	6
6	5	1	4	2	9	8	3	7
4	3	8	5	6	7	1	2	9

Solution # 851

6	2	7	3	8	4	5	9	1
4	9	5	1	2	6	7	8	3
1	3	8	5	7	9	6	4	2
7	1	6	9	5	3	4	2	8
9	4	3	2	6	8	1	7	5
5	8	2	7	4	1	3	6	9
3	7	4	8	9	5	2	1	6
2	5	9	6	1	7	8	3	4
8	6	1	4	3	2	9	5	7

Solution # 852

8	3	9	7	2	1	5	6	4
1	7	4	5	6	3	2	8	9
6	5	2	9	4	8	1	3	7
5	4	6	1	8	7	3	9	2
7	1	3	4	9	2	8	5	6
9	2	8	3	5	6	4	7	1
3	6	1	8	7	4	9	2	5
4	9	7	2	3	5	6	1	8
2	8	5	6	1	9	7	4	3

Solution # 853

4	1	3	7	5	8	9	2	6
5	2	8	9	6	1	4	7	3
7	6	9	3	4	2	1	8	5
3	4	7	2	9	5	8	6	1
9	8	6	1	3	4	7	5	2
1	5	2	6	8	7	3	9	4
6	9	4	5	7	3	2	1	8
8	7	1	4	2	6	5	3	9
2	3	5	8	1	9	6	4	7

Solution # 854

6	5	1	8	4	7	9	3	2
9	8	7	2	3	6	5	4	1
2	3	4	5	9	1	7	6	8
3	7	5	6	8	4	1	2	9
8	4	9	1	5	2	6	7	3
1	2	6	9	7	3	4	8	5
4	6	2	3	1	9	8	5	7
5	1	3	7	6	8	2	9	4
7	9	8	4	2	5	3	1	6

Solution # 855

3	6	4	1	9	2	7	8	5
2	7	8	5	6	3	1	4	9
1	9	5	8	4	7	2	6	3
8	1	6	4	3	5	9	7	2
9	3	7	2	8	6	4	5	1
5	4	2	9	7	1	6	3	8
4	2	9	7	5	8	3	1	6
6	5	1	3	2	4	8	9	7
7	8	3	6	1	9	5	2	4

Solution # 856

3	6	9	5	1	7	2	4	8
2	7	1	4	9	8	3	6	5
4	8	5	2	3	6	9	7	1
1	3	8	6	2	9	4	5	7
5	2	7	1	8	4	6	9	3
6	9	4	3	7	5	1	8	2
9	5	2	8	6	1	7	3	4
8	1	6	7	4	3	5	2	9
7	4	3	9	5	2	8	1	6

Solution # 857

5	4	9	1	2	7	6	3	8
6	3	1	4	8	5	9	2	7
2	7	8	6	9	3	1	5	4
3	6	4	2	7	8	5	1	9
1	9	2	5	3	4	7	8	6
7	8	5	9	1	6	3	4	2
9	5	7	8	4	1	2	6	3
4	1	3	7	6	2	8	9	5
8	2	6	3	5	9	4	7	1

Solution # 858

2	5	9	6	4	3	7	8	1
7	4	1	8	9	2	5	6	3
3	8	6	7	5	1	2	4	9
6	7	3	9	2	5	4	1	8
4	9	8	1	3	7	6	2	5
5	1	2	4	8	6	9	3	7
8	3	7	2	6	9	1	5	4
9	6	5	3	1	4	8	7	2
1	2	4	5	7	8	3	9	6

Solution # 859

3	5	1	6	8	7	2	4	9
4	6	7	2	9	5	1	8	3
9	8	2	4	1	3	6	5	7
8	2	6	1	7	4	9	3	5
5	9	3	8	2	6	7	1	4
1	7	4	5	3	9	8	6	2
2	3	8	9	5	1	4	7	6
7	4	9	3	6	8	5	2	1
6	1	5	7	4	2	3	9	8

Solution # 860

9	3	1	2	7	5	6	4	8
4	5	8	9	3	6	7	1	2
2	7	6	4	8	1	5	9	3
8	1	5	3	9	4	2	7	6
3	6	2	7	1	8	9	5	4
7	9	4	5	6	2	3	8	1
6	4	9	1	5	3	8	2	7
5	2	3	8	4	7	1	6	9
1	8	7	6	2	9	4	3	5

Solution # 861

8	7	3	9	4	5	2	1	6
4	9	1	2	6	8	3	5	7
5	6	2	1	7	3	9	8	4
9	1	8	5	3	4	7	6	2
2	5	4	6	9	7	8	3	1
7	3	6	8	2	1	5	4	9
1	2	5	7	8	6	4	9	3
6	4	9	3	5	2	1	7	8
3	8	7	4	1	9	6	2	5

Solution # 862

7	8	4	1	9	6	2	5	3
1	6	2	8	3	5	9	4	7
3	5	9	7	2	4	6	8	1
2	3	5	9	7	1	4	6	8
4	1	6	5	8	2	7	3	9
8	9	7	4	6	3	1	2	5
5	2	8	6	1	9	3	7	4
6	7	1	3	4	8	5	9	2
9	4	3	2	5	7	8	1	6

Solution # 863

7	3	5	9	1	2	4	6	8
8	2	4	6	3	5	7	1	9
9	6	1	8	7	4	5	3	2
1	8	7	3	5	6	9	2	4
2	9	3	1	4	8	6	5	7
5	4	6	7	2	9	1	8	3
4	7	8	2	6	1	3	9	5
6	5	2	4	9	3	8	7	1
3	1	9	5	8	7	2	4	6

Solution # 864

4	3	8	6	5	7	2	1	9
5	9	6	3	2	1	7	8	4
7	2	1	9	4	8	6	5	3
9	7	5	8	1	3	4	6	2
1	4	3	7	6	2	8	9	5
6	8	2	5	9	4	1	3	7
8	6	7	4	3	5	9	2	1
3	1	4	2	8	9	5	7	6
2	5	9	1	7	6	3	4	8

Solution # 865

3	5	1	8	2	7	6	4	9
4	7	8	1	6	9	2	3	5
9	6	2	5	3	4	8	7	1
8	3	5	6	4	1	7	9	2
1	9	7	3	8	2	5	6	4
6	2	4	7	9	5	1	8	3
5	8	6	9	1	3	4	2	7
2	1	3	4	7	8	9	5	6
7	4	9	2	5	6	3	1	8

Solution # 866

5	2	7	4	8	3	1	6	9
6	1	3	7	2	9	4	5	8
9	8	4	1	6	5	2	3	7
2	3	1	8	5	6	7	9	4
7	4	5	3	9	1	6	8	2
8	6	9	2	4	7	3	1	5
3	9	2	6	7	8	5	4	1
4	5	6	9	1	2	8	7	3
1	7	8	5	3	4	9	2	6

Solution # 867

8	1	4	2	5	6	7	3	9
9	5	7	4	3	8	1	2	6
6	3	2	9	7	1	4	5	8
4	2	3	5	1	9	6	8	7
7	8	1	6	4	3	5	9	2
5	6	9	7	8	2	3	1	4
3	4	6	8	9	5	2	7	1
1	7	8	3	2	4	9	6	5
2	9	5	1	6	7	8	4	3

Solution # 868

6	4	1	7	8	3	5	9	2
2	9	5	6	4	1	7	8	3
8	3	7	9	5	2	1	4	6
7	1	4	5	9	6	3	2	8
3	2	8	1	7	4	6	5	9
9	5	6	2	3	8	4	7	1
1	7	2	8	6	5	9	3	4
4	6	9	3	2	7	8	1	5
5	8	3	4	1	9	2	6	7

Solution # 869

7	3	5	4	6	1	2	8	9
6	9	1	7	2	8	4	5	3
8	2	4	9	3	5	7	6	1
9	7	2	3	8	4	5	1	6
3	5	6	2	1	9	8	7	4
1	4	8	5	7	6	9	3	2
2	8	9	6	5	3	1	4	7
5	6	7	1	4	2	3	9	8
4	1	3	8	9	7	6	2	5

Solution # 870

9	2	8	6	7	4	3	5	1
6	7	1	2	5	3	8	4	9
5	3	4	9	8	1	7	6	2
2	6	5	4	3	8	1	9	7
3	8	7	1	9	5	4	2	6
4	1	9	7	6	2	5	3	8
8	9	6	5	4	7	2	1	3
7	4	2	3	1	6	9	8	5
1	5	3	8	2	9	6	7	4

Solution # 871

4	8	5	3	9	6	7	1	2
9	6	1	7	2	5	4	8	3
3	7	2	8	4	1	5	9	6
7	1	3	2	5	8	9	6	4
8	5	9	4	6	7	2	3	1
6	2	4	9	1	3	8	7	5
2	3	8	6	7	4	1	5	9
1	9	6	5	8	2	3	4	7
5	4	7	1	3	9	6	2	8

Solution # 872

7	1	9	8	6	4	5	3	2
5	6	2	7	9	3	4	1	8
4	3	8	2	5	1	7	6	9
8	4	3	1	2	6	9	5	7
9	5	1	4	3	7	2	8	6
6	2	7	9	8	5	1	4	3
1	8	5	3	7	2	6	9	4
3	7	4	6	1	9	8	2	5
2	9	6	5	4	8	3	7	1

Solution # 873

2	3	7	5	6	8	1	9	4
5	8	1	7	4	9	3	2	6
9	6	4	2	1	3	8	5	7
7	5	3	9	8	6	2	4	1
8	1	6	4	2	5	7	3	9
4	9	2	1	3	7	6	8	5
6	4	9	8	7	2	5	1	3
1	7	8	3	5	4	9	6	2
3	2	5	6	9	1	4	7	8

Solution # 874

7	4	9	6	3	5	2	1	8
3	1	2	7	8	9	4	5	6
5	8	6	4	1	2	3	9	7
6	2	3	8	9	4	1	7	5
4	7	1	2	5	6	9	8	3
8	9	5	1	7	3	6	2	4
2	3	8	9	4	7	5	6	1
1	6	4	5	2	8	7	3	9
9	5	7	3	6	1	8	4	2

Solution # 875

5	6	3	7	4	1	2	8	9
1	8	9	5	3	2	4	7	6
7	2	4	9	8	6	3	1	5
2	4	6	1	5	8	7	9	3
9	1	7	2	6	3	8	5	4
8	3	5	4	9	7	6	2	1
4	7	2	6	1	5	9	3	8
3	9	1	8	2	4	5	6	7
6	5	8	3	7	9	1	4	2

Solution # 876

6	3	1	5	4	2	8	9	7
8	2	9	6	7	1	4	5	3
7	4	5	3	9	8	2	1	6
1	8	2	9	5	7	6	3	4
9	6	3	8	2	4	1	7	5
5	7	4	1	3	6	9	8	2
2	9	7	4	8	5	3	6	1
4	1	8	7	6	3	5	2	9
3	5	6	2	1	9	7	4	8

Solution # 877

6	8	7	9	5	2	1	4	3
4	5	3	1	8	6	2	9	7
1	9	2	7	3	4	8	6	5
2	1	6	8	4	5	7	3	9
3	7	5	6	2	9	4	1	8
8	4	9	3	1	7	6	5	2
5	6	1	2	7	3	9	8	4
9	2	4	5	6	8	3	7	1
7	3	8	4	9	1	5	2	6

Solution # 878

1	8	6	5	7	2	4	9	3
9	5	4	3	8	1	2	7	6
7	3	2	9	6	4	8	5	1
3	1	5	7	9	8	6	4	2
4	6	7	2	5	3	9	1	8
2	9	8	1	4	6	5	3	7
6	2	1	4	3	5	7	8	9
5	7	3	8	2	9	1	6	4
8	4	9	6	1	7	3	2	5

Solution # 879

8	7	6	9	5	1	3	4	2
4	5	9	2	7	3	1	8	6
1	2	3	4	6	8	5	7	9
7	8	2	1	3	9	4	6	5
9	4	5	6	8	2	7	1	3
6	3	1	7	4	5	9	2	8
5	9	7	8	1	6	2	3	4
3	1	8	5	2	4	6	9	7
2	6	4	3	9	7	8	5	1

Solution # 880

7	4	8	3	2	6	1	9	5
6	9	2	1	5	7	3	8	4
5	1	3	4	9	8	2	6	7
9	6	1	7	8	2	4	5	3
3	7	5	6	4	9	8	2	1
8	2	4	5	3	1	9	7	6
4	8	6	9	1	5	7	3	2
2	3	7	8	6	4	5	1	9
1	5	9	2	7	3	6	4	8

Solution # 881

7	3	1	9	8	2	4	6	5
9	8	6	4	5	3	1	2	7
2	4	5	7	6	1	8	3	9
6	5	9	8	1	4	3	7	2
4	7	2	6	3	5	9	1	8
8	1	3	2	7	9	5	4	6
5	9	4	3	2	6	7	8	1
1	2	8	5	4	7	6	9	3
3	6	7	1	9	8	2	5	4

Solution # 882

4	3	7	8	5	2	6	9	1
9	8	2	6	4	1	5	3	7
5	1	6	7	3	9	2	8	4
7	2	5	1	9	3	8	4	6
6	4	8	2	7	5	9	1	3
3	9	1	4	8	6	7	2	5
1	7	9	5	2	4	3	6	8
2	5	4	3	6	8	1	7	9
8	6	3	9	1	7	4	5	2

Solution # 883

4	9	8	5	1	7	3	2	6
7	3	1	9	6	2	4	5	8
2	6	5	8	4	3	1	7	9
3	2	6	7	5	9	8	1	4
8	7	9	1	3	4	2	6	5
1	5	4	6	2	8	7	9	3
6	4	7	3	9	1	5	8	2
9	8	3	2	7	5	6	4	1
5	1	2	4	8	6	9	3	7

Solution # 884

4	9	6	7	1	5	8	3	2
8	1	3	2	6	9	7	4	5
7	5	2	3	4	8	1	9	6
1	6	5	4	2	3	9	7	8
3	7	9	8	5	1	6	2	4
2	4	8	9	7	6	5	1	3
9	3	1	5	8	2	4	6	7
6	8	7	1	3	4	2	5	9
5	2	4	6	9	7	3	8	1

Solution # 885

2	8	3	1	9	6	5	7	4
4	5	9	8	2	7	3	1	6
6	7	1	5	3	4	9	8	2
7	1	5	6	4	9	8	2	3
9	2	4	3	7	8	1	6	5
8	3	6	2	1	5	4	9	7
5	4	2	9	6	1	7	3	8
3	9	8	7	5	2	6	4	1
1	6	7	4	8	3	2	5	9

Solution # 886

7	3	8	1	5	6	2	9	4
4	5	1	2	7	9	6	3	8
6	2	9	4	8	3	7	1	5
1	4	6	5	3	7	8	2	9
3	8	2	6	9	1	5	4	7
9	7	5	8	4	2	1	6	3
2	9	7	3	6	5	4	8	1
5	1	4	9	2	8	3	7	6
8	6	3	7	1	4	9	5	2

Solution # 887

4	9	3	5	1	7	6	2	8
2	1	8	3	9	6	5	7	4
7	5	6	2	4	8	3	9	1
3	4	1	7	2	5	9	8	6
6	8	7	1	3	9	2	4	5
5	2	9	6	8	4	1	3	7
8	3	2	4	5	1	7	6	9
1	6	4	9	7	3	8	5	2
9	7	5	8	6	2	4	1	3

Solution # 888

2	9	6	3	5	1	8	7	4
3	1	4	8	6	7	9	5	2
8	5	7	9	2	4	3	6	1
4	7	9	5	1	2	6	3	8
6	2	3	7	4	8	5	1	9
1	8	5	6	3	9	2	4	7
9	3	2	4	7	6	1	8	5
5	4	8	1	9	3	7	2	6
7	6	1	2	8	5	4	9	3

Solution # 889

5	8	3	4	2	7	1	9	6
7	6	4	1	9	3	5	8	2
2	1	9	5	8	6	7	4	3
4	5	7	6	3	2	8	1	9
6	2	1	8	4	9	3	5	7
9	3	8	7	1	5	6	2	4
1	4	6	9	7	8	2	3	5
3	9	5	2	6	1	4	7	8
8	7	2	3	5	4	9	6	1

Solution # 890

4	6	7	5	9	2	3	1	8
5	9	3	8	1	7	4	2	6
1	8	2	3	6	4	5	9	7
8	1	6	7	3	5	2	4	9
3	5	4	9	2	8	6	7	1
2	7	9	6	4	1	8	3	5
9	3	8	2	7	6	1	5	4
6	2	1	4	5	9	7	8	3
7	4	5	1	8	3	9	6	2

Solution # 891

8	7	5	1	3	9	6	2	4
4	6	3	5	7	2	9	1	8
9	2	1	8	4	6	3	7	5
5	3	6	2	9	4	7	8	1
7	8	2	6	1	3	4	5	9
1	9	4	7	8	5	2	6	3
3	5	7	4	2	8	1	9	6
2	4	8	9	6	1	5	3	7
6	1	9	3	5	7	8	4	2

Solution # 892

8	1	9	4	7	6	5	3	2
3	4	7	5	9	2	6	1	8
6	2	5	1	3	8	7	9	4
7	6	4	9	5	3	2	8	1
2	5	1	8	6	4	3	7	9
9	8	3	7	2	1	4	5	6
4	9	2	3	8	5	1	6	7
1	3	8	6	4	7	9	2	5
5	7	6	2	1	9	8	4	3

Solution # 893

7	4	1	5	8	3	9	2	6
8	5	3	9	2	6	7	4	1
2	9	6	1	7	4	3	8	5
4	2	9	7	6	8	5	1	3
5	1	8	3	4	9	6	7	2
3	6	7	2	1	5	8	9	4
1	3	5	8	9	2	4	6	7
9	7	4	6	5	1	2	3	8
6	8	2	4	3	7	1	5	9

Solution # 894

9	7	4	6	2	8	5	1	3
1	8	5	3	9	7	4	2	6
2	3	6	4	5	1	7	9	8
5	1	7	2	3	6	8	4	9
4	6	3	1	8	9	2	7	5
8	9	2	7	4	5	3	6	1
3	2	8	9	1	4	6	5	7
6	5	1	8	7	2	9	3	4
7	4	9	5	6	3	1	8	2

Solution # 895

7	8	9	3	2	1	5	6	4
6	2	3	9	4	5	8	1	7
1	4	5	8	7	6	3	9	2
8	9	1	6	5	7	4	2	3
3	6	7	2	8	4	1	5	9
4	5	2	1	3	9	6	7	8
5	1	4	7	9	8	2	3	6
9	3	6	4	1	2	7	8	5
2	7	8	5	6	3	9	4	1

Solution # 896

3	6	5	2	4	9	1	8	7
8	1	9	7	5	3	6	2	4
7	2	4	1	8	6	3	9	5
4	5	6	8	3	2	9	7	1
2	9	3	6	7	1	5	4	8
1	7	8	4	9	5	2	3	6
6	8	1	3	2	4	7	5	9
5	4	2	9	6	7	8	1	3
9	3	7	5	1	8	4	6	2

Solution # 897

6	9	3	8	2	1	5	7	4
4	7	2	5	6	9	3	1	8
5	1	8	7	4	3	6	9	2
3	8	9	6	1	5	4	2	7
2	6	7	9	8	4	1	5	3
1	4	5	2	3	7	9	8	6
7	5	6	3	9	2	8	4	1
8	2	1	4	5	6	7	3	9
9	3	4	1	7	8	2	6	5

Solution # 898

2	3	5	7	6	8	9	1	4
8	4	1	3	2	9	6	7	5
6	7	9	4	1	5	8	2	3
5	1	8	2	3	4	7	9	6
9	2	7	6	5	1	3	4	8
3	6	4	8	9	7	2	5	1
4	9	3	5	8	2	1	6	7
7	8	2	1	4	6	5	3	9
1	5	6	9	7	3	4	8	2

Solution # 899

5	9	2	3	8	4	1	6	7
4	8	7	1	5	6	2	9	3
3	6	1	2	7	9	4	5	8
8	5	4	6	3	7	9	1	2
1	3	6	5	9	2	7	8	4
2	7	9	4	1	8	6	3	5
6	4	8	9	2	3	5	7	1
9	1	3	7	4	5	8	2	6
7	2	5	8	6	1	3	4	9

Solution # 900

3	6	4	8	1	5	7	2	9
8	7	2	3	6	9	5	4	1
1	5	9	4	2	7	8	3	6
2	9	7	5	8	4	1	6	3
4	3	5	1	9	6	2	8	7
6	1	8	2	7	3	4	9	5
9	4	3	7	5	8	6	1	2
5	2	6	9	4	1	3	7	8
7	8	1	6	3	2	9	5	4

Solution # 901

1	2	9	4	6	3	7	5	8
6	8	7	1	5	9	2	3	4
4	5	3	7	8	2	1	9	6
9	1	4	5	3	7	8	6	2
2	3	8	6	1	4	5	7	9
5	7	6	9	2	8	3	4	1
3	4	5	8	9	1	6	2	7
7	6	1	2	4	5	9	8	3
8	9	2	3	7	6	4	1	5

Solution # 902

8	9	1	7	5	3	6	4	2
4	3	2	6	1	9	7	5	8
7	5	6	4	8	2	1	3	9
1	2	3	5	4	6	8	9	7
5	4	9	8	2	7	3	1	6
6	8	7	9	3	1	4	2	5
3	6	5	1	9	8	2	7	4
9	1	8	2	7	4	5	6	3
2	7	4	3	6	5	9	8	1

Solution # 903

5	3	7	8	4	9	6	1	2
2	6	8	3	1	7	4	9	5
9	1	4	6	2	5	8	7	3
1	9	6	5	7	4	3	2	8
8	5	2	9	3	1	7	4	6
7	4	3	2	8	6	1	5	9
4	8	5	7	9	3	2	6	1
3	7	9	1	6	2	5	8	4
6	2	1	4	5	8	9	3	7

Solution # 904

4	8	1	7	3	2	9	6	5
7	6	9	5	1	4	2	8	3
5	3	2	9	8	6	7	1	4
6	7	4	2	9	3	1	5	8
1	9	5	8	4	7	3	2	6
8	2	3	6	5	1	4	7	9
3	4	7	1	6	8	5	9	2
2	5	6	3	7	9	8	4	1
9	1	8	4	2	5	6	3	7

Solution # 905

6	1	9	4	7	5	8	3	2
4	3	5	2	1	8	6	9	7
7	2	8	3	9	6	5	4	1
9	6	2	7	3	1	4	8	5
5	7	4	6	8	2	9	1	3
3	8	1	9	5	4	2	7	6
1	5	3	8	6	9	7	2	4
8	4	7	5	2	3	1	6	9
2	9	6	1	4	7	3	5	8

Solution # 906

5	2	7	8	3	6	9	1	4
8	9	6	2	1	4	7	5	3
4	1	3	7	9	5	8	6	2
3	4	8	9	5	2	6	7	1
1	7	2	4	6	8	3	9	5
6	5	9	1	7	3	2	4	8
2	8	1	6	4	9	5	3	7
9	3	4	5	2	7	1	8	6
7	6	5	3	8	1	4	2	9

Solution # 907

1	6	8	9	5	2	3	4	7
9	3	5	7	6	4	1	2	8
4	2	7	8	3	1	5	9	6
5	1	9	4	8	3	6	7	2
2	8	3	5	7	6	9	1	4
7	4	6	1	2	9	8	3	5
3	5	1	6	4	7	2	8	9
6	9	4	2	1	8	7	5	3
8	7	2	3	9	5	4	6	1

Solution # 908

6	9	5	7	4	3	8	1	2
4	2	3	6	8	1	7	5	9
1	8	7	2	9	5	6	3	4
2	7	8	3	6	4	1	9	5
9	6	1	8	5	2	3	4	7
3	5	4	9	1	7	2	6	8
7	4	9	1	2	6	5	8	3
8	1	2	5	3	9	4	7	6
5	3	6	4	7	8	9	2	1

Solution # 909

3	6	1	8	7	5	2	9	4
8	5	9	2	4	3	6	7	1
7	2	4	9	6	1	3	5	8
4	8	2	6	5	7	1	3	9
6	9	5	1	3	8	7	4	2
1	3	7	4	9	2	8	6	5
2	7	6	5	8	4	9	1	3
5	1	3	7	2	9	4	8	6
9	4	8	3	1	6	5	2	7

Solution # 910

8	4	5	1	2	3	7	9	6
7	6	1	9	5	4	8	3	2
9	2	3	8	7	6	1	4	5
1	3	4	7	8	5	2	6	9
5	7	2	6	1	9	3	8	4
6	9	8	3	4	2	5	1	7
4	8	6	5	3	7	9	2	1
2	1	7	4	9	8	6	5	3
3	5	9	2	6	1	4	7	8

Solution # 911

2	3	1	6	8	9	5	4	7
6	5	7	3	2	4	9	8	1
4	8	9	7	5	1	2	6	3
9	4	6	1	3	5	7	2	8
8	2	3	4	7	6	1	5	9
1	7	5	8	9	2	4	3	6
3	9	8	5	4	7	6	1	2
7	1	4	2	6	3	8	9	5
5	6	2	9	1	8	3	7	4

Solution # 912

1	5	2	4	7	9	6	8	3
6	8	9	5	2	3	4	7	1
3	4	7	1	8	6	9	2	5
4	2	1	7	9	5	3	6	8
7	3	5	8	6	4	2	1	9
8	9	6	2	3	1	5	4	7
5	6	8	3	1	2	7	9	4
9	1	4	6	5	7	8	3	2
2	7	3	9	4	8	1	5	6

Solution # 913

9	7	6	8	2	1	4	3	5
1	5	8	4	3	6	2	9	7
2	3	4	7	5	9	6	1	8
7	4	5	2	1	8	3	6	9
6	8	2	3	9	5	7	4	1
3	9	1	6	7	4	8	5	2
5	6	3	9	8	2	1	7	4
4	2	9	1	6	7	5	8	3
8	1	7	5	4	3	9	2	6

Solution # 914

9	5	8	7	4	3	2	1	6
7	2	4	9	1	6	8	3	5
3	6	1	2	5	8	7	9	4
6	4	9	3	8	7	5	2	1
5	1	3	4	6	2	9	7	8
2	8	7	5	9	1	6	4	3
8	7	6	1	2	4	3	5	9
1	3	5	6	7	9	4	8	2
4	9	2	8	3	5	1	6	7

Solution # 915

3	7	4	1	6	5	8	2	9
8	9	1	4	2	3	5	7	6
6	2	5	7	8	9	3	1	4
4	5	8	6	9	7	2	3	1
9	3	2	8	1	4	6	5	7
7	1	6	3	5	2	4	9	8
5	8	7	2	4	1	9	6	3
1	4	9	5	3	6	7	8	2
2	6	3	9	7	8	1	4	5

Solution # 916

2	4	1	3	9	8	6	5	7
8	7	3	5	6	4	9	2	1
5	6	9	1	7	2	4	8	3
6	9	8	7	1	5	3	4	2
3	5	4	2	8	6	7	1	9
1	2	7	4	3	9	8	6	5
4	1	6	9	5	7	2	3	8
7	8	5	6	2	3	1	9	4
9	3	2	8	4	1	5	7	6

Solution # 917

3	7	8	1	9	5	6	4	2
9	4	5	6	3	2	1	7	8
2	1	6	4	8	7	5	9	3
1	2	7	5	6	9	8	3	4
6	8	4	3	7	1	9	2	5
5	3	9	2	4	8	7	6	1
4	5	2	9	1	6	3	8	7
7	9	1	8	2	3	4	5	6
8	6	3	7	5	4	2	1	9

Solution # 918

1	7	9	3	6	5	8	2	4
2	3	5	4	8	7	1	9	6
4	8	6	9	2	1	7	5	3
5	9	4	1	7	2	3	6	8
7	2	8	6	3	9	4	1	5
6	1	3	8	5	4	9	7	2
8	5	2	7	9	3	6	4	1
3	4	7	2	1	6	5	8	9
9	6	1	5	4	8	2	3	7

Solution # 919

3	6	9	4	2	7	5	1	8
2	8	4	1	5	6	9	7	3
5	1	7	3	9	8	6	2	4
7	9	5	8	3	1	4	6	2
6	3	1	5	4	2	8	9	7
4	2	8	6	7	9	3	5	1
9	4	3	2	1	5	7	8	6
8	7	2	9	6	4	1	3	5
1	5	6	7	8	3	2	4	9

Solution # 920

7	1	9	2	5	3	4	6	8
3	6	2	4	8	9	7	5	1
5	8	4	1	7	6	2	9	3
4	7	5	8	3	1	9	2	6
9	2	1	5	6	7	8	3	4
6	3	8	9	2	4	5	1	7
8	9	3	7	1	2	6	4	5
2	5	6	3	4	8	1	7	9
1	4	7	6	9	5	3	8	2

Solution # 921

5	9	7	4	1	3	8	2	6
1	8	3	6	2	7	9	4	5
6	2	4	9	8	5	1	7	3
7	4	1	5	3	2	6	9	8
3	5	9	7	6	8	2	1	4
2	6	8	1	4	9	5	3	7
8	3	6	2	7	1	4	5	9
9	7	2	8	5	4	3	6	1
4	1	5	3	9	6	7	8	2

Solution # 922

8	9	6	5	7	4	1	3	2
4	1	3	2	9	8	5	6	7
5	2	7	3	1	6	8	4	9
3	7	8	6	2	5	9	1	4
9	4	2	1	8	3	7	5	6
1	6	5	9	4	7	2	8	3
2	3	9	4	5	1	6	7	8
7	5	4	8	6	2	3	9	1
6	8	1	7	3	9	4	2	5

Solution # 923

7	5	9	2	4	8	1	3	6
4	6	3	1	5	7	8	9	2
1	2	8	9	3	6	4	7	5
3	8	6	7	2	9	5	1	4
9	4	2	6	1	5	3	8	7
5	1	7	3	8	4	2	6	9
6	3	1	4	7	2	9	5	8
2	7	5	8	9	3	6	4	1
8	9	4	5	6	1	7	2	3

Solution # 924

2	6	9	4	3	1	5	7	8
7	5	1	8	9	2	6	3	4
8	3	4	6	5	7	9	1	2
4	7	6	3	8	5	1	2	9
3	9	8	2	1	6	7	4	5
5	1	2	7	4	9	8	6	3
9	2	5	1	6	3	4	8	7
6	8	7	5	2	4	3	9	1
1	4	3	9	7	8	2	5	6

Solution # 925

3	5	6	1	9	4	2	7	8
8	1	7	5	2	6	4	3	9
2	4	9	8	7	3	1	6	5
5	6	2	9	4	7	8	1	3
9	8	3	2	6	1	5	4	7
4	7	1	3	5	8	6	9	2
7	3	5	6	1	2	9	8	4
1	9	8	4	3	5	7	2	6
6	2	4	7	8	9	3	5	1

Solution # 926

7	1	4	6	5	3	9	8	2
2	3	6	8	9	7	1	5	4
8	5	9	4	2	1	7	6	3
3	9	5	2	6	4	8	1	7
4	7	2	1	8	5	6	3	9
6	8	1	3	7	9	2	4	5
1	2	7	5	4	6	3	9	8
9	4	3	7	1	8	5	2	6
5	6	8	9	3	2	4	7	1

Solution # 927

1	4	6	3	8	9	7	5	2
5	8	3	7	1	2	4	6	9
9	7	2	4	5	6	1	8	3
6	3	8	2	4	5	9	1	7
4	5	9	8	7	1	2	3	6
2	1	7	6	9	3	8	4	5
8	6	5	1	2	7	3	9	4
3	2	1	9	6	4	5	7	8
7	9	4	5	3	8	6	2	1

Solution # 928

1	8	5	9	6	2	3	4	7
4	3	2	5	1	7	6	9	8
9	7	6	8	4	3	2	1	5
2	1	8	3	7	4	5	6	9
5	6	4	1	8	9	7	2	3
3	9	7	6	2	5	4	8	1
8	4	9	7	3	6	1	5	2
7	2	1	4	5	8	9	3	6
6	5	3	2	9	1	8	7	4

Solution # 929

9	7	2	4	8	3	5	6	1
3	6	5	7	2	1	4	9	8
8	1	4	6	9	5	2	7	3
2	8	9	3	6	4	7	1	5
7	5	3	8	1	2	9	4	6
6	4	1	9	5	7	8	3	2
5	3	6	2	4	9	1	8	7
1	9	8	5	7	6	3	2	4
4	2	7	1	3	8	6	5	9

Solution # 930

6	3	8	7	4	5	2	1	9
4	9	7	2	1	6	5	3	8
2	5	1	3	8	9	7	6	4
8	4	6	5	7	3	1	9	2
3	7	9	8	2	1	6	4	5
5	1	2	9	6	4	3	8	7
1	6	5	4	9	2	8	7	3
9	8	3	6	5	7	4	2	1
7	2	4	1	3	8	9	5	6

Solution # 931

7	6	2	3	8	5	9	4	1
3	4	5	6	9	1	8	2	7
9	1	8	2	4	7	6	5	3
5	8	6	7	1	9	4	3	2
2	9	1	8	3	4	5	7	6
4	3	7	5	6	2	1	8	9
6	7	4	1	5	3	2	9	8
8	2	9	4	7	6	3	1	5
1	5	3	9	2	8	7	6	4

Solution # 932

5	8	1	6	9	3	2	7	4
2	6	3	7	4	8	5	9	1
4	7	9	5	2	1	6	8	3
7	2	5	9	3	6	1	4	8
6	9	8	4	1	2	3	5	7
1	3	4	8	5	7	9	6	2
8	4	2	1	6	9	7	3	5
9	1	7	3	8	5	4	2	6
3	5	6	2	7	4	8	1	9

Solution # 933

2	6	8	9	1	3	4	5	7
7	9	3	5	8	4	6	2	1
4	5	1	6	2	7	8	9	3
9	4	5	1	7	8	2	3	6
1	2	7	3	9	6	5	4	8
8	3	6	4	5	2	7	1	9
3	8	4	2	6	9	1	7	5
5	7	9	8	4	1	3	6	2
6	1	2	7	3	5	9	8	4

Solution # 934

8	6	5	2	1	4	9	3	7
2	3	1	6	7	9	8	5	4
7	4	9	8	3	5	1	6	2
6	2	7	4	8	1	5	9	3
4	1	8	5	9	3	2	7	6
9	5	3	7	2	6	4	8	1
3	7	2	9	4	8	6	1	5
1	9	6	3	5	2	7	4	8
5	8	4	1	6	7	3	2	9

Solution # 935

1	6	4	3	7	8	9	2	5
7	9	5	2	4	1	6	8	3
2	8	3	6	9	5	1	7	4
9	5	8	1	3	6	2	4	7
6	3	7	5	2	4	8	9	1
4	1	2	9	8	7	5	3	6
5	2	9	7	1	3	4	6	8
3	4	1	8	6	9	7	5	2
8	7	6	4	5	2	3	1	9

Solution # 936

6	5	7	8	9	3	1	2	4
9	4	3	6	1	2	8	7	5
2	1	8	4	5	7	6	3	9
4	8	1	3	6	9	2	5	7
5	6	2	1	7	4	3	9	8
3	7	9	5	2	8	4	6	1
8	9	5	2	4	6	7	1	3
1	2	4	7	3	5	9	8	6
7	3	6	9	8	1	5	4	2

Solution # 937

7	3	2	8	6	9	4	5	1
6	4	5	1	2	3	9	7	8
8	1	9	7	5	4	3	6	2
5	9	4	3	8	7	2	1	6
2	8	1	6	9	5	7	3	4
3	7	6	4	1	2	8	9	5
1	6	7	2	3	8	5	4	9
9	2	3	5	4	6	1	8	7
4	5	8	9	7	1	6	2	3

Solution # 938

3	8	1	7	2	6	5	9	4
4	7	5	1	8	9	3	2	6
9	6	2	3	5	4	7	1	8
2	4	3	5	1	7	8	6	9
6	9	7	8	4	3	2	5	1
5	1	8	9	6	2	4	3	7
1	3	9	2	7	8	6	4	5
8	5	4	6	3	1	9	7	2
7	2	6	4	9	5	1	8	3

Solution # 939

2	8	5	1	4	9	6	7	3
4	3	1	2	7	6	5	8	9
6	9	7	3	5	8	1	2	4
3	2	4	8	1	7	9	5	6
1	7	6	9	2	5	4	3	8
9	5	8	6	3	4	7	1	2
8	6	3	7	9	1	2	4	5
5	1	2	4	6	3	8	9	7
7	4	9	5	8	2	3	6	1

Solution # 940

7	2	9	5	6	8	3	4	1
1	8	4	3	2	7	5	9	6
5	3	6	1	9	4	8	2	7
9	4	7	2	5	1	6	8	3
8	6	2	9	7	3	4	1	5
3	5	1	4	8	6	9	7	2
4	1	8	6	3	2	7	5	9
6	7	5	8	1	9	2	3	4
2	9	3	7	4	5	1	6	8

Solution # 941

7	4	1	3	8	9	5	6	2
9	8	6	2	5	1	3	7	4
3	2	5	4	7	6	8	1	9
5	9	7	6	4	2	1	3	8
2	1	8	7	9	3	4	5	6
4	6	3	5	1	8	9	2	7
8	3	9	1	2	7	6	4	5
1	7	4	8	6	5	2	9	3
6	5	2	9	3	4	7	8	1

Solution # 942

8	1	3	5	6	2	7	9	4
4	7	2	9	1	8	5	6	3
5	9	6	4	7	3	2	8	1
3	5	9	6	2	4	1	7	8
2	4	1	7	8	5	6	3	9
6	8	7	1	3	9	4	2	5
9	6	4	8	5	7	3	1	2
7	2	5	3	9	1	8	4	6
1	3	8	2	4	6	9	5	7

Solution # 943

8	3	9	5	2	4	6	7	1
5	6	2	7	1	9	3	4	8
7	1	4	8	3	6	9	5	2
2	4	6	1	7	5	8	3	9
1	9	8	3	4	2	7	6	5
3	7	5	6	9	8	1	2	4
4	2	3	9	8	7	5	1	6
6	8	1	2	5	3	4	9	7
9	5	7	4	6	1	2	8	3

Solution # 944

3	7	9	2	6	4	5	1	8
2	1	5	9	7	8	4	3	6
6	8	4	1	5	3	7	2	9
9	5	2	7	4	1	6	8	3
1	3	6	5	8	2	9	7	4
8	4	7	6	3	9	1	5	2
5	6	8	4	2	7	3	9	1
7	2	1	3	9	6	8	4	5
4	9	3	8	1	5	2	6	7

Solution # 945

1	4	2	9	6	7	8	3	5
6	5	7	8	2	3	1	9	4
8	3	9	5	4	1	6	2	7
3	1	4	7	8	5	2	6	9
2	8	5	1	9	6	7	4	3
9	7	6	4	3	2	5	8	1
4	9	1	2	7	8	3	5	6
5	2	3	6	1	9	4	7	8
7	6	8	3	5	4	9	1	2

Solution # 946

2	4	8	9	5	7	6	1	3
1	5	3	6	2	4	7	8	9
7	6	9	1	3	8	2	5	4
6	1	7	8	4	5	9	3	2
5	9	2	3	7	6	8	4	1
8	3	4	2	9	1	5	6	7
4	7	1	5	6	2	3	9	8
9	8	6	7	1	3	4	2	5
3	2	5	4	8	9	1	7	6

Solution # 947

1	2	3	5	6	9	8	7	4
4	6	7	3	8	2	5	1	9
5	9	8	7	1	4	6	3	2
2	3	6	1	7	5	4	9	8
7	5	1	9	4	8	2	6	3
8	4	9	6	2	3	1	5	7
3	1	4	8	9	6	7	2	5
6	8	5	2	3	7	9	4	1
9	7	2	4	5	1	3	8	6

Solution # 948

5	2	3	7	6	4	9	1	8
1	9	4	8	3	2	7	5	6
7	8	6	5	9	1	4	2	3
8	4	1	2	7	9	3	6	5
6	3	9	1	4	5	2	8	7
2	7	5	6	8	3	1	4	9
4	6	8	3	2	7	5	9	1
3	5	2	9	1	6	8	7	4
9	1	7	4	5	8	6	3	2

Solution # 949

8	1	6	5	7	2	9	3	4
7	5	4	9	8	3	6	1	2
2	3	9	6	4	1	7	8	5
3	9	1	8	2	6	4	5	7
5	8	7	3	9	4	2	6	1
4	6	2	7	1	5	8	9	3
6	7	3	4	5	8	1	2	9
1	4	8	2	3	9	5	7	6
9	2	5	1	6	7	3	4	8

Solution # 950

9	7	3	6	5	8	4	2	1
2	1	6	7	9	4	3	8	5
4	5	8	2	3	1	7	9	6
1	6	7	4	8	3	9	5	2
5	3	2	9	1	6	8	7	4
8	4	9	5	2	7	6	1	3
3	9	5	8	4	2	1	6	7
7	8	1	3	6	5	2	4	9
6	2	4	1	7	9	5	3	8

Solution # 951

9	6	2	8	4	3	5	7	1
1	5	4	2	7	6	9	3	8
3	7	8	9	5	1	4	6	2
5	9	7	1	2	4	3	8	6
4	8	1	6	3	7	2	9	5
2	3	6	5	9	8	1	4	7
7	1	9	3	8	5	6	2	4
6	4	3	7	1	2	8	5	9
8	2	5	4	6	9	7	1	3

Solution # 952

6	2	9	8	1	7	5	3	4
3	4	7	2	5	9	6	1	8
1	8	5	3	4	6	9	2	7
9	3	2	4	8	1	7	6	5
8	7	6	5	9	2	3	4	1
4	5	1	6	7	3	8	9	2
2	6	4	7	3	5	1	8	9
5	9	8	1	6	4	2	7	3
7	1	3	9	2	8	4	5	6

Solution # 953

5	9	2	8	1	7	6	4	3
6	1	8	5	4	3	2	7	9
7	3	4	9	2	6	1	8	5
4	8	1	7	5	9	3	2	6
3	5	9	2	6	8	7	1	4
2	6	7	4	3	1	5	9	8
9	7	3	1	8	5	4	6	2
8	2	5	6	7	4	9	3	1
1	4	6	3	9	2	8	5	7

Solution # 954

8	2	9	7	5	1	6	3	4
3	4	7	6	2	8	9	5	1
1	6	5	4	9	3	2	7	8
2	5	3	8	6	9	4	1	7
6	7	1	5	3	4	8	2	9
9	8	4	2	1	7	5	6	3
5	3	8	1	4	2	7	9	6
7	9	6	3	8	5	1	4	2
4	1	2	9	7	6	3	8	5

Solution # 955

3	1	6	7	8	5	2	4	9
7	5	4	9	2	3	1	6	8
8	9	2	4	1	6	7	3	5
1	2	9	6	7	8	3	5	4
6	7	3	5	4	9	8	1	2
5	4	8	2	3	1	6	9	7
9	6	7	1	5	2	4	8	3
2	8	1	3	9	4	5	7	6
4	3	5	8	6	7	9	2	1

Solution # 956

9	1	6	7	4	8	2	3	5
4	5	2	6	1	3	7	8	9
8	3	7	2	9	5	1	4	6
2	7	5	4	3	9	8	6	1
3	4	1	8	7	6	5	9	2
6	9	8	5	2	1	4	7	3
7	2	3	9	5	4	6	1	8
5	8	9	1	6	7	3	2	4
1	6	4	3	8	2	9	5	7

Solution # 957

7	2	4	1	3	5	6	9	8
6	8	3	2	4	9	5	7	1
5	9	1	8	7	6	4	2	3
1	3	2	7	6	8	9	4	5
9	7	8	4	5	3	2	1	6
4	6	5	9	1	2	3	8	7
8	1	6	3	2	4	7	5	9
2	5	7	6	9	1	8	3	4
3	4	9	5	8	7	1	6	2

Solution # 958

8	3	6	5	9	2	7	4	1
4	2	1	3	7	8	5	6	9
7	9	5	4	6	1	3	8	2
5	4	8	1	2	6	9	7	3
9	6	2	7	3	4	8	1	5
3	1	7	9	8	5	6	2	4
6	8	9	2	4	3	1	5	7
1	7	4	8	5	9	2	3	6
2	5	3	6	1	7	4	9	8

Solution # 959

7	1	6	2	4	8	5	3	9
8	3	4	1	9	5	7	6	2
2	9	5	3	7	6	1	4	8
9	6	7	8	5	4	2	1	3
3	8	1	6	2	7	4	9	5
4	5	2	9	3	1	6	8	7
6	7	8	5	1	9	3	2	4
5	2	9	4	6	3	8	7	1
1	4	3	7	8	2	9	5	6

Solution # 960

8	3	6	4	9	7	1	5	2
5	4	1	6	2	8	9	3	7
9	7	2	1	5	3	6	4	8
4	1	9	2	7	5	8	6	3
6	8	7	3	1	4	2	9	5
3	2	5	9	8	6	7	1	4
7	9	3	8	4	1	5	2	6
2	5	4	7	6	9	3	8	1
1	6	8	5	3	2	4	7	9

Solution # 961

6	7	8	1	9	4	5	3	2
1	9	5	3	6	2	8	4	7
2	4	3	8	5	7	6	9	1
9	8	6	7	2	1	4	5	3
7	1	4	5	3	6	9	2	8
3	5	2	4	8	9	7	1	6
8	2	7	9	4	3	1	6	5
5	3	9	6	1	8	2	7	4
4	6	1	2	7	5	3	8	9

Solution # 962

5	4	3	6	8	7	9	2	1
6	9	7	1	5	2	4	3	8
8	1	2	3	9	4	7	6	5
1	2	5	4	6	9	8	7	3
3	6	9	5	7	8	1	4	2
4	7	8	2	3	1	5	9	6
7	8	1	9	2	6	3	5	4
9	5	6	8	4	3	2	1	7
2	3	4	7	1	5	6	8	9

Solution # 963

5	6	3	9	4	1	7	8	2
7	9	1	8	2	6	5	4	3
2	8	4	5	7	3	9	6	1
8	7	5	4	1	9	3	2	6
9	4	6	2	3	7	1	5	8
3	1	2	6	5	8	4	9	7
6	2	7	3	9	4	8	1	5
4	3	8	1	6	5	2	7	9
1	5	9	7	8	2	6	3	4

Solution # 964

8	6	4	5	9	3	2	7	1
9	5	7	2	6	1	4	3	8
2	1	3	8	7	4	9	5	6
6	9	8	1	5	7	3	2	4
1	3	5	9	4	2	8	6	7
7	4	2	6	3	8	1	9	5
3	7	1	4	2	6	5	8	9
5	8	6	3	1	9	7	4	2
4	2	9	7	8	5	6	1	3

Solution # 965

5	1	4	6	8	2	3	9	7
8	7	2	5	3	9	1	6	4
3	9	6	7	1	4	2	5	8
4	5	1	9	2	6	7	8	3
6	8	7	1	4	3	5	2	9
2	3	9	8	7	5	4	1	6
1	4	8	2	6	7	9	3	5
9	2	3	4	5	8	6	7	1
7	6	5	3	9	1	8	4	2

Solution # 966

9	7	1	5	3	8	4	2	6
2	3	6	7	4	9	1	8	5
8	5	4	2	1	6	9	7	3
7	4	8	3	9	5	6	1	2
5	6	2	4	7	1	8	3	9
3	1	9	6	8	2	5	4	7
6	2	7	8	5	4	3	9	1
4	9	3	1	6	7	2	5	8
1	8	5	9	2	3	7	6	4

Solution # 967

6	2	8	3	9	7	4	1	5
4	3	9	1	8	5	6	2	7
1	5	7	2	6	4	8	3	9
7	8	1	6	2	3	9	5	4
3	6	5	7	4	9	1	8	2
2	9	4	5	1	8	7	6	3
8	7	3	4	5	6	2	9	1
9	4	2	8	3	1	5	7	6
5	1	6	9	7	2	3	4	8

Solution # 968

2	1	8	4	9	5	3	7	6
6	7	4	1	2	3	9	5	8
5	9	3	7	6	8	2	1	4
3	4	2	9	1	6	7	8	5
9	6	5	8	4	7	1	2	3
1	8	7	3	5	2	6	4	9
4	5	6	2	7	9	8	3	1
7	3	9	5	8	1	4	6	2
8	2	1	6	3	4	5	9	7

Solution # 969

8	6	1	2	4	9	7	3	5
3	2	9	5	7	8	1	6	4
7	5	4	1	3	6	2	8	9
9	8	5	3	1	7	6	4	2
2	1	6	9	5	4	3	7	8
4	3	7	8	6	2	5	9	1
5	7	8	6	9	1	4	2	3
1	4	2	7	8	3	9	5	6
6	9	3	4	2	5	8	1	7

Solution # 970

3	7	9	4	1	6	8	5	2
6	1	2	5	8	7	4	9	3
8	5	4	2	9	3	1	7	6
7	9	5	3	6	8	2	1	4
2	6	8	9	4	1	7	3	5
1	4	3	7	2	5	9	6	8
5	8	7	1	3	2	6	4	9
4	2	1	6	5	9	3	8	7
9	3	6	8	7	4	5	2	1

Solution # 971

6	1	4	9	8	7	5	2	3
3	9	8	1	2	5	6	7	4
2	7	5	6	3	4	9	8	1
8	3	1	4	9	6	7	5	2
4	6	7	2	5	8	3	1	9
9	5	2	7	1	3	4	6	8
5	4	3	8	6	2	1	9	7
1	8	6	3	7	9	2	4	5
7	2	9	5	4	1	8	3	6

Solution # 972

2	7	6	8	5	9	1	3	4
8	3	9	1	7	4	2	6	5
1	4	5	2	3	6	7	9	8
3	8	1	4	9	5	6	7	2
6	2	7	3	8	1	4	5	9
9	5	4	6	2	7	8	1	3
5	1	8	9	6	2	3	4	7
7	6	3	5	4	8	9	2	1
4	9	2	7	1	3	5	8	6

Solution # 973

5	1	3	2	6	4	7	8	9
6	4	9	7	1	8	2	3	5
7	8	2	9	3	5	6	4	1
9	2	5	6	7	3	8	1	4
8	6	1	4	2	9	5	7	3
4	3	7	5	8	1	9	2	6
2	9	6	3	4	7	1	5	8
1	5	4	8	9	2	3	6	7
3	7	8	1	5	6	4	9	2

Solution # 974

9	5	8	1	3	7	2	6	4
1	6	3	5	4	2	9	8	7
4	7	2	9	6	8	3	5	1
5	1	4	7	9	3	8	2	6
7	3	9	8	2	6	1	4	5
2	8	6	4	1	5	7	9	3
6	4	1	2	7	9	5	3	8
3	2	5	6	8	1	4	7	9
8	9	7	3	5	4	6	1	2

Solution # 975

2	1	8	7	5	4	3	9	6
3	7	4	9	6	2	5	8	1
5	6	9	3	1	8	7	4	2
6	9	1	2	8	7	4	5	3
7	5	2	6	4	3	9	1	8
4	8	3	1	9	5	6	2	7
8	4	7	5	2	6	1	3	9
9	2	6	4	3	1	8	7	5
1	3	5	8	7	9	2	6	4

Solution # 976

4	2	6	8	5	7	9	1	3
7	8	1	4	9	3	2	6	5
9	5	3	2	6	1	8	7	4
6	3	8	9	2	4	1	5	7
5	4	2	7	1	8	3	9	6
1	7	9	6	3	5	4	2	8
2	9	4	3	7	6	5	8	1
3	6	5	1	8	2	7	4	9
8	1	7	5	4	9	6	3	2

Solution # 977

1	5	3	2	4	7	9	8	6
2	9	6	8	3	1	4	7	5
7	4	8	5	6	9	1	3	2
6	8	5	3	9	2	7	4	1
4	3	1	6	7	8	5	2	9
9	7	2	1	5	4	3	6	8
5	1	4	7	2	6	8	9	3
8	6	9	4	1	3	2	5	7
3	2	7	9	8	5	6	1	4

Solution # 978

6	8	1	4	5	3	7	9	2
3	9	4	8	2	7	6	5	1
2	7	5	9	1	6	3	4	8
1	6	7	3	8	9	5	2	4
4	2	3	1	7	5	8	6	9
9	5	8	2	6	4	1	3	7
7	4	9	6	3	8	2	1	5
5	1	6	7	4	2	9	8	3
8	3	2	5	9	1	4	7	6

Solution # 979

3	1	2	5	4	9	7	6	8
9	7	4	6	1	8	5	3	2
6	8	5	3	2	7	1	9	4
1	6	7	8	3	5	4	2	9
2	9	3	7	6	4	8	5	1
5	4	8	1	9	2	6	7	3
7	2	9	4	5	1	3	8	6
4	5	6	9	8	3	2	1	7
8	3	1	2	7	6	9	4	5

Solution # 980

6	3	1	8	9	7	5	4	2
4	5	7	2	6	1	9	8	3
8	9	2	3	4	5	1	7	6
2	1	4	6	8	3	7	5	9
3	7	5	9	1	2	8	6	4
9	8	6	7	5	4	3	2	1
5	4	3	1	7	6	2	9	8
7	2	8	4	3	9	6	1	5
1	6	9	5	2	8	4	3	7

Solution # 981

5	2	4	8	3	6	9	7	1
1	7	6	2	9	4	3	5	8
8	3	9	5	1	7	2	6	4
6	4	7	1	5	2	8	9	3
2	1	8	3	7	9	6	4	5
3	9	5	6	4	8	1	2	7
4	5	2	9	8	1	7	3	6
7	6	1	4	2	3	5	8	9
9	8	3	7	6	5	4	1	2

Solution # 982

7	9	3	2	5	4	6	1	8
4	6	2	9	1	8	3	5	7
1	8	5	7	3	6	9	2	4
6	1	9	4	8	3	2	7	5
8	2	4	6	7	5	1	3	9
3	5	7	1	9	2	8	4	6
2	3	6	5	4	9	7	8	1
9	4	1	8	2	7	5	6	3
5	7	8	3	6	1	4	9	2

Solution # 983

5	1	4	2	8	3	9	6	7
6	3	2	9	7	4	8	1	5
8	7	9	5	1	6	4	2	3
7	8	3	1	2	9	5	4	6
9	4	6	3	5	8	2	7	1
2	5	1	6	4	7	3	9	8
3	2	7	4	6	5	1	8	9
1	6	5	8	9	2	7	3	4
4	9	8	7	3	1	6	5	2

Solution # 984

4	3	5	8	1	2	7	6	9
2	1	9	6	7	5	4	8	3
8	6	7	9	3	4	1	2	5
7	2	3	1	4	8	9	5	6
6	8	1	5	2	9	3	4	7
5	9	4	3	6	7	2	1	8
1	7	6	4	5	3	8	9	2
9	4	2	7	8	6	5	3	1
3	5	8	2	9	1	6	7	4

Solution # 985

5	9	3	1	7	4	8	2	6
4	2	8	6	3	5	7	9	1
6	7	1	2	9	8	4	5	3
9	5	6	8	2	1	3	4	7
8	1	4	3	5	7	9	6	2
2	3	7	9	4	6	5	1	8
1	4	9	7	8	2	6	3	5
3	8	2	5	6	9	1	7	4
7	6	5	4	1	3	2	8	9

Solution # 986

3	4	1	9	6	7	8	2	5
7	9	8	3	2	5	6	1	4
5	2	6	4	8	1	3	7	9
9	3	5	7	1	2	4	6	8
4	1	2	8	9	6	7	5	3
6	8	7	5	4	3	1	9	2
2	7	4	1	5	8	9	3	6
1	5	9	6	3	4	2	8	7
8	6	3	2	7	9	5	4	1

Solution # 987

9	4	8	2	7	6	3	5	1
1	3	5	4	9	8	2	7	6
7	2	6	5	3	1	8	9	4
5	7	3	1	4	9	6	2	8
4	8	1	6	5	2	7	3	9
6	9	2	3	8	7	1	4	5
3	1	9	7	6	5	4	8	2
2	5	7	8	1	4	9	6	3
8	6	4	9	2	3	5	1	7

Solution # 988

3	5	8	2	9	6	1	7	4
2	4	6	1	7	3	9	8	5
9	1	7	8	5	4	3	6	2
7	3	9	4	6	1	5	2	8
5	2	1	7	3	8	4	9	6
6	8	4	9	2	5	7	3	1
4	7	2	6	1	9	8	5	3
8	9	3	5	4	2	6	1	7
1	6	5	3	8	7	2	4	9

Solution # 989

2	1	6	3	5	8	7	9	4
9	5	7	1	6	4	2	3	8
3	4	8	2	9	7	5	6	1
4	7	2	9	1	6	8	5	3
1	8	9	4	3	5	6	2	7
6	3	5	7	8	2	1	4	9
7	6	1	5	4	3	9	8	2
8	2	4	6	7	9	3	1	5
5	9	3	8	2	1	4	7	6

Solution # 990

9	7	1	3	8	2	4	5	6
6	8	5	7	9	4	1	2	3
3	4	2	6	5	1	9	7	8
8	6	3	9	2	5	7	4	1
2	9	7	4	1	6	8	3	5
1	5	4	8	3	7	2	6	9
4	1	6	5	7	8	3	9	2
7	3	8	2	6	9	5	1	4
5	2	9	1	4	3	6	8	7

Solution # 991

5	6	3	1	7	2	8	9	4
9	1	2	4	8	5	3	6	7
7	4	8	9	3	6	5	2	1
2	7	1	8	9	4	6	5	3
4	9	5	7	6	3	2	1	8
8	3	6	5	2	1	4	7	9
3	5	7	2	4	9	1	8	6
1	8	4	6	5	7	9	3	2
6	2	9	3	1	8	7	4	5

Solution # 992

7	9	1	6	8	5	4	2	3
6	4	8	2	7	3	5	1	9
5	3	2	4	1	9	8	7	6
9	2	7	3	4	8	1	6	5
8	1	3	5	6	7	9	4	2
4	6	5	1	9	2	3	8	7
3	8	4	7	5	6	2	9	1
2	7	9	8	3	1	6	5	4
1	5	6	9	2	4	7	3	8

Solution # 993

7	8	4	3	9	2	1	6	5
9	2	5	1	6	7	8	4	3
1	6	3	8	4	5	7	9	2
8	1	6	2	5	4	3	7	9
3	5	7	6	1	9	4	2	8
4	9	2	7	3	8	5	1	6
5	3	9	4	2	1	6	8	7
6	7	1	9	8	3	2	5	4
2	4	8	5	7	6	9	3	1

Solution # 994

9	5	3	4	1	7	8	6	2
7	2	6	8	9	3	4	1	5
4	1	8	5	2	6	3	7	9
5	6	4	1	8	2	9	3	7
8	9	1	3	7	5	2	4	6
3	7	2	6	4	9	1	5	8
2	3	9	7	5	1	6	8	4
6	8	5	2	3	4	7	9	1
1	4	7	9	6	8	5	2	3

Solution # 995

1	7	2	8	5	3	4	6	9
4	3	6	9	2	7	8	1	5
5	9	8	1	6	4	2	7	3
9	1	5	7	8	6	3	2	4
8	6	4	3	1	2	5	9	7
3	2	7	4	9	5	6	8	1
2	8	1	5	4	9	7	3	6
7	4	9	6	3	8	1	5	2
6	5	3	2	7	1	9	4	8

Solution # 996

5	3	8	2	6	4	7	9	1
9	6	2	8	7	1	5	4	3
4	7	1	5	3	9	8	2	6
7	4	6	3	2	8	1	5	9
1	9	3	4	5	7	2	6	8
2	8	5	9	1	6	3	7	4
8	2	9	1	4	5	6	3	7
6	5	4	7	8	3	9	1	2
3	1	7	6	9	2	4	8	5

Solution # 997

4	6	1	5	9	2	8	7	3
8	9	2	7	1	3	5	6	4
7	3	5	8	6	4	1	2	9
9	7	4	6	2	5	3	8	1
1	2	3	9	8	7	4	5	6
6	5	8	3	4	1	2	9	7
3	1	6	2	5	9	7	4	8
2	8	7	4	3	6	9	1	5
5	4	9	1	7	8	6	3	2

Solution # 998

2	5	6	7	8	3	9	1	4
9	1	4	5	2	6	7	8	3
3	8	7	4	9	1	6	5	2
7	4	9	6	3	8	1	2	5
8	6	3	2	1	5	4	9	7
1	2	5	9	4	7	3	6	8
6	7	1	3	5	2	8	4	9
5	9	8	1	7	4	2	3	6
4	3	2	8	6	9	5	7	1

Solution # 999

3	8	2	9	5	6	1	7	4
9	5	4	7	8	1	2	3	6
6	7	1	2	3	4	5	8	9
8	6	5	3	1	2	9	4	7
7	2	3	4	9	8	6	1	5
1	4	9	5	6	7	3	2	8
4	3	7	6	2	9	8	5	1
2	1	6	8	7	5	4	9	3
5	9	8	1	4	3	7	6	2

Solution # 1000

8	1	9	7	2	5	6	4	3
6	2	4	3	9	1	5	7	8
7	5	3	4	8	6	9	2	1
3	4	1	6	5	8	2	9	7
9	7	5	2	1	4	8	3	6
2	6	8	9	3	7	4	1	5
1	9	7	5	6	2	3	8	4
5	8	2	1	4	3	7	6	9
4	3	6	8	7	9	1	5	2

Solution # 1001

9	8	1	6	2	3	4	5	7
4	6	5	1	7	9	2	8	3
2	7	3	8	5	4	1	9	6
5	2	6	7	9	1	8	3	4
7	1	4	5	3	8	6	2	9
3	9	8	2	4	6	5	7	1
8	3	9	4	6	5	7	1	2
1	4	2	3	8	7	9	6	5
6	5	7	9	1	2	3	4	8

Solution # 1002

5	6	1	3	7	8	9	2	4
9	2	4	6	1	5	3	7	8
7	3	8	2	9	4	6	1	5
6	4	2	1	8	3	7	5	9
3	5	9	7	4	2	8	6	1
8	1	7	5	6	9	4	3	2
2	8	5	9	3	6	1	4	7
4	7	6	8	2	1	5	9	3
1	9	3	4	5	7	2	8	6

Solution # 1003

3	6	4	7	5	8	1	9	2
7	5	1	4	9	2	3	8	6
2	9	8	3	1	6	7	5	4
1	3	6	5	4	7	9	2	8
4	8	7	1	2	9	5	6	3
9	2	5	8	6	3	4	7	1
6	4	2	9	3	5	8	1	7
8	1	9	6	7	4	2	3	5
5	7	3	2	8	1	6	4	9

Solution # 1004

3	4	5	7	1	9	8	2	6
9	8	1	6	2	4	3	5	7
2	6	7	5	8	3	9	4	1
7	5	8	4	9	2	6	1	3
6	1	2	3	7	5	4	8	9
4	3	9	8	6	1	5	7	2
1	9	6	2	4	8	7	3	5
5	2	4	9	3	7	1	6	8
8	7	3	1	5	6	2	9	4

Solution # 1005

5	1	6	4	7	2	8	9	3
9	7	3	5	8	6	4	1	2
4	8	2	3	1	9	5	6	7
2	4	5	7	9	1	6	3	8
1	6	9	8	3	5	7	2	4
8	3	7	6	2	4	9	5	1
6	9	1	2	4	7	3	8	5
7	5	8	1	6	3	2	4	9
3	2	4	9	5	8	1	7	6

Solution # 1006

9	3	6	8	4	2	7	1	5
5	2	4	7	1	3	8	9	6
7	1	8	9	5	6	4	3	2
2	5	7	3	6	4	9	8	1
3	8	1	2	9	5	6	7	4
4	6	9	1	8	7	5	2	3
6	9	2	5	7	1	3	4	8
8	4	3	6	2	9	1	5	7
1	7	5	4	3	8	2	6	9

Solution # 1007

6	3	8	9	2	4	7	5	1
2	7	1	3	8	5	4	9	6
5	4	9	1	6	7	8	3	2
7	6	2	8	1	9	5	4	3
9	5	3	2	4	6	1	7	8
1	8	4	5	7	3	6	2	9
4	2	5	6	3	1	9	8	7
8	9	6	7	5	2	3	1	4
3	1	7	4	9	8	2	6	5

Solution # 1008

5	1	8	7	3	9	2	6	4
6	9	4	5	8	2	3	1	7
2	3	7	6	4	1	9	8	5
7	4	2	8	9	3	1	5	6
9	6	5	2	1	7	8	4	3
1	8	3	4	5	6	7	9	2
3	5	9	1	7	4	6	2	8
8	7	6	9	2	5	4	3	1
4	2	1	3	6	8	5	7	9

Solution # 1009

9	3	1	6	4	5	8	2	7
7	6	2	8	9	1	4	5	3
5	4	8	3	2	7	6	9	1
1	8	7	4	6	2	9	3	5
4	2	9	5	8	3	1	7	6
3	5	6	1	7	9	2	4	8
8	9	4	7	3	6	5	1	2
2	1	3	9	5	8	7	6	4
6	7	5	2	1	4	3	8	9

Solution # 1010

2	7	9	8	5	3	6	1	4
8	1	6	4	2	9	3	5	7
3	5	4	7	1	6	8	9	2
9	3	1	2	6	8	7	4	5
6	2	5	1	4	7	9	3	8
7	4	8	3	9	5	2	6	1
4	6	3	5	7	2	1	8	9
1	9	2	6	8	4	5	7	3
5	8	7	9	3	1	4	2	6

Solution # 1011

3	8	1	5	7	6	4	2	9
6	2	9	1	3	4	8	5	7
7	4	5	2	9	8	3	6	1
5	1	6	7	2	3	9	8	4
2	3	8	9	4	1	5	7	6
4	9	7	6	8	5	1	3	2
1	5	4	8	6	2	7	9	3
8	7	2	3	1	9	6	4	5
9	6	3	4	5	7	2	1	8

Solution # 1012

2	3	8	4	9	1	5	6	7
4	9	5	7	2	6	1	3	8
7	6	1	5	8	3	4	2	9
5	7	6	8	4	2	3	9	1
8	2	9	1	3	5	7	4	6
1	4	3	6	7	9	2	8	5
6	8	2	3	1	7	9	5	4
3	5	7	9	6	4	8	1	2
9	1	4	2	5	8	6	7	3

Solution # 1013

4	1	9	7	2	8	5	3	6
3	2	6	4	9	5	1	7	8
7	8	5	1	6	3	2	9	4
8	9	7	2	4	6	3	1	5
5	6	2	3	7	1	4	8	9
1	3	4	8	5	9	7	6	2
6	7	8	5	3	2	9	4	1
9	5	3	6	1	4	8	2	7
2	4	1	9	8	7	6	5	3

Solution # 1014

7	6	2	8	1	5	3	9	4
8	9	3	2	4	7	1	6	5
5	1	4	3	9	6	2	8	7
2	8	5	6	3	9	4	7	1
4	3	6	7	2	1	9	5	8
9	7	1	5	8	4	6	2	3
1	4	7	9	5	2	8	3	6
3	5	9	1	6	8	7	4	2
6	2	8	4	7	3	5	1	9

Solution # 1015

3	2	9	7	5	4	8	6	1
4	7	1	9	6	8	5	2	3
6	5	8	3	1	2	7	4	9
9	3	4	1	8	7	6	5	2
2	1	5	4	3	6	9	7	8
7	8	6	2	9	5	3	1	4
8	6	2	5	4	3	1	9	7
5	9	7	8	2	1	4	3	6
1	4	3	6	7	9	2	8	5

Solution # 1016

1	5	4	9	8	3	6	7	2
2	6	9	7	5	1	8	4	3
8	3	7	2	4	6	9	1	5
7	8	5	4	6	2	1	3	9
3	4	2	8	1	9	5	6	7
9	1	6	5	3	7	4	2	8
5	2	3	1	9	4	7	8	6
4	7	8	6	2	5	3	9	1
6	9	1	3	7	8	2	5	4

Solution # 1017

9	8	4	1	5	6	3	7	2
7	6	1	3	9	2	8	4	5
5	3	2	8	7	4	9	1	6
8	4	3	7	6	5	2	9	1
2	5	6	9	1	3	7	8	4
1	9	7	2	4	8	6	5	3
4	1	8	6	3	9	5	2	7
3	7	9	5	2	1	4	6	8
6	2	5	4	8	7	1	3	9

Solution # 1018

7	2	6	4	8	5	9	3	1
1	3	4	6	9	2	7	8	5
8	9	5	1	3	7	2	6	4
4	6	8	7	1	9	3	5	2
5	7	9	3	2	4	6	1	8
2	1	3	5	6	8	4	9	7
6	8	2	9	4	1	5	7	3
3	5	1	2	7	6	8	4	9
9	4	7	8	5	3	1	2	6

Solution # 1019

9	4	6	2	3	1	5	8	7
7	8	3	9	5	4	2	6	1
2	1	5	8	7	6	3	9	4
5	3	1	7	2	9	8	4	6
8	6	9	5	4	3	7	1	2
4	7	2	6	1	8	9	5	3
3	9	7	1	6	5	4	2	8
1	2	8	4	9	7	6	3	5
6	5	4	3	8	2	1	7	9

Solution # 1020

5	1	2	9	7	4	3	6	8
3	8	4	1	5	6	2	7	9
6	9	7	3	8	2	1	5	4
8	3	1	2	9	5	7	4	6
2	4	9	6	3	7	8	1	5
7	5	6	8	4	1	9	2	3
1	7	8	4	6	9	5	3	2
4	2	3	5	1	8	6	9	7
9	6	5	7	2	3	4	8	1

Solution # 1021

6	2	4	7	8	1	9	5	3
8	3	5	4	6	9	1	2	7
7	9	1	2	5	3	6	4	8
3	8	2	1	7	4	5	6	9
4	1	9	5	3	6	8	7	2
5	7	6	8	9	2	3	1	4
1	5	7	3	2	8	4	9	6
2	6	3	9	4	5	7	8	1
9	4	8	6	1	7	2	3	5

Solution # 1022

3	1	9	8	5	7	2	6	4
4	6	7	1	3	2	8	9	5
2	8	5	4	9	6	1	3	7
7	5	6	9	8	3	4	2	1
8	2	3	6	4	1	5	7	9
1	9	4	7	2	5	3	8	6
9	4	1	2	6	8	7	5	3
5	7	2	3	1	9	6	4	8
6	3	8	5	7	4	9	1	2

Solution # 1023

2	8	3	6	9	5	1	7	4
7	4	5	1	3	8	6	9	2
1	9	6	2	4	7	8	5	3
5	7	4	3	8	2	9	6	1
3	1	8	5	6	9	4	2	7
9	6	2	7	1	4	5	3	8
8	2	9	4	5	3	7	1	6
4	3	1	9	7	6	2	8	5
6	5	7	8	2	1	3	4	9

Solution # 1024

9	6	4	1	2	8	7	5	3
8	3	1	7	5	4	2	9	6
7	5	2	3	9	6	8	1	4
5	7	9	6	1	2	3	4	8
1	4	6	8	3	7	5	2	9
3	2	8	5	4	9	1	6	7
6	1	7	9	8	5	4	3	2
2	9	5	4	7	3	6	8	1
4	8	3	2	6	1	9	7	5

Solution # 1025

1	6	7	2	4	8	3	9	5
4	9	2	5	3	6	8	1	7
3	8	5	1	9	7	2	6	4
5	7	6	9	8	1	4	2	3
9	3	4	6	7	2	5	8	1
2	1	8	3	5	4	6	7	9
7	5	9	8	6	3	1	4	2
8	2	3	4	1	9	7	5	6
6	4	1	7	2	5	9	3	8

Solution # 1026

9	6	8	5	3	4	1	7	2
3	4	7	2	6	1	5	9	8
5	1	2	9	7	8	3	4	6
1	3	6	7	5	9	8	2	4
4	2	9	8	1	6	7	5	3
8	7	5	4	2	3	9	6	1
6	9	4	1	8	5	2	3	7
7	5	1	3	4	2	6	8	9
2	8	3	6	9	7	4	1	5

Solution # 1027

3	1	2	9	4	7	6	5	8
9	7	5	3	6	8	2	4	1
6	4	8	2	1	5	3	7	9
2	9	4	5	3	1	7	8	6
1	8	3	4	7	6	5	9	2
7	5	6	8	2	9	1	3	4
8	2	1	7	9	3	4	6	5
4	3	9	6	5	2	8	1	7
5	6	7	1	8	4	9	2	3

Solution # 1028

1	2	6	8	3	5	4	9	7
9	5	4	2	7	6	3	8	1
7	8	3	9	1	4	6	5	2
2	9	5	4	6	1	7	3	8
6	3	1	7	5	8	2	4	9
4	7	8	3	9	2	5	1	6
5	4	9	6	8	7	1	2	3
3	6	2	1	4	9	8	7	5
8	1	7	5	2	3	9	6	4

Solution # 1029

9	8	4	5	1	2	6	3	7
7	2	1	3	9	6	8	4	5
3	5	6	8	7	4	9	2	1
5	1	3	6	4	7	2	9	8
6	9	8	2	5	3	1	7	4
4	7	2	9	8	1	3	5	6
8	4	7	1	3	9	5	6	2
2	3	5	4	6	8	7	1	9
1	6	9	7	2	5	4	8	3

Solution # 1030

7	1	2	4	9	3	5	8	6
8	5	3	6	7	2	9	4	1
4	9	6	1	8	5	2	3	7
6	2	8	7	4	1	3	5	9
1	3	5	9	2	6	4	7	8
9	4	7	3	5	8	1	6	2
5	8	1	2	3	7	6	9	4
3	6	4	8	1	9	7	2	5
2	7	9	5	6	4	8	1	3

Solution # 1031

4	9	5	8	6	7	1	2	3
1	3	7	2	4	9	6	8	5
6	2	8	3	1	5	9	4	7
5	1	4	6	3	2	7	9	8
7	6	9	5	8	4	2	3	1
2	8	3	7	9	1	5	6	4
9	4	6	1	7	3	8	5	2
8	5	1	4	2	6	3	7	9
3	7	2	9	5	8	4	1	6

Solution # 1032

4	3	8	2	1	9	7	6	5
2	1	7	5	3	6	9	8	4
9	6	5	7	4	8	2	3	1
6	2	4	3	8	5	1	7	9
5	8	9	6	7	1	4	2	3
3	7	1	4	9	2	8	5	6
7	5	6	1	2	4	3	9	8
8	4	2	9	6	3	5	1	7
1	9	3	8	5	7	6	4	2

Solution # 1033

6	8	7	4	1	3	5	9	2
1	9	2	5	8	7	3	6	4
3	4	5	6	9	2	1	8	7
2	7	1	9	5	6	4	3	8
5	3	9	8	7	4	6	2	1
4	6	8	3	2	1	7	5	9
7	1	3	2	6	8	9	4	5
8	5	4	7	3	9	2	1	6
9	2	6	1	4	5	8	7	3

Solution # 1034

8	7	6	2	9	3	1	4	5
3	5	1	7	6	4	2	9	8
9	2	4	5	1	8	3	6	7
2	6	7	1	3	5	4	8	9
5	1	8	4	7	9	6	3	2
4	3	9	8	2	6	5	7	1
6	4	2	9	8	1	7	5	3
7	9	5	3	4	2	8	1	6
1	8	3	6	5	7	9	2	4

Solution # 1035

6	9	2	3	4	1	7	5	8
3	4	8	5	2	7	1	6	9
1	5	7	9	6	8	2	3	4
7	3	5	4	9	6	8	1	2
9	8	1	2	7	5	3	4	6
4	2	6	1	8	3	9	7	5
8	6	3	7	5	9	4	2	1
2	7	9	6	1	4	5	8	3
5	1	4	8	3	2	6	9	7

Solution # 1036

9	4	1	3	5	2	7	6	8
6	8	2	7	4	9	5	1	3
7	5	3	6	8	1	4	2	9
4	3	8	2	9	7	6	5	1
5	7	9	1	6	8	2	3	4
2	1	6	4	3	5	9	8	7
8	6	7	9	2	3	1	4	5
3	9	4	5	1	6	8	7	2
1	2	5	8	7	4	3	9	6

Solution # 1037

8	2	6	1	7	4	3	9	5
5	9	7	6	8	3	2	1	4
1	3	4	5	2	9	6	8	7
7	5	8	4	6	2	1	3	9
4	6	3	8	9	1	7	5	2
2	1	9	7	3	5	8	4	6
3	4	2	9	1	7	5	6	8
9	8	1	2	5	6	4	7	3
6	7	5	3	4	8	9	2	1

Solution # 1038

6	1	9	2	3	7	8	5	4
5	4	3	1	8	6	9	2	7
8	2	7	5	4	9	3	6	1
3	9	1	8	5	4	6	7	2
7	8	5	6	2	1	4	3	9
4	6	2	7	9	3	1	8	5
9	5	6	3	1	2	7	4	8
1	3	8	4	7	5	2	9	6
2	7	4	9	6	8	5	1	3

Solution # 1039

5	7	6	9	4	3	2	8	1
3	2	9	1	7	8	6	5	4
1	4	8	5	6	2	9	7	3
4	5	3	8	2	6	7	1	9
9	1	7	3	5	4	8	6	2
8	6	2	7	9	1	3	4	5
2	8	5	4	3	7	1	9	6
7	3	4	6	1	9	5	2	8
6	9	1	2	8	5	4	3	7

Solution # 1040

5	3	8	9	1	7	2	4	6
2	4	9	6	8	3	7	5	1
1	7	6	4	2	5	3	8	9
9	2	5	7	3	6	4	1	8
3	6	1	5	4	8	9	7	2
4	8	7	2	9	1	6	3	5
8	9	2	3	5	4	1	6	7
6	1	3	8	7	9	5	2	4
7	5	4	1	6	2	8	9	3

Solution # 1041

7	3	8	1	9	2	5	6	4
2	9	4	5	7	6	3	1	8
1	6	5	3	4	8	2	9	7
6	1	2	8	3	7	4	5	9
8	4	3	9	6	5	1	7	2
5	7	9	2	1	4	6	8	3
3	5	7	4	8	1	9	2	6
4	2	6	7	5	9	8	3	1
9	8	1	6	2	3	7	4	5

Solution # 1042

5	4	9	6	8	3	7	1	2
7	1	6	4	2	9	3	8	5
2	8	3	1	5	7	6	4	9
3	5	8	2	1	4	9	6	7
9	7	4	3	6	5	8	2	1
1	6	2	7	9	8	5	3	4
6	3	5	9	4	2	1	7	8
8	2	7	5	3	1	4	9	6
4	9	1	8	7	6	2	5	3

Solution # 1043

4	2	6	7	1	5	9	8	3
9	5	3	2	4	8	1	7	6
8	1	7	9	3	6	2	4	5
5	9	4	3	2	7	8	6	1
6	3	8	4	5	1	7	2	9
1	7	2	8	6	9	3	5	4
7	6	9	1	8	4	5	3	2
2	4	1	5	7	3	6	9	8
3	8	5	6	9	2	4	1	7

Solution # 1044

7	1	2	8	3	5	6	9	4
6	5	4	2	9	7	1	3	8
3	9	8	1	6	4	2	5	7
2	4	9	5	1	3	8	7	6
5	6	3	7	8	2	9	4	1
1	8	7	6	4	9	3	2	5
8	2	5	9	7	1	4	6	3
4	7	1	3	2	6	5	8	9
9	3	6	4	5	8	7	1	2

Solution # 1045

4	7	6	8	1	2	3	5	9
2	9	5	3	7	6	8	1	4
8	3	1	5	4	9	7	6	2
9	1	7	6	2	4	5	8	3
6	5	8	1	9	3	2	4	7
3	2	4	7	8	5	6	9	1
5	4	9	2	6	7	1	3	8
7	8	3	9	5	1	4	2	6
1	6	2	4	3	8	9	7	5

Solution # 1046

4	2	7	8	1	3	5	6	9
5	1	9	6	7	4	3	2	8
6	3	8	2	5	9	7	1	4
7	4	5	3	6	1	9	8	2
9	6	3	4	8	2	1	5	7
2	8	1	7	9	5	4	3	6
8	9	2	1	3	7	6	4	5
1	5	6	9	4	8	2	7	3
3	7	4	5	2	6	8	9	1

Solution # 1047

7	4	8	2	1	9	3	5	6
6	3	1	4	5	8	9	7	2
5	9	2	7	3	6	8	4	1
9	7	4	5	6	2	1	8	3
1	8	5	9	4	3	2	6	7
3	2	6	8	7	1	5	9	4
2	1	7	6	8	5	4	3	9
8	6	3	1	9	4	7	2	5
4	5	9	3	2	7	6	1	8

Solution # 1048

1	2	9	5	4	3	8	6	7
4	6	5	8	1	7	3	9	2
8	7	3	2	6	9	5	1	4
6	5	1	7	2	8	4	3	9
9	3	7	1	5	4	2	8	6
2	8	4	3	9	6	1	7	5
5	9	2	6	8	1	7	4	3
7	4	8	9	3	2	6	5	1
3	1	6	4	7	5	9	2	8

Solution # 1049

8	3	2	4	5	6	1	7	9
1	5	6	7	8	9	2	4	3
9	4	7	2	1	3	6	5	8
4	8	3	6	2	7	5	9	1
5	2	9	1	3	8	7	6	4
7	6	1	5	9	4	3	8	2
2	7	4	9	6	1	8	3	5
6	1	8	3	4	5	9	2	7
3	9	5	8	7	2	4	1	6

Solution # 1050

3	2	5	4	7	9	6	8	1
8	7	4	6	1	2	9	3	5
6	9	1	8	3	5	4	7	2
9	5	2	7	6	1	8	4	3
7	4	8	5	9	3	2	1	6
1	6	3	2	8	4	7	5	9
4	3	7	1	2	6	5	9	8
5	1	6	9	4	8	3	2	7
2	8	9	3	5	7	1	6	4

www.ingramcontent.com/pod-product-compliance
Lightning Source LLC
Chambersburg PA
CBHW080826220526
45467CB00008B/2208
* 9 7 9 8 3 9 1 0 8 6 4 1 3 *